Modelling and Applications of Transport Phenomena in Porous Media

Theory and Applications of Transport in Porous Media

Series Editor:
Jacob Bear, *Technion – Israel Institute of Technology, Haifa, Israel*

Volume 5

The titles published in this series are listed at the end of this volume.

Contents

4 HEAT TRANSFER IN SELF-HEATED PARTICLE BEDS SUBMERGED IN LIQUID COOLANT

KENT MEHR and JORGEN WÜRTZ
Commission of the European Communities
Joint Research Centre, Ispra, Italy. 277

PREFACE

Transport phenomena in porous media are encountered in various disciplines, e.g., civil engineering, chemical engineering, reservoir engineering, agricultural engineering and soil science. In these disciplines, problems are encountered in which various extensive quantities, e.g., mass and heat, are transported through a porous material domain. Often, the void space of the porous material contains two or three fluid phases, and the various extensive quantities are transported simultaneously through the multiphase system. In all these disciplines, decisions related to a system's development and its operation have to be made. To do so a tool is needed that will provide a forecast of the system's response to the implementation of proposed decisions. This response is expressed in the form of spatial and temporal distributions of the state variables that describe the system's behavior. Examples of such state variables are pressure, stress, strain, density, velocity, solute concentration, temperature, etc., for each phase in the system,

The tool that enables the required predictions is the model. A model may be defined as a simplified version of the real porous medium system and the transport phenomena that occur in it. Because the model is a simplified version of the real system, no unique model exists for a given porous medium system. Different sets of simplifying assumptions, each suitable for a particular task, will result in different models.

By analyzing a problem on hand and the objectives of modeling, a conceptual model is constructed, which is subsequently represented as a mathematical one. The solution of the mathematical model yields the required forecast. Usually, the solution is obtained numerically.

Because of the complexity associated with the presence of a solid and a void space that is filled with one or more fluid phase, and the fact that it is impossible to describe in detail the geometry of the interphase boundaries, and to observe and measure variables at points within each phase, models of transport phenomena are formulated, presented and solved at the macroscopic, or continuum level.

The main objectives of this short course are (a) to present a methodology for constructing conceptual and mathematical models of problems of transport in porous medium domains on the basis of the continuum approach, and (b) to present and discuss a number of applications of the theory to important problems of practical interest, especially problems related to the nuclear industry.

We hope that the material presented in this volume will enable participants in the course, as well as other readers of this volume, to formulate problems of transport of any extensive quantity in any porous medium domain which they may encounter.

The frontiers of knowledge related to modeling transport in porous media are continuously being pushed forward. Pressed by urgent needs to provide solutions for problems of importance to society, a large number of scientists, in a variety of disciplines, are currently engaged in research that continuously contributes to the understanding and quantitative description of transport phenomena in porous media and to the solution of models that describe them. Examples of such problems are groundwater pollution by hazardous wastes, produced by industry and by agricultural activities, contamination from repositories for radioactive wastes, improved techniques for enhanced oil and gas production and storage of energy in aquifers. Concurrently, research continues on methodologies of utilizing field data for solving field problems. These include design of observation networks, regionalization of point data, system identification (including parameter estimation), taking into account the spatial heterogeneity of the porous medium and stochasticity of the processes involved. Efforts are also made to improve model solving techniques. Much effort is devoted to methods for coping with the various aspects of uncertainty associated with models, from uncertainty in the processes that take place, through uncertainty in the spatial distribution of model coefficients and scale effects. It is our intention at the Von Karman Institute for Fluid Dynamics to present many of these advanced topics and research results in subsequent short courses.

Jacob Bear
Jean-Marie Buchlin
1989

Chapter 1

EIGHT LECTURES ON MATHEMATICAL MODELLING OF TRANSPORT IN POROUS MEDIA

J. BEAR
Albert and Anne Mansfield
Chair in Water Resources
Technion - Israel Institute of Technology
Haifa, Israel 32000.

1.1   Lecture One: Introduction

The objective of this lecture is to set the stage and define the goals for this series of lectures. Accordingly, we shall start by defining a porous medium, understand what models are and why we need them to describe transport in porous media and discuss the modelling process. We shall concentrate on the class of models that visualize a porous medium domain as a *continuum*, or as a set of *overlapping continua*.

---

[0]A series of eight lectures delivered in the 1987/88 Lecture Series Programme on Modelling and Application of Transport in Porous Media at the von Karman Institute for Fluid Dynamics, Belgium, Nov. 30 - Dec. 3, 1987. The lectures and the lecture notes are based on the book INTRODUCTION TO MODELING OF TRANSPORT PHENOMENA IN POROUS MEDIA, by J. Bear and Y. Bachmat, KLUWER ACADEMIC PUBLISHERS, Dordrecht, The Netherlands, 1990.

1.1.1   Porous medium

Porous materials and the transport in them of *extensive quantities*, such as mass, momentum, and energy of a phase, or mass of a component of a phase, are encountered in a large number of scientific and engineering disciplines. Flow of oil, water and gas in a petroleum reservoir, transport of water and contaminants in the unsaturated zone and in aquifers, storage of heat in the unsaturated zone and in aquifers, storage of energy in the form of compressed air injected into and produced from anticlinal geological formations, the possible migration of radionuclides from a radioactive waste repository in deep geological formations, mass, heat and chemical component transport in packed-bed reactors, and drying processes in industry, may serve as examples of transport problems of practical interest encountered in civil engineering, agricultural engineering, reservoir engineering, chemical engineering and even biomedical engineering (where the movement of fluids and solutes through such organs as lungs and kidneys is considered).

In these cases, one or more extensive quantities are transported through the solid and/or through the fluid phases that together occupy a porous medium domain. The term *extensive quantity* is used here as a quantity that is additive over volume (e.g., mass, energy).

Sand, sandstone, soil, fissured rock, Karstic limestone, ceramic, foam rubber, industrial filters, wicks, bread and lungs, are just a few examples of porous materials encountered in practice. Common to all of them is the presence of both a persistent *solid matrix* and a persistent *void space*. The latter is occupied by one or more fluid phases, e.g., air, water, oil. A *phase* is defined as a chemically homogeneous portion of space that is separated from other such portions by a well defined physical boundary (= *interface*, or *interphase boundary*). There can be only a single gaseous phase in the void space, as all gaseous phases are completely *miscible* and do not maintain a distinct interface between them. However, we may have a number of (*immiscible*) liquid phases, each occupying a well defined portion of the void space.

A phase may be composed of more than one component. A *component* is a part of a phase that is made up of an identifiable, homogeneous chemical constituent, or an assembly of constituents, e.g., ions, or molecules. A component is characterized by the fact that its behavior can be described by a unique set of variables. The number of components comprising a phase is the minimum number of *independent* chemical constituents required in order to completely describe the composition of the phase.

Another common characteristic of a porous medium domain is that the void space (and, therefore, also the solid matrix) is distributed throughout it. This implies that wherever a sufficiently large sample is taken around a point within a porous medium domain, it will always contain both a solid phase and a void space. It is obvious that if this sample is to represent the behavior of the porous medium in the vicinity of the sampling point (say, the centroid of the sample), it must be sufficiently small. However, the very size of the sample may influence the description of this behavior, presented in terms of average values of state variables of the phases present in the sample and of the characteristics of the void space configuration. For example, if this sample is too small, it may contain *only* one phase, say a solid. If it is too large, the average taken over it may smear details, or variations of interest. For the time being, let us refer to the volume of a sample that provides a distinct, unique description of the behavior at a point, as a *Representative Elementary Volume* (abbreviated REV).. Later we shall discuss this concept in more detail.

Transport of a considered extensive quantity through a porous medium domain may take place within a single (fluid, or solid) phase, through a number of phases (possibly including the solid phase), or through all of them. In the first case, at least part of the domain occupied by that phase must be *connected*, whereas in the last two cases, the transfer of a considered quantity may also take place across the (microscopic) interphase boundaries separating the phases through which the transport occurs.

With these considerations, and for the purpose of this series of lectures, we define a porous medium as a *multiphase material body* characterized by the following features:

(a) A Representative Elementary Volume (abbreviated REV) can be determined for it, such that no matter where we place it within a porous medium domain, it will always contain *both* a a solid phase and a void space. Similarly, a Representative Elementary Area (abbreviated REA) can be determined, such that no matter where we place it within a porous medium domain, it will always contain both a solid phase and a void space. *If such an REV and REA cannot be determined for a given domain, the latter cannot qualify as a porous medium domain.*

(b) The size of an REV, or an REA, is such that parameters that represent the distributions of void and of the solid within it, are statistically meaningful. The quantification of this requirement is discussed in detail below.

In the discussion so far, each phase (gas, liquid, or solid) is regarded as a *continuum*. This continuum is obtained by averaging the behavior *of the molecules* that comprise the phase over a volume of a size such that the resulting averages represent phase behavior.

A similar methodology will be followed below, in order to pass from the *microscopic* (phase continuum) level to the *macroscopic* one.

1.1.2 Modelling process

Our starting point is that management decisions have to be made with respect to an investigated system–here, a given porous medium domain. In order to make such decisions, or to select some 'best' mode of operation, from amongst various possible alternative modes, we need to know the response of the system to the implementation of the contemplated decisions.

Our objective in investigating *processes*, or phenomena of transport of extensive quantities in a porous medium domain (= system) is, therefore, to predict the response of the system of interest as a whole, of a phase, or of a component of the latter, to excitation(s) imposed on it. The response takes the form of spatial and temporal variations of values of state variables (e.g., pressure, density, temperature) that describe the behavior of the system. In a particular case, we may be interested only in those variables that are relevant to a considered practical (management, or decision making) problem. The excitation may take the form of sink/source phenomena, (e.g., injection, or production of a fluid phase), at points within the domain, or of imposed changes in conditions (say, of flux, or pressure) on the domain's boundary.

The tool that provides the required predictions is a *model*. A model may be defined as a *selected simplified* version of the real system, that satisfactorily (e.g., from the point of view of details and accuracy) approximately simulates those excitation - response relations of the latter, that are relevant to the problem on hand.

The real system and its behavior is very complex, with the degree of complexity depending on the strength of the 'magnifying glass' through which we observe the porous medium and the phenomena that occur in it. For example, is the configuration of each individual soil grain required to describe the flow? The need for simplification requires no further justification. The simplification is stated in the form of a *set of assumptions, subjectively selected by the modeller.* They express, *in words,* his, or her's understanding and approximation of the real system under consideration, and the processes that take place in it, for the purpose of providing information on the future behavior of the system in a particular case of interest, in response to specified excitations.. From the very definition of a model, it follows that there exists no unique model for a given porous medium domain, or even for a given problem. Different objectives (in terms of the type and accuracy of the required information) will lead to different models. The selection of the appropriate model to be used in each case depends, therefore, on the objective of the investigations (and, hence, on the type of predictions required) and on the available resources (and that includes field, or laboratory data).

We shall use the term *conceptual model* for the set of selected assumptions that *verbally* describe the system's composition, the processes of transport that take place in it, the mechanisms that govern them and the relevant medium properties, all as envisioned, or approximated by the modeller, for the purpose of constructing a model intended to provide information for a specific problem.

The first step in the modelling process is the construction of a conceptual model, which should relate to the following features:

- Configuration of the boundary of the domain of interest.

- Dimensionality of the model (one, two or three dimensions).

- Steady-state, or time dependent.

- Materials (solid, fluid phases and components) comprising the domain, and their *relevant* properties (density, viscosity, compressibility).

- Isothermal, or nonisothermal conditions.

- Presence of assumed sharp fluid-fluid (macroscopic) interfaces (whether fluids are miscible, or not).

- Processes of interest that take place within the domain.

- State variables that describe these processes.

- Sources/sinks of the considered extensive quantities, with reference to their type and to whether they are point, or distributed ones.

- Initial conditions within the domain.

- Conditions on the domain's boundary.

Selecting the appropriate conceptual model for a given problem is, perhaps, the most important step in the modelling process. Undersimplification may lead to a model that does not provide the required information, while oversimplification may result in lack of measurements (data) required for *model calibration* and *parameter estimation* (discussed below).

As a second step, the (verbal) conceptual model is expressed in the form of a *mathematical model*. The solution of the mathematical model yields the required predictions for the considered transport phenomena.

In principle, transport phenomena that take place in a porous medium domain, may be treated at the *microscopic level*, at which attention is focused on the behavior of a phase (or of a component of it) at points *within* that phase, considered as a *continuum*. As mentioned above, this continuum description is obtained from the molecular one by averaging over volumes that represent the behavior of the phase at points within the latter.

Following the modelling process, once a (microscopic) conceptual model has been stated, it should be translated into a mathematical one, in order to enable a solution for particular cases. As will be explained below, an intrinsic element of every mathematical model of a process, is the detailed description of the geometrical configuration of the boundary of the domain throughout which the considered process takes place. Obviously, it is impossible to describe the configuration of the surface that bounds the considered phase (say, solid surface, in the case of a single fluid phase that occupies the entire void space), at the microscopic level of description. Hence, it is impossible to state and solve transport problems at that level. Moreover, even if we could construct a complete model, and solve it for values of state variables (e.g., pressure) at that level, it would be impossible to take measurements in the real system at that level, in order to *validate* the model and to estimate model coefficients (see below).

To circumvent these difficulties, the transport problem is transformed from the microscopic level to a *macroscopic* one, at which quantities can be measured and boundary value problems can be fully stated and solved. Accordingly, the real porous medium domain, comprised of two, or more phases (each of which already regarded as a continuum), that together completely occupy disjoint subdomains within the porous medium domain, is replaced by a *model* in which each phase is assumed to behave as a continuum that fills up the *entire* domain. We speak of *overlapping continua*, each corresponding to one of the phases. If the individual phases interact with each other, so do these continua. In fact, every extensive quantity of every phase, or of a component of a phase, is modelled as a continuum that fills up the entire domain.

For every point within these continua, average values of phase (or component) variables are taken over elementary volumes (abbreviated EV) centered at the point, regardless of whether, in the real domain, this point falls within the considered phase, or not. The averaged values are referred to as *macroscopic* values of the considered variables. By traversing the entire porous medium domain with a moving EV, we obtain for each phase and for each variable, a *field* that is a function of the space coordinates.

Later in this lecture, we shall discuss the size of the EV, and suggest a special kind of EV, to which we shall refer as a Representative Elementary Volume (abbreviated, REV).

Thus, by introducing the macroscopic level of description, we have

(a) circumvented the need to know the exact configuration of (microscopic) interphase boundaries, and

(b) obtained a description of processes in terms of measurable quantities.

Transport problems of practical interest can, thus, be described, solved and validated by measurements. All this at the expense of losing information concerning the microscopic configuration of interphase boundaries and the variation of quantities within each phase, information that we do not have, anyway.

As we shall see below, the lack of this information is compensated for by the introduction of various *coefficients* that reflect, at the macroscopic level, the effects of the configuration of the microscopic interphase boundaries. The detailed structure of each coefficient depends on the model (e.g., statistical) that is introduced to represent the *microscopic* reality. In cases of specific porous media, *the numerical values* of these coefficients must be determined experimentally, in the laboratory, or in the field. Actually, these coefficients are *coefficients of the model* that we have selected for the problem on hand. This means that their interpretation and numerical values for a given domain, may differ from one model to the next. Nevertheless, following common practice, we refer to these coefficients as ones of the porous medium (and not as ones of the model). Obviously, no model can be used for solving a particular problem, unless the numerical values of all the coefficients appearing it are known for that problem.

In what follows, we shall present a methodology for passing from the microscopic level to the macroscopic one. The ultimate objective is to derive mathematical models of transport phenomena of interest at the macroscopic level, which is the level at which measurements are taken and problems of practical interest are stated and solved. In developing such models, the underlying (often implicit) assumption is that the use of the continuum approach is permitted. In fact, our very use of the term 'porous medium', already implies that we consider a continuum called *porous medium.*

Having stated the conceptual model for a given problem, including in it the assumption that the continuum approach is permissible and will be employed, this model is translated into a mathematical one (this time, at the macroscopic level). This macroscopic mathematical model consists of the following items:

- A definition of the surface that bounds the considered porous medium domain.

- Equations that express the balances of the extensive quantities that are relevant to the problem.

- Flux equations, sometimes referred to as *dynamic constitutive equations*, that express the fluxes of the relevant extensive quantities, in terms of the state variables of the problem.

- Constitutive equations that define the behavior (e.g., pressure–density relations) of the particular phases and components involved.

- Source functions for the considered extensive quantities, expressed in terms of the relevant state variables.

- Initial conditions that describe the *known* state of the considered porous medium system at some initial time.

- Boundary conditions on the domain's boundary.

This standard content of a mathematical model is the same whether we construct a model at the microscopic level, or at the macroscopic one. However, in this series of lectures, our objective is to develop transport models at the macroscopic level. The considered extensive quantities will be: mass, mass of a component, momentum and energy, all of a phase, in a single, or multiphase system. Each fluid phase may be composed of more than one component. The balance equations, constituting the core of a transport model, take the form of partial differential equations, written in terms of *macroscopic state variables,* each of which is an average taken over the Representative Elementary Volume of the considered domain.

A special case, that deserves a comment, is that of the *momentum balance equation.* In the continuum approach, subject to certain simplifying assumptions (included in the conceptual model) as to the solid-fluid interaction, negligible friction in the fluid, and negligible inertial effects, the average momentum balance equation reduces to the linear motion equation, known as *Darcy's law*, used as a flux equation for fluids in porous media. With certain modifications, it is also applicable to multiphase flow systems. Any flow model, in which the transported extensive quantity is the mass of a fluid phase, always involves also the transport of another such quantity–the momentum of a phase, the density of which is the velocity. This requires the statement of a momentum balance equation as part of the model. However, subject to certain simplifying assumptions, in a macroscopic model, the balance of momentum usually appears as a flux expression–Darcy's law– and not as a balance equation. This topic will be discussed in detail in LECTURE 4.

We have emphasized above that no model can be used in practice for a given domain, unless the model coefficients pertinent to that domain are known, and that these coefficients can be obtained only by experiments conducted in the considered domain. A typical experiment consists of exciting the system and observing its response, while applying the same excitation (and obviously the same initial and boundary conditions) to the model of the system. The values of the coefficients are obtained by comparing the response predicted by the model with that actually observed in the real system (in the field, or in the laboratory). The sought values of the coefficients are those that will make the two sets of values identical. Obviously, because the model is only an approximation of the real system, we should never expect the two sets of values to be *identical.* Instead, we search for the 'best fit' between them, according to some criterion. Various techniques exist for determining the 'best', or 'optimal' values of the coefficients, i.e., values that will bring the observed values and the predicted ones sufficiently close to each other.

Once a mathematical model has been constructed in terms of relevant state variables, it has to be solved for cases of practical interest. The solution takes the form of spatial and temporal distributions of the state variables within the space and time domains of interest. The preferable method of solution is the analytical one, because once such a solution has been derived, it can be employed for a variety of planned, or anticipated situations. However, in most cases of practical interest, this method of solution is not feasible because of the irregular shape of the domain's boundaries, the heterogeneity of the domain, as expressed by the spatial distributions of its coefficients, and the irregular temporal and spatial distributions of the various excitations, and sink/source functions. In addition, many problems of transport are nonlinear, and it is impossible to derive analytical solutions for them for most cases of practical interest. Approximate analytical solutions have been derived, primarily for one-dimensional domains. Although they serve an important purpose by providing an insight into internal relations affecting the model's results, and by helping in model validation and verification, their use is, naturally, limited. Instead, *numerical methods* are usually employed for solving the mathematical model.

The main features of the various numerical methods for solving models of transport in porous media are:

- The solution is sought for the numerical values of state variables only at specified points in the space and time domains defined for the problem, rather than for the continuous variations of the variables within these domains. Interpolation is used to obtain values at other points.

- The partial differential equations, that represent balances of the considered extensive quantities, are replaced by a set of algebraic equations that are written in terms of the sought, discrete values of the state variables at the discrete points in space and time mentioned above.

- The solution is obtained for a specified set of numerical values of the various model coefficients, rather than as general relationships in terms of these coefficients.

- Because of the very large number of algebraic equations that have to be solved simultaneously, a computer program must be prepared and used to obtain a solution by means of a digital computer.

Sometimes, the term *numerical model,* or *computer model,* is used, rather than speaking of a *numerical method of solution* of the mathematical model. This is justified on the ground that a number of assumptions are introduced *in addition* to those underlying the mathematical model. This makes the numerical model a model in its own right. It represents a different approximate version of the real system. It is sometimes possible to pass directly from the conceptual model to the numerical one, without first constructing a separate mathematical model. The numerical model has its own set of coefficients that have to be identified before the model can be used for any particular problem.

Another important feature of modelling, closely associated with the problem of parameter identification, is *uncertainty*. We are uncertain whether the selected conceptual model (i.e., our set of assumptions) indeed represents what happens in the real domain, albeit to the accepted degree of approximation. Furthermore, even when employing some identification technique, we are uncertain about the values of the coefficients to be used in the model. Possible errors in observed data used for parameter identification also contribute to uncertainty in the values of model parameters. In many cases, data for model calibration is insufficient, or even non-existent. As a consequence, we should also expect uncertainty in the values of the state variables predicted by the model. These considerations pave the way for the development of *stochastic models*. In the latter, the information on coefficients appears in the form of probability distributions of values, rather than as deterministic ones. These probability distributions are derived by appropriate methods of solving the inverse problem, where the input data also appear in probabilistic forms. Probabilistic values of model coefficients will yield probabilistic predicted values of state variables. Management decisions will have to take into account this probabilistic nature of the predicted behavior of the system.

In view of these uncertainties and, most often, the lack of sufficient data for validating a model and determining its coefficients, the use of models should, be redefined beyond the standard approach, mentioned above, that envisages the model only as a tool for predicting the response of an investigated system. Models may be used also to identify trends and directions of change and ranges of probable responses, and, through a *sensitivity analysis*, (or *parametric analysis*) to deepen our understanding of the system and its behavior and help to design observation networks.

1.1.3   Selecting the size of an REV

We have discussed the process of developing and using continuum models. We have still to discuss the size of the EV that should be used for the passage to the (macroscopic) continuum level.

In principle, any arbitrarily selected *Elementary Volume* (EV) may be used as an averaging volume for passing from the microscopic level of description to the macroscopic one. Obviously, different EV's will yield different average values for each quantity of interest, and there is no sense in asking which of them is more 'correct'. The selection of an averaging volume in any particular case depends only on the model's objectives. Also, the size of the 'window' of the instrument that measures an averaged value should correspond to that of the selected EV. In this way, within the range of error introduced by the conceptual model of the problem, the predicted and measured values will always be the same. However, the main drawback of this approach is that since every averaged value may strongly depend on the size of the selected EV, it must be 'labelled' by the size of the EV over which it was taken.

To circumvent this difficulty, rather than select the volume of averaging arbitrarily, we need a universal criterion which is based on measurable characteristics and which determines, for any given porous medium, a *range* of averaging volumes within which these characteristics remain, more or less, constant. As long as the instrument 'window' is in that range, observed and computed values will be close, within a prescribed level of error. An averaging volume which belongs to that range will be referred to as a *Representative Elementary Volume* (abbreviated REV).

Our next objective is to attempt to quantify the REV concept by proposing how to determine its size range.

We start by requiring, as part of the very definition of an REV, that the value of any averaged (= macroscopic) characteristic of the microstructure of the void space, at any point in the porous medium domain, be a *single valued function* of that point and of time. Denoting the volume of an REV by U_o, and selecting *porosity*, ϕ, as a typical geometrical characteristic of the void space, this requirement means that

$$\left.\frac{\partial\phi(\mathbf{x}_o, U)}{\partial U}\right|_{U=U_o} = 0. \tag{1.1.1}$$

Let us introduce the *characteristic function* of the void space, $\gamma(\mathbf{x})$, as a means of describing the spatial distribution of the void space, with

$$\gamma(\mathbf{x}) = \begin{cases} 1 & \text{when } \mathbf{x} \text{ is in the void space,} \\ \\ 0 & \text{when } \mathbf{x} \text{ is in the solid matrix,} \end{cases} \tag{1.1.2}$$

where $\mathbf{x}$ denotes a position vector of a point (Fig. 1.1.1).

Making use of $\gamma(\mathbf{x})$, we may define various geometrical characteristics of the void space. For example

$$\begin{aligned}
\overline{\gamma}(\mathbf{x}_o) &= \frac{1}{U(\mathbf{x}_o)} \int_{U(\mathbf{x}_o)} \gamma(\mathbf{x})\, dU(\mathbf{x}) \\ \\
&= \left. \mathrm{E}(\gamma|\mathbf{x}_o)\right|_U = \left. \phi(\mathbf{x}_o)\right|_U,
\end{aligned} \tag{1.1.3}$$

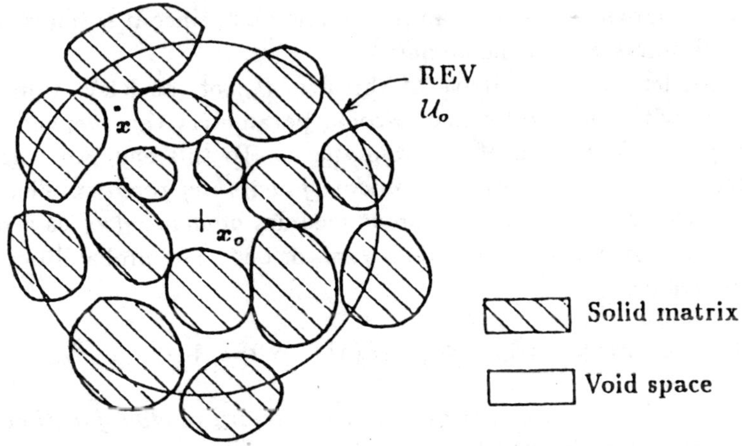

Figure 1.1.1: Planner section through REV.

where $\mathcal{U}$ denotes the averaging domain of volume U. Equation (1.1.3) relates the average of γ to the porosity, ϕ.

Another geometrical characteristic is the *covariance* of γ

$$\overline{\mathring{\gamma}(\mathbf{x})\mathring{\gamma}(\mathbf{x}+\mathbf{h})}\bigg|_{\mathbf{x}_o,U,\mathbf{h}} = \frac{1}{U}\int_{\mathcal{U}(\mathbf{x}_o)} \mathring{\gamma}(\mathbf{x})\mathring{\gamma}(\mathbf{x}+\mathbf{h})\, dU(\mathbf{x})$$

$$= \left. \mathrm{Cov}_\gamma(\mathbf{h})\right|_{\mathbf{x}_o,U}. \tag{1.1.4}$$

Thus, for an appropriately selected volume $U = U_o$, we should have

$$\left.\frac{\partial \mathrm{Cov}_\gamma(\mathbf{h})|_{\mathbf{x}_o,U}}{\partial U}\right|_{U=U_o} = 0. \tag{1.1.5}$$

In principle, for every point $\mathbf{x}_o$, one can visualize an experiment within a given domain, consisting of a succession of gradually increasing volumes $U_1 < U_2 < U_3 < \ldots$, all centered at $\mathbf{x}_o$, and a concurrent determination of $\overline{\gamma}$ and $\overline{\mathring{\gamma}(\mathbf{x})\mathring{\gamma}(\mathbf{x}+\mathbf{h})}$, for each such volume, hoping that a volume $U = U_o$, which satisfies both (1.1.1) and (1.1.5), will be found.

After repeating this procedure for determining U_o at all points $\mathbf{x}_o$ within a considered porous medium domain, one can replace the actual domain by a model of a fictitious continuum, provided U_o is uniform throughout the domain. Obviously, this is an impossible task, since it is impractical to observe all points within the domain.

Instead, let us try to arrive at the size, U_o, of the REV, from its relationships with *measurable* macroscopic parameters that depend on the microscopic configuration of the void space. To this end, we regard the characteristic function, $\gamma(\mathbf{x})$, as a *random function of position*. This means that at any point, $\mathbf{x} = \mathbf{x}_p$, in a porous medium domain, $\gamma(\mathbf{x}_p)$ is a random variable which may attain the values zero, or one. The probabilities, θ and $1 - \theta$ are defined by

$$P(\gamma|_{\mathbf{x}_p} = 1) = \theta_{\mathbf{x}_p}, \quad P(\gamma|_{\mathbf{x}_p} = 0) = 1 - \theta_{\mathbf{x}_p}.$$

Let us further assume that $\gamma(\mathbf{x})$ is a *stationary random function* within the considered domain, which means that

(a) The expected value of γ, given $\mathbf{x}$, satisfies

$$E[\gamma(\mathbf{x})] = \theta = \text{const.}$$

(b) The *covariance* of γ-values at any two points, $\mathbf{x}_p$ and $\mathbf{x}_q$, satisfies

$$\text{Cov}[\gamma(\mathbf{x}_p), \gamma(\mathbf{x}_q)] \equiv E\{[\gamma(\mathbf{x}_p) - \theta][\gamma(\mathbf{x}_q) - \theta]\} = f(\mathbf{h}_{pq}),$$

where $\mathbf{h}_{pq} = \mathbf{x}_p - \mathbf{x}_q$ is the oriented distance between points $\mathbf{x}_p$ and $\mathbf{x}_q$. As a consequence

(c) The *variance* of γ satisfies

$$\text{Var}[\gamma] = E\{[\gamma(\mathbf{x}) - \theta]^2\} = f(0) = \text{const.}$$

We shall refer to a domain for which (a), (b) and (c) hold as *macroscopically homogeneous* with respect to $\gamma(\mathbf{x})$. It will be called *isotropic* with respect to γ, if

$$\text{Cov}_\gamma[\mathbf{h}_{pq}] = \text{Cov}_\gamma[h], \qquad h = |\mathbf{h}_{pq}|,$$

i.e., the *correlation function* between the values of γ at different points within the considered domain depends only on the distance, h, between them, and not on their relative orientation.

Next we *assume* that $\gamma(\mathbf{x})$ possesses the *ergodic property* within the considered domain. A *stationary random function* is said to be *ergodic*, if the average of any statistical characteristic of the function, taken over a sufficiently large domain of its argument in a single realization, is an *unbiased* and consistent estimate of that characteristic over the entire set of possible realizations of the function. An estimate of a population parameter is said to be *unbiased* if its expected value is equal to the value of the parameter. An estimate is said to be *consistent* if it approaches, probabilistically, the value of the parameter as the sample size increases.

Under the *ergodic hypothesis*

$$\overline{\gamma}(\mathbf{x}_o) \equiv \frac{1}{U_o} \int_{\mathcal{U}_o} \gamma(\mathbf{x})dU = \phi(\mathbf{x}_o) \simeq E(\gamma|_{\mathbf{x}_o}) = \theta|_{\mathbf{x}_o}, \tag{1.1.6}$$

and

$$\left.\overline{\mathring{\gamma}(\mathbf{x})\mathring{\gamma}(\mathbf{x}+\mathbf{h})}\right|_{U_o,\mathbf{h}} = \frac{1}{U_o}\int_{\mathcal{U}_o} \mathring{\gamma}(\mathbf{x})\mathring{\gamma}(\mathbf{x}+\mathbf{h})dU$$

$$\simeq \left.\mathrm{Cov}_\gamma(\mathbf{h})\right|_{\mathbf{x}_o} = \mathrm{Var}_\gamma(\mathbf{x}_o)\tau_\gamma(\mathbf{h})$$

$$= \phi(1-\phi)\tau_\gamma(\mathbf{h}), \tag{1.1.7}$$

where $\mathrm{Cov}_\gamma(\mathbf{h})|_{\mathbf{x}_o}$ is the covariance of γ in $\mathcal{U}_o$ for points spaced an oriented distance $\mathbf{h}$ apart, $\mathrm{Var}_\gamma(\mathbf{x}_o) = \phi(1-\phi)$, $\tau_\gamma(\mathbf{h})$ is the *correlation coefficient* at $\mathbf{x}_o$, between values of γ at points spaced an oriented distance $\mathbf{h}$ apart, and $\phi(= U_{ov}/U_o)$ is the porosity of the porous medium. By definition

$$\tau_\gamma(0) = 1. \tag{1.1.8}$$

In fact, the volume U_o of an REV should be sufficiently large so that the volumetric averages can be considered as satisfactory estimates of all relevant population parameters of the void space configuration at $\mathbf{x}_o$, i.c., estimates which are free of errors caused by the size of the sample and its random choice.

In the present case, a sufficient condition for (1.1.6) and (1.1.7) to hold is that

$$\left| \int_o^\infty \tau_\gamma(h)dh \right| < \infty.$$

As shown by Debye *et al.* (1957), for an isotropic porous medium and any function $\tau_\gamma(h)$, the relation

$$\left.\frac{\partial \tau_\gamma}{\partial h}\right|_{h=0} = -\frac{1}{4\Delta_v(1-\phi)} \tag{1.1.9}$$

holds, where $\Delta_v = U_{ov}/S_{vs}$ is the *hydraulic radius* of the void space (of volume U_{ov} and area of contact with the solid, S_{vs}). An example of an approximate expression for $\tau_\gamma(h)$ for an isotropic porous medium, with a random distribution of void and solid spaces, is given by Debye *et al.* (1957), in the form

$$\tau_\gamma(h) \cong \exp\left\{-\frac{h}{4\Delta_v(1-\phi)}\right\}, \qquad h = |\mathbf{h}|. \qquad (1.1.10)$$

It follows that a necessary condition for obtaining nonrandom estimates of the geometrical characteristics of the void space at any point, $\mathbf{x}_o$, which serves as a centroid of a sphere of volume U_o, and diameter ℓ, i.e., ones that are not subject to sampling error, is

$$h_{\max} = \ell_{\min} \gg \Delta_v. \qquad (1.1.11)$$

Bachmat and Bear (1986) consider an example in which $\ell_{\min}$ is determined on the basis of porosity. They consider a volume U_o centered at $\mathbf{x}_o$, and divide this volume into N disjoint elementary subdomains, $\delta U = U_o/N$, such that in each of them one may find (more or less) *either* solid, *or* void. The average of γ over the N samples is taken as an estimate, $\hat{\phi}$, of the porosity, ϕ. By employing (1.1.10), they relate N to the variance, $\sigma_{\hat{\phi}}^2$, and to ϕ (Fig. 1.1.2). For example, for $\phi = 0.4$ and $\sigma_{\hat{\phi}}^2 = 0.0032$, they find $N = 8000$. For a cubical REV, they find

$$\ell_{\min}^{(\phi)} = \{N(\phi, \sigma_{\hat{\phi}}^2)\}^{\frac{1}{3}} C_\Delta \Delta_v, \qquad C_\Delta = \left.\frac{\ell_{\min}^{(\phi)}}{\Delta_v}\right|_{N=1},$$

where C_Δ is a numerical coefficient, and superscript ϕ indicates that we are considering the porosity in determining $\ell_{\min}$.

The requirement of ergodicity also sets an upper bound on the size of the REV, namely, $\ell < \ell_{\max}$, where $\ell_{\max}$ is the distance between points in the porous medium domain beyond which the domain of averaging ceases to be *statistically homogeneous* with respect to the moments of $\gamma(\mathbf{x})$.

In reality, the requirement of statistical homogeneity is seldom satisfied, as the macroscopic parameters of the void geometry usually vary from point to point. However, even for a domain that is heterogeneous with respect to these parameters, one can define around every point a sufficiently small subdomain within which these parameters may still be considered uniform, *up to a prescribed error level.* The size of such a subdomain around a given point serves as the upper bound for the size of the REV at that point.

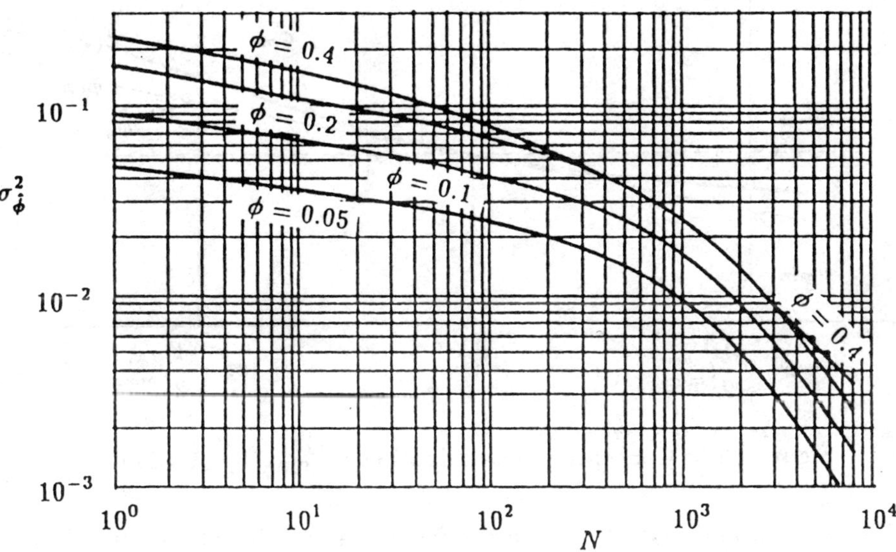

Figure 1.1.2: Variance of the estimate, $\hat{\phi}$, of porosity as a function of number, N, of elementary subdomains (Bachmat and Bear, 1986).

In order to determine this upper bound, let us consider a domain centered at a point $\mathbf{x}_o$ as homogeneous if for all points $\mathbf{x} \subset \mathcal{U}$, we have

$$E_\gamma(\mathbf{x}) \equiv \phi(\mathbf{x}) = \text{const.} = \phi_o,$$

$$\text{Cov}[\gamma(\mathbf{x} + \mathbf{h}), \gamma(\mathbf{x})] = f(\mathbf{h}),$$

i.e., a function of $\mathbf{h}$ only for all $\mathbf{x} \subset \mathcal{U}$. Then

$$\text{Var}\,\gamma(\mathbf{x}) = f(0) = \text{const.} = \phi_o(1 - \phi_o). \qquad (1.1.12)$$

For a heterogeneous domain, $\mathcal{U}$, we have $\phi = \phi(\mathbf{x})$. We shall refer to the domain $\mathcal{U}$ as *approximately homogeneous*, here with respect to porosity, if within it

$$\frac{\phi_{\max} - \phi_{\min}}{\overline{\phi}} \equiv \delta \ll 1, \qquad (1.1.13)$$

where $\phi_{\max}$, $\phi_{\min}$ and $\overline{\phi}$ are the largest, the smallest and the average values of ϕ, respectively, within $\mathcal{U}$, and δ (with $0 < \delta \le 1$) is an arbitrarily selected small number.

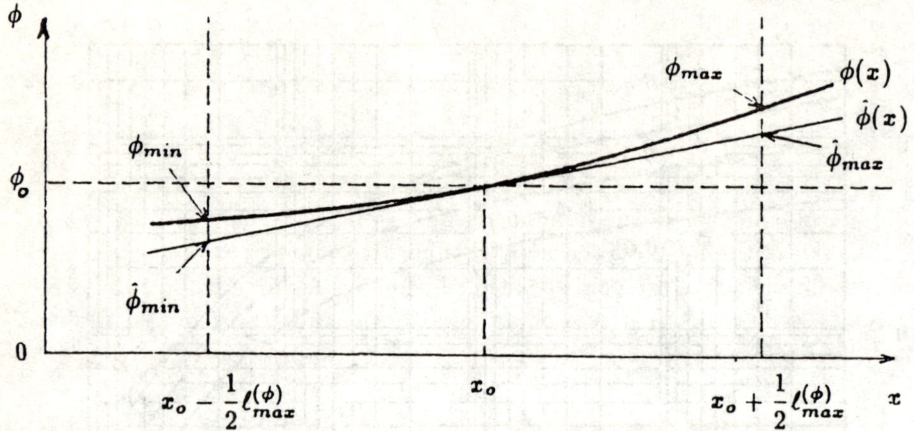

Figure 1.1.3: Conceptual determination of $\ell_{\max}^{(\phi)}$ by (1.1.17).

For a sufficiently small domain around $\mathbf{x}_o$, any differentiable function, $\phi(\mathbf{x})$, can be approximated by its linear part (Fig. 1.1.3), i.e.

$$\hat{\phi}(\mathbf{x}) = \phi_o + (\text{grad}\phi)|_{\mathbf{x}_o}\cdot(\mathbf{x} - \mathbf{x}_o), \qquad (1.1.14)$$

where $\phi_o = \phi|_{\mathbf{x}_o}$.

By introducing the definition $\ell_{\max}^{(\phi)} = 2\text{Max}|\mathbf{x}-\mathbf{x}_o|$, and employing (1.1.13) and (1.1.14), we obtain

$$\ell_{\max}^{(\phi)} = \frac{\phi_o}{|\text{grad }\phi|_{\mathbf{x}_o}}\hat{\delta}, \qquad (1.1.15)$$

where $\hat{\delta} = (\hat{\phi}_{\max} - \hat{\phi}_{\min})/\overline{\phi}$.

The distance $\ell_{\max}^{(\phi)}$ (based on porosity) is thus the upper limit for the size of the REV at a point $\mathbf{x}_o$ within a porous medium domain, at the selected error level. We have to scan all points $\mathbf{x}_o$ within the given domain, in order to determine the smallest value of $\ell_{\max}^{(\phi)}$.

Altogether, $\ell^{(\phi)}$ has to satisfy the condition

$$\ell_{\min}^{(\phi)} \ll \ell^{(\phi)} \ll \ell_{\max}^{(\phi)}. \qquad (1.1.16)$$

If $\ell_{\max}^{(\phi)} \leq \ell_{\min}^{(\phi)}$ at $\mathbf{x}_o$, an REV cannot be defined for the considered domain. On the other hand, only if a non-zero range of $\ell^{(\phi)}$ can be found, which is common to all points within a given spatial porous medium domain, one can adopt the continuum model for the porous medium within that domain.

Finally, we have to relate $\ell^{(\phi)}$ to the size of the considered domain. If L^* is a characteristic length of the domain, we require that

$$\ell^{(\phi)} \ll L^*, \qquad (1.1.17)$$

in order to ensure that the boundary region of the domain, which has a width $\ell^{(\phi)}$, and in which the continuum approach is not applicable, be small compared to the size of the domain itself.

The size of the REV in a domain is, thus, determined by the porosity and the specific surface of the void space in the domain, by prescribed acceptable reliability and error levels in estimating ϕ, by the size of the domain, by the spatial variation of ϕ within the REV and by a prescribed tolerable deviation of ϕ from uniformity within it.

So far, the concept and size of the REV have been related to porosity as a geometrical porous medium property. We have indicated this fact by using the superscript ϕ. Whenever additional characteristics of the porous medium appear in the macroscopic model describing a transport problem, e.g., permeability, a range for REV has to be determined for each of them. If a common REV range can be found, a continuum model of the porous medium can be employed.

One of the requirements for the range of the REV is that $\partial\phi/\partial U = 0$ within it, as defined by (1.1.1). This does not necessarily imply that $\phi(\mathbf{x})$ is uniform within $\mathcal{U}_o$. To illustrate this point, consider the ratio $U_v(\mathbf{x}_o)/U(\mathbf{x}_o)$, where $U(\mathbf{x}_o)$ is the volume of a sphere centered at an arbitrary point $\mathbf{x}_o$, within the domain, and $U_v(\mathbf{x}_o)$ is the volume of the void space within $U(\mathbf{x}_o)$.

Figure 1.1.4 shows the variation of the ratio U_v/U as $U(\mathbf{x}_o)$ increases. For very small values of $U(\mathbf{x}_o)$, the above ratio is one, or zero, depending on whether $\mathbf{x}_o$ happens to fall in the void space, or in the solid. As $U(\mathbf{x}_o)$ increases, we note large fluctuations in the ratio U_v/U. However, as U continues to grow, these fluctuations are gradually attenuated, until, above some value $U = U_{\min}$, they decay, leaving only small amplitude fluctuations around some constant value.

Bachmat and Bear (1986) show that in the vicinity of $\mathbf{x}_o$, the function $\phi(U)$ has a plateau and $\phi(\mathbf{x}|_U)$ is a linear function of $\mathbf{x}$.

If $U(\mathbf{x}_o)$ is further increased, say beyond some value $U = U_{\max}$, we may observe a trend in the considered ratio U_v/U, due to a systematic variation in the latter. The representative elementary volume is that volume, $U_o(\mathbf{x}_o)$,

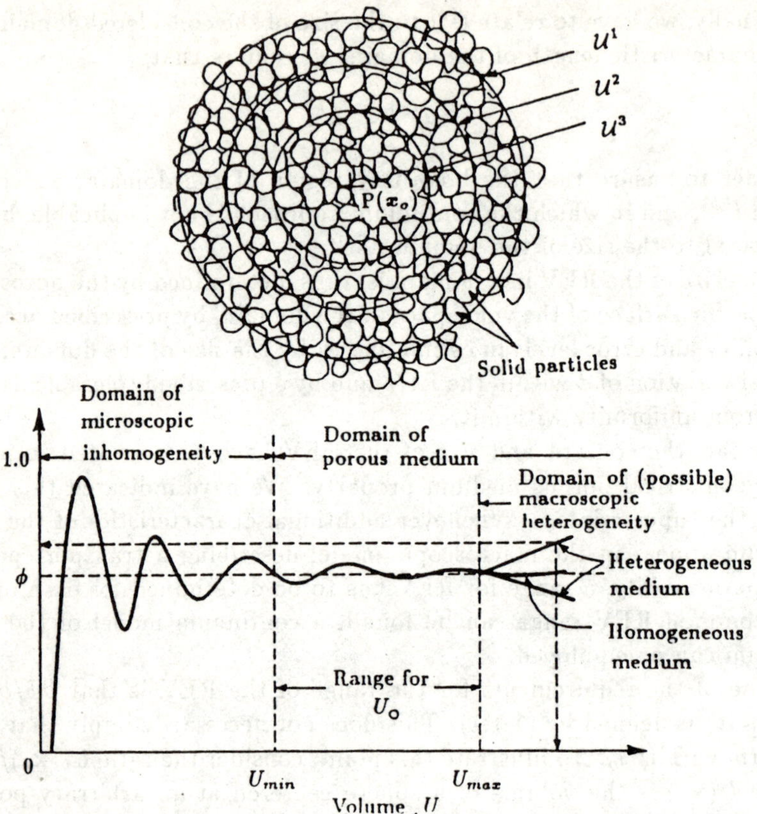

Figure 1.1.4: Variation of porosity in the neighborhood of a point as a function of the average volume(Bachmat and Bear, 1986).

within the range of $U_{min} < U < U_{max}$ that will make the ratio U_v/U independent of U, and hence a single valued function of $\mathbf{x}_o$ only. For $U = U_o$, the ratio U_v/U represents the porous medium's porosity, at $\mathbf{x}_o$. By definition, for the REV, the volumetric fraction of the solid, $(1 - \phi)$, is also a single valued function of $\mathbf{x}_o$.

By (1.1.15), the upper limit of $\ell^{(\phi)}$, for a given δ, is defined by $\ell^{(\phi)}_{max}$, which, in turn, depends on U_{max}, that indicates the point of deviation from the plateau, produced by the linearity of $\phi(\mathbf{x})$ in the vicinity of $\mathbf{x}$.

Once an REV has been (conceptually) selected for a given porous medium domain, each extensive quantity (i.e., mass, momentum and energy) of each phase and each component within that domain, as well as the porous medium as a whole, can be regarded as a continuum. Models describing the transport of these extensive quantities can then be written at the continuum level.

This concludes the discussion on the (conceptual) selection of the REV size.

To summarize this lecture:

- We have defined what we mean by a 'transport problem' and a 'porous medium', and why we need models to solve problems of transport in porous media.

- We have explained what we mean by a model, why we need models, what modelling objectives are and what the modelling process is.

- We have emphasized the importance of the conceptual model, as the first step in the modelling process, to be followed by its transformation into a mathematical model.

- We have introduced the continuum approach and discussed the concept of the Representative Elementary Volume (REV).

- We have discussed how the size of an REV can be (conceptually) selected.

1.2   Lecture Two: Microscopic Balance Equations

In the previous lecture, we have introduced the continuum approach to modelling transport in porous media. The core of a transport model consists of a balance equation of the considered extensive quantity. In this lecture, we shall start by presenting the general microscopic balance equation for any extensive quantity, followed by a number of examples of balance equations for particular extensive quantities. Then, we shall present a number of so called *'averaging rules'* which will be employed in LECTURE 3 in order to transform microscopic models into macroscopic (continuum) ones.

In preparation for constructing the balance equation of an extensive quantity, E, we shall introduce the definitions of velocity and fluxes of E.

1.2.1   Velocity and flux

The velocity, $\mathbf{V}^E$, of an E-continuum particle is defined by

$$\mathbf{V}^E = \left. \frac{\partial \mathbf{x}^E}{\partial t} \right|_{\boldsymbol{\xi}^E = \text{const.}}, \tag{1.2.1}$$

where $\mathbf{x}^E$ is the position in space of a fixed E-particle and t is time. The relation expressed by (1.2.1) is a *Lagrangian* concept, as it follows what happens to a fixed particle as it moves through space and undergoes changes (e.g., in its density, or temperature). The symbol $\boldsymbol{\xi}^E$ represents the location of the considered E-particle at some initial time. This location serves as the *label* of the particle. As an example

$$\mathbf{V}^m = \left. \frac{\partial \mathbf{x}^m}{\partial t} \right|_{\boldsymbol{\xi}^m = \text{const.}}, \tag{1.2.2}$$

is the velocity of a mass particle, i.e., the velocity of a particle of a mass continuum.

Consider a phase that is composed of a number (say, N) of γ-components, $\gamma = 1,2,3,...,N$. Here a component is made up of identical molecules, or ions. The density of a γ-component (= mass of component per unit volume of phase) is denoted by ρ^γ, with $\sum_{(\gamma)} \rho^\gamma = \rho = (m/U) = $ the density of the phase. A number of weighted velocities may be defined for such a a multicomponent phase. The more useful ones for us are

(a) Mass weighted velocity, $\mathbf{V}^m$

$$\mathbf{V}^m = \frac{1}{m}\sum_{\gamma=1}^{N}\mathbf{V}^\gamma m^\gamma = \sum_{\gamma=1}^{N}\omega^\gamma\mathbf{V}^\gamma, \qquad (1.2.3)$$

where $\omega^\gamma = m^\gamma/m = \rho^\gamma/\rho$ is the mass fraction of the γ-component in the phase, with $\sum_{(\gamma)}\omega^\gamma = 1$. The mass weighted velocity is often called the *barycentric velocity*.

(b) Volume weighted velocity, $\mathbf{V}$

$$\begin{aligned}
\mathbf{V} &= \frac{1}{U}\sum_{\gamma=1}^{N}\mathbf{V}^\gamma U^\gamma = \sum_{\gamma=1}^{N}\frac{\partial U}{\partial m^\gamma}\frac{m^\gamma}{U}\mathbf{V}^\gamma \\
&= \sum_{\gamma=1}^{N}\frac{\partial U}{\partial m^\gamma}\rho^\gamma\mathbf{V}^\gamma, \qquad (1.2.4)
\end{aligned}$$

where $\partial U/\partial m^\gamma$ is the partial specific volume of the γ-component in the phase, and $U = \sum_{(\gamma)}(\partial U/\partial m^\gamma)\,m^\gamma$.

For a single component phase, i.e., a phase that is comprised of identical molecules, $N = 1$, and $\mathbf{V}^m = \mathbf{V}$.

The total flux of E, at point $\mathbf{x}$ and time t, with respect to a fixed coordinate system, i.e., the quantity of E passing through a fixed unit area normal to the velocity, $\mathbf{V}^E$, per unit time, is expressed by

$$\mathbf{j}^{tE}(\mathbf{x},t) = (e\mathbf{V}^E)\Big|_{(\mathbf{x},t)}. \qquad (1.2.5)$$

Examples are: $\mathbf{j}^{tm} = \rho\mathbf{V}^m$, $\mathbf{j}^{tM} = \rho\mathbf{V}^m\mathbf{V}^M$, for the fluxes of mass and momentum, respectively.

The total flux of E, $\mathbf{j}^{tE}$, can be rewritten in the form

$$\mathbf{j}^{tE} \equiv e\mathbf{V}^E = e\mathbf{V} + \mathbf{j}^{EU}, \qquad \mathbf{j}^{EU} \equiv e(\mathbf{V}^E - \mathbf{V}), \qquad (1.2.6)$$

or

$$\mathbf{j}^{tE} \equiv e\mathbf{V}^E = e\mathbf{V}^m + \mathbf{j}^{Em}, \qquad \mathbf{j}^{Em} = e(\mathbf{V}^E - \mathbf{V}^m), \qquad (1.2.7)$$

where $e\mathbf{V}$ (or $e\mathbf{V}^m$), called the *advective flux* of E, expresses the quantity of E carried by the volume weighted (or the mass weighted) velocity of the phase, with respect to fixed coordinates, and $\mathbf{j}^{EU}$ (or $\mathbf{j}^{Em}$) is the *diffusive flux* of E, relative to the advective one.

For example, for $E = m^\gamma$

$$\mathbf{j}^{tm^\gamma} = \rho^\gamma \mathbf{V}^m + \rho^\gamma(\mathbf{V}^{m^\gamma} - \mathbf{V}^m), \qquad (1.2.8)$$

or

$$\mathbf{j}^{tm^\gamma} = \rho^\gamma \mathbf{V} + \rho^\gamma(\mathbf{V}^{m^\gamma} - \mathbf{V}), \qquad (1.2.9)$$

where $\rho^\gamma(\mathbf{V}^{m^\gamma} - \mathbf{V}^m)$ denotes the *molecular diffusion of the γ-component, relative to the mass weighted velocity of the phase*, and $\rho^\gamma(\mathbf{V}^{m^\gamma} - \mathbf{V})$ denotes the *molecular diffusion of the γ-component, relative to the volume weighted velocity of the phase.* We note that

$$\sum_{(\gamma)} \rho^\gamma(\mathbf{V}^{m^\gamma} - \mathbf{V}^m) = 0. \qquad \sum_{(\gamma)} \rho^\gamma(\mathbf{V}^{m^\gamma} - \mathbf{V}) \neq 0.$$

1.2.2   The general balance equation

Let E denote an extensive quantity of a phase, (e.g., $E = m$ for mass, or $E = M$ for momentum, and let e denote its density, i.e.

$$e = \frac{dE}{dU},$$

where U denotes the volume of the phase. In what follows, we shall develop the microscopic balance equation of E in the vicinity of a point within a phase, by employing the *Eulerian approach*. According to this approach, we focus our attention not on a fixed particle, as in the *Lagrangian approach*, but on a fixed point within the considered domain, and examine changes (e.g., in density of a fluid) that take place at that point. We recall that in order to visualize E of a phase as a continuum, we had to average the behavior of molecules in the vicinity of every point, over some Representative Elementary Volume centered at the point. This means that when we speak about the 'behavior at a point' in a phase continuum, we refer to the 'averaged behavior in the vicinity of that point'.

Consider a quantity E within a domain $\mathcal{U}$ of volume U, bounded by a surface $\mathcal{S}$ of area S (Fig. 1.2.1). The balance of E within $\mathcal{U}$ can be expressed, verbally, by

$$\left\{ \begin{array}{c} \text{rate of} \\ \text{accumulation of} \\ E \text{ within } \mathcal{U} \end{array} \right\} = \left\{ \begin{array}{c} \text{Net influx of} \\ E \text{ into } \mathcal{U} \\ \text{through } \mathcal{S} \end{array} \right\} + \left\{ \begin{array}{c} \text{Net rate of} \\ \text{production of} \\ E \text{ within } \mathcal{U} \end{array} \right\}.$$

$$\quad\quad (a) \qquad\qquad\qquad (b) \qquad\qquad\qquad (c)$$

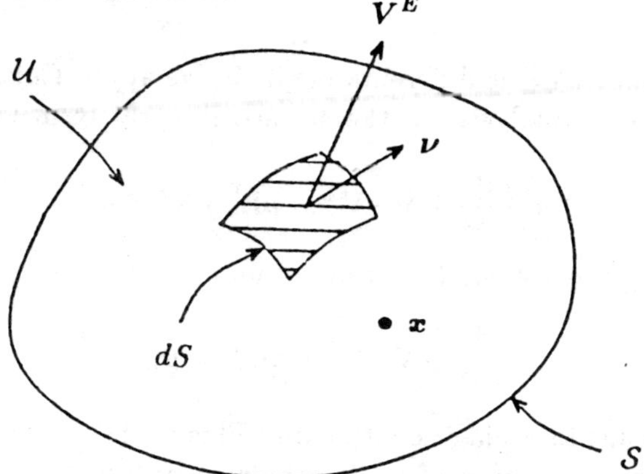

Figure 1.2.1: A control volume $\mathcal{U}$ bounded by a surface S.

Let us consider these balance components in detail.

(a) The rate of accumulation of the amount of E within $\mathcal{U}$, is expressed by

$$\frac{\partial}{\partial t}\int_{\mathcal{U}} e\, dU \left(=\int_{\mathcal{U}}\frac{\partial e}{\partial t}\, dU\right),$$

where the exchange of integration and differentiation is permitted in view of the fact that the boundary of the domain, $\mathcal{U}$, is fixed.

(b) The net influx (= total influx minus total efflux) of E into $\mathcal{U}$, through S, is given by

$$-\int_{S} e\mathbf{V}^E\cdot\boldsymbol{\nu}\, dS,$$

where $\boldsymbol{\nu}$ is the outward normal unit vector on the elemental area dS.

(c) The net rate of production of E from sources within $\mathcal{U}$, is expressed by

$$\int_{\mathcal{U}}\rho\Gamma^E\, dU,$$

where ρ is the mass density of the phase and Γ^E denotes the rate of internal production of E, per unit mass of the phase.

Altogether, the balance of E in $\mathcal{U}$ takes the form

$$\int_{\mathcal{U}}\frac{\partial e}{\partial t}\, dU = -\int_{S} e\mathbf{V}^E\cdot\boldsymbol{\nu}\, dS + \int_{\mathcal{U}}\rho\Gamma^E\, dU. \tag{1.2.10}$$

Assuming that $e\mathbf{V}^E$ is differentiable in $\mathcal{U}$, we apply *Gauss theorem* (see any text on vector analysis) to the surface integral term in (1.2.10), and obtain

$$\int_{\mathcal{U}} \left(\frac{\partial e}{\partial t} + \nabla \cdot e\mathbf{V}^E - \rho \Gamma^E \right) dU = 0.$$

By shrinking the volume U to zero around an arbitrary point, we obtain

$$\frac{\partial e}{\partial t} + \nabla \cdot e\mathbf{V}^E - \rho \Gamma^E = 0, \tag{1.2.11}$$

valid for any point in a phase continuum. This is the *Eulerian microscopic differential balance equation of any extensive quantity, E.*

It may be of interest to have a physical interpretation of the term $\nabla \cdot e\mathbf{V}^E$ appearing in (1.2.11).

Consider a parallelpiped control volume $U(= \Delta x \Delta y \Delta z)$ (Fig. 1.2.2), and a total flux, $\mathbf{j}^{tE}$, that is differentiable at all points of the considered domain. The net rate of influx into $\mathcal{U}$, per unit volume, as this volume shrinks to zero, is given by

$$\lim_{U \to 0} \frac{1}{U} \int_S \mathbf{j}^{tE} \cdot \boldsymbol{\nu} \, dS$$

$$= \lim_{\Delta x, \Delta y, \Delta z \to 0} \frac{1}{\Delta x \Delta y \Delta z} \left\{ \left(j_x^{tE} \big|_{x - \frac{\Delta x}{2}, y, z} - j_x^{tE} \big|_{x + \frac{\Delta x}{2}, y, z} \right) \Delta y \Delta z \right.$$

$$+ \left(j_y^{tE} \big|_{x, y - \frac{\Delta y}{2}, z} - j_y^{tE} \big|_{x, y + \frac{\Delta y}{2}, z} \right) \Delta x \Delta z$$

$$\left. + \left(j_z^{tE} \big|_{x, y, z - \frac{\Delta z}{2}} - j_z^{tE} \big|_{x, y, z + \frac{\Delta z}{2}} \right) \Delta x \Delta y \right\}$$

$$= \lim_{\Delta x, \Delta y, \Delta z \to 0} \left\{ \frac{j_x^{tE} \big|_{x - \frac{\Delta x}{2}, y, z} - j_x^{tE} \big|_{x + \frac{\Delta x}{2}, y, z}}{\Delta x} \right.$$

$$\left. + \frac{j_y^{tE} \big|_{x, y - \frac{\Delta y}{2}, z} - j_y^{tE} \big|_{x, y + \frac{\Delta y}{2}, z}}{\Delta y} + \frac{j_z^{tE} \big|_{x, y, z - \frac{\Delta z}{2}} - j_z^{tE} \big|_{x, y, z + \frac{\Delta z}{2}}}{\Delta z} \right\}$$

$$= -\left(\frac{\partial j_x^{tE}}{\partial x} + \frac{\partial j_y^{tE}}{\partial y} + \frac{\partial j_z^{tE}}{\partial z} \right).$$

Hence

$$\operatorname{div} \mathbf{j}^{tE} (\equiv \nabla \cdot \mathbf{j}^{tE}) = \frac{\partial j_x^{tE}}{\partial x} + \frac{\partial j_y^{tE}}{\partial y} + \frac{\partial j_z^{tE}}{\partial z}. \tag{1.2.12}$$

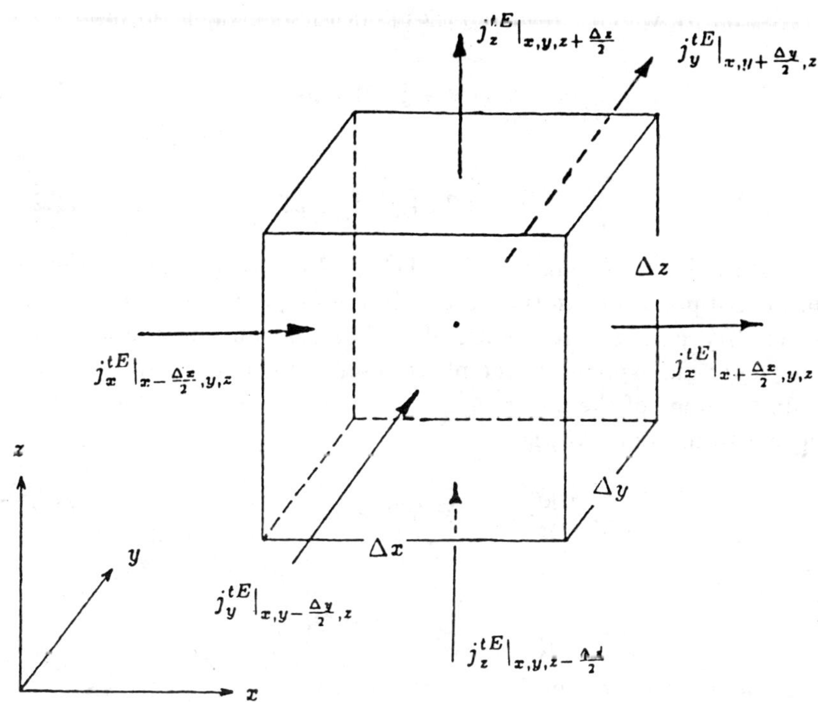

Figure 1.2.2: A control box, in a rectangular cartesian coordinate system, for deriving an expression for the divergence of a flux.

Thus, the physical interpretation of the *divergence of a flux of an exten-sive quantity* at a point is the *excess of outflow over inflow (of that quantity), per unit volume around the point and per unit time.* This interpretation should enable us to easily construct balance equations for any extensive quantity.

In the absence of sources and sinks, Gauss theorem

$$\int_U \nabla\cdot\mathbf{j}\, dU = \int_S \mathbf{j}\cdot\boldsymbol{\nu}\, dS, \tag{1.2.13}$$

may also lead to the same interpretation of $\nabla\cdot\mathbf{j}$, viz.

$$\nabla\cdot\mathbf{j} = \lim_{U\to 0}\frac{1}{U}\int_U \nabla\cdot\mathbf{j}\, dU = \lim_{U\to 0}\int_S \mathbf{j}\cdot\boldsymbol{\nu}\, dS. \tag{1.2.14}$$

Following the decomposition of the total flux into advective and diffusive fluxes (see above), we may rewrite (1.2.11) in the equivalent forms

$$\frac{\partial e}{\partial t} = -\nabla \cdot (e\mathbf{V} + \mathbf{j}^{EU}) + \rho\Gamma^E, \qquad (1.2.15)$$

and

$$\frac{\partial e}{\partial t} = -\nabla \cdot (e\mathbf{V}^m + \mathbf{j}^{Em}) + \rho\Gamma^E. \qquad (1.2.16)$$

Each of the balance equations (1.2.11), (1.2.15), or (1.2.16), states that at a considered point, i.e., in the close vicinity of a point, the rate of increase in the quantity E, per unit volume of a phase, is equal to the net rate of influx of E per unit volume of the phase, plus the net rate of production of E, per unit volume of the phase.

Another form of (1.2.15) is

$$\frac{D_E e}{Dt} = -e\nabla \cdot \mathbf{V}^E + \rho\Gamma^E, \qquad (1.2.17)$$

with

$$\frac{D_E(..)}{Dt} \equiv \frac{\partial(..)}{\partial t} + \mathbf{V}^E \cdot \nabla(..)$$

denoting the material, or total derivative of $(..)$.

While (1.2.15) is an Eulerian form of the balance equation, (1.2.17) is written as a Lagrangian equation. Its l.h.s. represents the rate of change in e as a particle moves, while its r.h.s., which is an Eulerian expression, represents the sources that cause this change. They include the effects of dilatation and production.

1.2.3   Particular balance equations

(a)   Mass of a phase

Here, $E = m$, $e = \rho$, $\mathbf{V}^E = \mathbf{V}^m$, $\mathbf{j}^{mm} = 0$. Assuming that the total mass of a phase cannot be created, $\Gamma^m = 0$. Hence, we obtain from (1.2.16)

$$\frac{\partial \rho}{\partial t} = -\nabla \cdot \rho\mathbf{V}^m, \qquad (1.2.18)$$

and from (1.2.15)

$$\frac{\partial \rho}{\partial t} = -\nabla \cdot (\rho\mathbf{V} + \mathbf{j}^{mU}), \qquad (1.2.19)$$

where $\mathbf{j}^{mU} = \rho(\mathbf{V}^m - \mathbf{V})$ is the diffusive mass flux (= flux due to molecular diffusion) of the mass of the phase.

Equation (1.2.18) is the *differential balance equation for the mass of a phase*.

With (1.2.17), equation (1.2.18) can be rewritten in the form

$$\frac{1}{\rho}\frac{\mathrm{D}_m\rho}{\mathrm{D}t} = -\nabla\cdot\mathbf{V}^m. \tag{1.2.20}$$

This equation says that as the particle moves, its density, ρ, changes due to its expansion. When ρ does not vary as the particle moves, i.e., $\mathrm{D}_m\rho/\mathrm{D}t = 0$, equation (1.2.20) reduces to $\nabla\cdot\mathbf{V}^m = 0$. We have here an interpretation of the material derivative of ρ, since

$$\nabla\cdot\mathbf{V}^m = -\frac{1}{\rho}\frac{\mathrm{D}_m\rho}{\mathrm{D}t} = \frac{1}{v}\frac{\mathrm{D}_m v}{\mathrm{D}t}, \tag{1.2.21}$$

where $v(= 1/\rho)$ is the *specific volume* (of the mass continuum). Accordingly, $\nabla\cdot\mathbf{V}^m$ gives the rate of *dilatation* of a conservative mass.

(b) Mass of a γ-component of a phase

Now, $E = m^\gamma$, $e = \rho^\gamma \equiv c^\gamma$ is the concentration (= mass per unit volume) of the γ-component in a fluid phase. We obtain from (1.2.15)

$$\frac{\partial c^\gamma}{\partial t} = -\nabla\cdot(c^\gamma\mathbf{V} + \mathbf{j}^{m\gamma U}) + \rho\Gamma^{m\gamma}, \tag{1.2.22}$$

or, from (1.2.16)

$$\frac{\partial c^\gamma}{\partial t} = -\nabla\cdot(c^\gamma\mathbf{V}^m + \mathbf{j}^{m\gamma m}) + \rho\Gamma^{m\gamma}. \tag{1.2.23}$$

Equation (1.2.22) is the *differential balance equation for the component of a phase*.

By summing (1.2.23) over all γ-components, we obtain

$$\frac{\partial \sum_{(\gamma)} c^\gamma}{\partial t} = -\nabla\cdot\left(\mathbf{V}^m\sum_{(\gamma)} c^\gamma + \sum_{(\gamma)}\mathbf{j}^{m\gamma m}\right) + \rho\sum_{(\gamma)}\Gamma^{m\gamma}, \tag{1.2.24}$$

which reduces to the mass balance equation (1.2.18), since $\sum_{(\gamma)} c^\gamma = \rho$, the sum of diffusive fluxes, with respect to the mass weighted velocity, vanishes, i.e., $\sum_{(\gamma)} \mathbf{j}^{m^\gamma m} = \sum_{(\gamma)} c^\gamma(\mathbf{V}^{m^\gamma} - \mathbf{V}^m) = 0$, and the total mass production vanishes, i.e. $\sum_{(\gamma)} \Gamma^{m^\gamma} = \Gamma^m = 0$.

The component balance equation may also be written in the forms

$$\frac{D_m c^\gamma}{Dt} = -c^\gamma \nabla \cdot \mathbf{V}^m - \nabla \cdot \mathbf{j}^{m^\gamma m} + \rho \Gamma^{m^\gamma}, \qquad (1.2.25)$$

or, with $D_m \rho / Dt = 0$

$$\rho \frac{D_m(c^\gamma/\rho)}{Dt} = -\nabla \cdot \mathbf{j}^{m^\gamma m} + \rho \Gamma^{m^\gamma}. \qquad (1.2.26)$$

For an isochoric motion of a phase, i.e., a motion for which $\nabla \cdot \mathbf{V}^m \equiv 0$, equation (1.2.25) reduces to

$$\frac{D_m c^\gamma}{Dt} = -\nabla \cdot \mathbf{j}^{m^\gamma m} + \rho \Gamma^{m^\gamma}. \qquad (1.2.27)$$

The last three equations are written in a mixed Eulerian–Lagrangian form.

(c) Momentum of a phase

In this case, $E = \mathbf{M}$, $e = \rho \mathbf{V}^m$, $\Gamma^E = \mathbf{F} =$ the resultant body force per unit mass of the phase, $\mathbf{j}^{Em} = \mathbf{j}^{Mm} = \rho \mathbf{V}^m(\mathbf{V}^M - \mathbf{V}^m) = -\sigma$, where σ is the stress tensor, representing the *diffusive flux of momentum* carried by molecules of the phase, with respect to the mass weighted velocity. For this case, (1.2.16) becomes

$$\frac{\partial \rho \mathbf{V}^m}{\partial t} = -\nabla \cdot (\rho \mathbf{V}^m \mathbf{V}^m - \sigma) + \rho \mathbf{F}, \qquad (1.2.28)$$

where $\mathbf{V}^m \mathbf{V}^m$ is the *dyadic product*, $V_i^m V_j^m$, of the two vectors. In (1.2.28)

$$
\begin{aligned}
\partial \rho \mathbf{V}^m / \partial t \ &= \ \text{rate of accumulation of momentum,} \\
-\nabla \cdot \rho \mathbf{V}^m \mathbf{V}^m \ &= \ \text{rate of momentum gained by advection,} \\
+\nabla \cdot \sigma \ &= \ \text{rate of momentum gained by the diffusive flux} \\
&\qquad \text{of momentum, and} \\
\rho \mathbf{F} \ &= \ \text{rate of supply of momentum to the phase by} \\
&\qquad \text{the body force.}
\end{aligned}
$$

and all terms are per unit volume of the phase.

Equation (1.2.28) is the *differential balance equation for the linear momentum of a phase.*

When combined with the mass balance equation (1.2.18), equation (1.2.28) becomes

$$\rho\frac{\partial \mathbf{V}^m}{\partial t} = -\rho\mathbf{V}^m\cdot\nabla\mathbf{V}^m + \rho\mathbf{F} + \nabla\cdot\boldsymbol{\sigma}, \qquad (1.2.29)$$

or, in indicial notation, making use, as everywhere else in these lectures, of *Einstein's summation convention*

$$\rho\frac{\partial V_j^m}{\partial t} = \frac{\partial \sigma_{ij}}{\partial x_i} - \rho V_i^m\frac{\partial V_j^m}{\partial x_i} + \rho F_j. \qquad (1.2.30)$$

Finally, we may rewrite the momentum balance equation in the form

$$\rho\frac{D_m\mathbf{V}^m}{Dt} = \nabla\cdot\boldsymbol{\sigma} + \rho\mathbf{F}, \qquad (1.2.31)$$

also known as the *motion equation*.

(d) Energy of a phase.

Here, the energy density is given by the sum $\rho I + \frac{1}{2}\rho(V^m)^2$, where I is the specific *internal energy* (i.e., internal energy per unit mass of the phase), due to the thermal agitation and short range intermolecular forces, and $\frac{1}{2}\rho(V^m)^2$ is the *kinetic energy* (per unit volume of the phase).

Energy is supplied to the phase contained in a spatial domain, $\mathcal{U}$, through its surface, $\mathcal{S}$, by advection through $\mathcal{S}$, expressed by

$$-\int_{\mathcal{S}}\{\rho I + \tfrac{1}{2}\rho(V^m)^2\}\mathbf{V}^m\cdot\boldsymbol{\nu}\,dS$$

and by the (diffusive) flux of heat, $\mathbf{j}^H$, through $\mathcal{S}$, expressed by

$$\int_{\mathcal{S}}\mathbf{j}^H\,\boldsymbol{\nu}\,dS,$$

in which the minus sign results from the definition of $\boldsymbol{\nu}$ as the outward normal unit vector on $\mathcal{S}$.

Transforming the sum of the last two integrals into a volume one by employing the *divergence theorem*, we obtain

$$-\int_{\mathcal{U}}\nabla\cdot\{\rho(I + \tfrac{1}{2}(V^m)^2)\mathbf{V}^m + \mathbf{j}^H\}\,dU.$$

The rate of production of energy within $\mathcal{U}$, is expressed by

$$\int_{\mathcal{U}}\rho\Gamma^H\,dU,$$

where Γ^H is the rate of heat produced within $\mathcal{U}$ per unit mass, e.g., by chemical reactions.

Finally, energy is added to the domain $\mathcal{U}$ by the work of the forces acting on the phase contained in $\mathcal{U}$. These include:

(a) $\int_{\mathcal{U}} \mathbf{V}^m \cdot \rho \mathbf{F} \, dU$, where $\mathbf{F}$ represents body force per unit mass, expressing the rate of supply of kinetic energy by the body force acting on the phase contained in $\mathcal{U}$, and

(b) $- \int_{S} \mathbf{V}^m \cdot (-\boldsymbol{\sigma}) \cdot \boldsymbol{\nu} \, dS$, expressing the rate of work done by the surface force acting on the surface S of $\mathcal{U}$. Employing the divergence theorem, this term can be replaced by $\int_{\mathcal{U}} \nabla \cdot (\boldsymbol{\sigma} \cdot \mathbf{V}^m) \, dU$.

By combining all the above terms, dividing by U and passing to the limit as $U \to 0$, we obtain the differential *energy balance equation* in the form

$$\frac{\partial}{\partial t} \left\{ \rho \left(I + \tfrac{1}{2}(V^m)^2 \right) \right\}$$

$$= -\nabla \cdot \left\{ \rho \left(I + \frac{(V^m)^2}{2} \right) \mathbf{V}^m + \mathbf{j}^H \right\}$$

$$+ \{ \mathbf{V}^m \cdot \rho \mathbf{F} + \nabla \cdot (\boldsymbol{\sigma} \cdot \mathbf{V}^m) + \rho \Gamma^H \}. \tag{1.2.32}$$

Written in this form, we recognize the similarity between (1.2.32) and the general (microscopic) balance equation (1.2.16).

By combining (1.2.32) with the mass balance equation (1.2.16), and the momentum balance equation (1.2.31), we obtain

$$\rho \frac{\mathrm{D}_m I}{\mathrm{D} t} = -\nabla \cdot \mathbf{j}^H + \boldsymbol{\sigma} : \nabla \mathbf{V}^m + \rho \Gamma^H, \tag{1.2.33}$$

where all terms are per unit volume of the phase. In (1.2.33)

$$
\begin{aligned}
\rho \mathrm{D}_m I / \mathrm{D} t \;&=\; \text{material rate of growth of the internal energy,}\\
\nabla \cdot \mathbf{j}^H \;&=\; \text{net influx of internal energy by heat conduction,}\\
\boldsymbol{\sigma} : \nabla \mathbf{V}^m \;&=\; \text{rate of increase of energy by work done}\\
&\qquad \text{in producing strain, and}\\
\rho \Gamma^H \;&=\; \text{rate of increase of internal energy}\\
&\qquad \text{from internal sources.}
\end{aligned}
$$

In a fluid continuum, we replace the stress, $\boldsymbol{\sigma}$, by

$$\boldsymbol{\sigma} = \boldsymbol{\tau} - p\mathbf{I}, \tag{1.2.34}$$

where $\boldsymbol{\tau}$ (like $\boldsymbol{\sigma}$, positive for tension) is the *deviator stress*, p is the pressure (positive for compression) and $\mathbf{I}$ is the unit tensor.

With (1.2.34), equation (1.2.33) can be rewritten in the form

$$\rho \frac{D_m I}{Dt} = -\nabla \cdot \mathbf{j}^H + \boldsymbol{\tau} : \nabla \mathbf{V}^m - p \nabla \cdot \mathbf{V}^m + \rho \Gamma^H. \tag{1.2.35}$$

In this equation

$$
\begin{aligned}
\boldsymbol{\tau} : \nabla \mathbf{V}^m \;\; &= \;\; \text{irreversible rate of internal energy gain by shear,} \\
-p \nabla \cdot \mathbf{V}^m \;\; &= \;\; \text{reversible rate of internal energy gain by compression.}
\end{aligned}
$$

In order to rewrite the internal energy balance equation, say, (1.2.35) in terms of the absolute temperature, T, and the *heat capacity*, we note that according to a basic postulate of thermodynamics, the *specific internal energy* of a phase, I, is a single valued function of a set of specific values (per unit of mass) of extensive state variables, e.g.,

$$I = I(s, v, \omega^\gamma), \tag{1.2.36}$$

where s is the *specific entropy*, $v(= 1/\rho)$ is the specific volume, and $\omega^\gamma(= \rho^\gamma/\rho)$ is the mass fraction of a γ-component, $\gamma = 1, 2, \ldots$, all of a phase. Hence

$$dI = Tds - pdv + \sum_{(\gamma)} \mu^\gamma d\omega^\gamma, \tag{1.2.37}$$

where $\partial I/\partial s|_{v,\omega^\gamma} = T$, $\partial I/\partial v|_{s,\omega^\gamma} = -p$ and $\partial I/\partial \omega^\gamma|_{s,v,\omega^\delta, \delta \neq \gamma} = \mu^\gamma$ is the *chemical potential* of the γ-component.

From (1.2.37) it follows that

$$T = T(s, v, \omega^\gamma) \quad \text{or} \quad s = s(T, v, \omega^\gamma) \qquad \gamma = 1, 2, \ldots$$

Hence, neglecting changes in component concentration, we have

$$ds = \left. \frac{\partial s}{\partial T} \right|_{v,\omega^\gamma} dT + \left. \frac{\partial s}{\partial v} \right|_{T,\omega^\gamma} dv.$$

Since (see any text on reversible thermodynamics)

$$\left. \frac{\partial s}{\partial v} \right|_{T,\omega^\gamma} = \left. \frac{\partial p}{\partial T} \right|_{v,\omega^\gamma}, \quad C_V = \left. \frac{\partial I}{\partial T} \right|_{v,\omega^\gamma} = T \left. \frac{\partial s}{\partial T} \right|_{v,\omega^\gamma},$$

where C_V is the *specific heat* of the considered phase per unit mass at constant volume, we obtain

$$dI = \left(T \left. \frac{\partial p}{\partial T} \right|_{v,\omega^\gamma} - p \right) dv + C_V dT + \sum_{(\gamma)} \mu^\gamma d\omega^\gamma. \tag{1.2.38}$$

Hence, the l.h.s. of (1.2.35) becomes

$$\rho\frac{D_m I}{Dt} = \rho\left(T\frac{\partial p}{\partial T}\Big|_{v,\omega^\gamma} - p\right)\frac{D_m v}{Dt} + \rho C_V\frac{D_m T}{Dt} + \rho\sum_{(\gamma)}\mu^\gamma\frac{D_m \omega^\gamma}{Dt}. \qquad (1.2.39)$$

Since

$$\frac{1}{v}\frac{D_m v}{Dt} \equiv \rho\frac{D_m 1/\rho}{Dt} = -\frac{1}{\rho}\frac{D_m \rho}{Dt} = \nabla\cdot\mathbf{V}^m,$$

$$\rho C_V\frac{D_m T}{Dt} = \boldsymbol{\tau}:\nabla\mathbf{V}^m - \nabla\cdot\mathbf{j}^H - T\frac{\partial p}{\partial T}\Big|_{v,\omega^\gamma}\nabla\cdot\mathbf{V}^m - \rho\sum_{(\gamma)}\mu^\gamma\frac{D_m \omega^\gamma}{Dt} + \rho\Gamma^H.$$

$$(1.2.40)$$

This is the (microscopic) *heat balance equation.*

It is also possible to write the energy balance equation in terms of the *enthalpy*, h, defined by

$$h = I + pv \qquad (1.2.41)$$

The energy balance equation, then takes the form

$$\rho\frac{D_m h}{Dt} = -\nabla\cdot\mathbf{j}^H + \boldsymbol{\tau}:\nabla\mathbf{V}^m + \frac{D_m p}{Dt} + \rho\Gamma^H, \qquad (1.2.42)$$

or, when combined with the mass balance equation (1.2.18)

$$\frac{\partial\rho h}{\partial t} = -\nabla\cdot(\rho h\mathbf{V}^m + \mathbf{j}^H) + \boldsymbol{\tau}:\nabla\mathbf{V}^m + \frac{D_m p}{Dt} + \rho\Gamma^H. \qquad (1.2.43)$$

The use of the h as the dependent variable is especially convenient when dealing with fluids that undergo phase change.

Very often, since $|\boldsymbol{\tau}:\nabla\mathbf{V}^m| \ll |\nabla\cdot\mathbf{j}^H|$, the term $\boldsymbol{\tau}:\nabla\mathbf{V}^m$ is dropped. Under such conditions, another form of (1.2.40) is

$$\frac{\partial\rho C_V T}{\partial t} = -\nabla\cdot(\rho C_V T\mathbf{V}^m + \mathbf{j}^H) - T\frac{\partial p}{\partial T}\Big|_v\nabla\cdot\mathbf{V}^m - \rho\sum_{(\gamma)}\mu^\gamma\frac{D_m \omega^\gamma}{Dt}$$

$$+\rho\Gamma^H. - T\rho\frac{D_m C_V}{Dt}. \qquad (1.2.44)$$

For $C_V = $ const., the last term on the r.h.s. of (1.2.44) vanishes.

For an isotropic thermoelastic solid, the energy balance equation (1.2.33) becomes

$$\frac{\partial\rho_s C_s T_s}{\partial t} = -\nabla\cdot\left(\rho_s C_s\mathbf{V}_s T_s\mathbf{j}_s^H\right) - T_s\eta\frac{\partial\varepsilon_s}{\partial t}, \qquad (1.2.45)$$

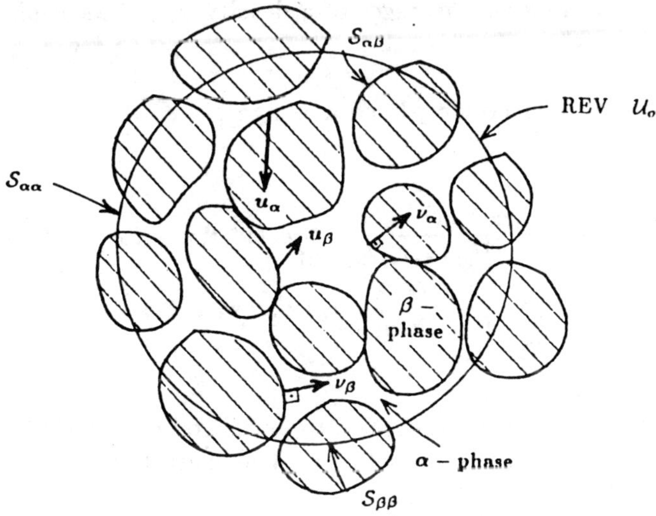

Figure 1.2.3: Nomenclature for averaging over an REV.

where η is a coefficient defined by $\eta\delta_{ij} = -(\partial\sigma_{sij}/\partial T_s)|_{\varepsilon_s}$, ε_s is the strain in the solid, and we have introduced subscript s to indicate solid phase values.

We have now the microscopic differential balance equations for mass, mass of a component, momentum and energy.

In order to average these equations, with the objective of deriving the corresponding macroscopic balance equations, we have to introduce laws that will express the average of spatial and temporal derivatives, in term of the derivatives of the averaged values. This topic is next on our agenda.

1.2.4 Averaging rules

We start by taking a look at the nomenclature shown on Fig. 1.2.3. In this figure, $\mathcal{U}_o$ is the REV domain of volume U_o, $\mathcal{S}_o$ is the surface that bounds the REV (of area S_o), α is a considered phase, β indicates all other phases within $\mathcal{U}_o$, $\mathbf{x}$ is the center of $\mathcal{U}_o$, $\mathbf{x}'$ is the position of a point within $\mathcal{U}_{o\alpha}$, which is the domain occupied by the α-phase within $\mathcal{U}_o$, $\mathcal{S}_{\alpha\beta}$ is the surface separating the α-phase from all other phases within $\mathcal{U}_o$, ν_α is the outward (from $\mathcal{U}_{o\alpha}$) normal unit vector on $\mathcal{S}_{\alpha\beta}$, $\mathbf{u}$ is the velocity of $\mathcal{S}_{\alpha\beta}$, and $\mathcal{S}_{\alpha\alpha}$ is the α-α portion of $\mathcal{S}_o$.

A number of different averages of a density, e_α, of an α-phase, can be defined.

(Volumetric) intrinsic phase average of e_α.

$$
\begin{aligned}
\overline{e_\alpha}^\alpha(\mathbf{x}, t) &= \frac{1}{U_{o\alpha}(\mathbf{x}, t)} \int_{\mathcal{U}_o(\mathbf{x},t)} e_\alpha(\mathbf{x}', t; \mathbf{x}) \gamma_\alpha(\mathbf{x}', t)\, dU(\mathbf{x}') \\
&\equiv \frac{1}{U_{o\alpha}} \int_{\mathcal{U}_{o\alpha}(\mathbf{x},t)} e_\alpha(\mathbf{x}', t; \mathbf{x})\, dU_\alpha(\mathbf{x}'),
\end{aligned}
\qquad (1.2.46)
$$

where $dU_\alpha = \gamma_\alpha dU$.

The $\mathbf{x}$ in the argument of e_α, is introduced to indicate that we consider points $\mathbf{x}'$ belonging to $\mathcal{U}_o$ centered at $\mathbf{x}$. The characteristic function, γ_x, is defined by

$$
\gamma_\alpha(\mathbf{x}') = \begin{cases} 1 & \text{if } \mathbf{x}' \text{ is in } \mathcal{U}_{o\alpha}, \\ 0 & \text{if } \mathbf{x}' \text{ is in } \mathcal{U}_{o\beta}. \end{cases}
\qquad (1.2.47)
$$

(Volumetric) phase average of e_α.

$$
\begin{aligned}
\overline{e_\alpha}(\mathbf{x}, t) &= \frac{1}{U_o} \int_{\mathcal{U}_o(\mathbf{x})} e_\alpha(\mathbf{x}', t; \mathbf{x}) \gamma_\alpha(\mathbf{x}', t; \mathbf{x})\, dU(\mathbf{x}') \\
&= \frac{1}{U_o} \int_{\mathcal{U}_{o\alpha}(\mathbf{x},t)} e_\alpha(\mathbf{x}', t; \mathbf{x})\, dU_\alpha(\mathbf{x}').
\end{aligned}
\qquad (1.2.48)
$$

Here, the total amount of E of the α-phase is averaged over the *entire* domain, $\mathcal{U}$, of the REV.

The relation between the two averages is given by

$$
\overline{e_\alpha} = \theta_\alpha \overline{e_\alpha}^\alpha,
\qquad (1.2.49)
$$

where θ_α is the volumetric fraction of the α-phase within the REV.

The selection of the kind of average to be used in each case depends on the measured quantity.

The adjective 'volumetric' was added to the names of the averages to indicate that the averages are taken over the volume of the REV. It is also possible to define areal averages in a similar way (Bachmat and Bear, 1986).

When the considered density, e, exists within all the phases present in $\mathcal{U}_o$, we may define a *volume average* of e

$$
\begin{aligned}
\bar{e}(\mathbf{x},t) &= \frac{1}{U_o(\mathbf{x},t)} \int_{\mathcal{U}_o} e \, dU \\
&\equiv \frac{1}{U_o} \sum_{(\alpha)} \int_{\mathcal{U}_{o\alpha}} e_\alpha \, dU_\alpha = \sum_{(\alpha)} \overline{e_\alpha}.
\end{aligned}
\tag{1.2.50}
$$

A *deviation*, $\mathring{e}_\alpha(\mathbf{x}',t;\mathbf{x})$, of $e_\alpha(\mathbf{x}',t;\mathbf{x})$ from the average, $\overline{e_\alpha}^\alpha(\mathbf{x},t)$, is defined by

$$
\mathring{e}_\alpha(\mathbf{x}',t;\mathbf{x}) = e_\alpha(\mathbf{x}',t;\mathbf{x}) - \overline{e_\alpha}^\alpha(\mathbf{x},t),
\tag{1.2.51}
$$

with

$$
\overline{\mathring{e}_\alpha}^\alpha = 0.
\tag{1.2.52}
$$

With the above definitions, we may now present a number of averaging rules (for proofs, see Bachmat and Bear, 1986 and Bear and Bachmat, 1990).

(a) Average of a sum

For the average of the sum of two quantities, $G_{1\alpha}$ and $G_{2\alpha}$ of an α-phase, we have

$$
\overline{G_{1\alpha} + G_{2\alpha}}^\alpha = \overline{G_{1\alpha}}^\alpha + \overline{G_{2\alpha}}^\alpha.
\tag{1.2.53}
$$

(b) Average of a product

The average of the product of the same two quantities, is given by

$$
\overline{G_{1\alpha}G_{2\alpha}}^\alpha = \overline{G_{1\alpha}}^\alpha \overline{G_{2\alpha}}^\alpha + \overline{\mathring{G}_{1\alpha}\mathring{G}_{2\alpha}}^\alpha.
\tag{1.2.54}
$$

We note that the average of a product *is not* equal to the product of the averages. A second term is added that is related to the variations of $G_{1\alpha}$ and $G_{2\alpha}$ within $\mathcal{U}_{o\alpha}$. In LECTURE 8, we shall interpret this term as a *dispersive flux*, when $G_{1\alpha}$ is a velocity, and $G_{2\alpha}$ is an intensive quantity.

(c) Average of a time derivative

The rule is

$$
\overline{\frac{\partial G_\alpha}{\partial t}} = \frac{\partial \overline{G_\alpha}}{\partial t} + \overline{G_\alpha \mathbf{u} \cdot \boldsymbol{\nu}}^{\alpha\beta} \Sigma_{\alpha\beta},
\tag{1.2.55}
$$

in which

$$\overline{G_\alpha \mathbf{u}\cdot\boldsymbol{\nu}}^{\alpha\beta}\, \Sigma_{\alpha\beta} \equiv \frac{1}{U_o} \int_{\mathcal{S}_{\alpha\beta}} G_\alpha \mathbf{u}\cdot\boldsymbol{\nu}\, dS, \qquad \overline{(\ \ )}^{\alpha\beta} \equiv \frac{1}{S_{\alpha\beta}} \int_{\mathcal{S}_{\alpha\beta}} (\ \ )\, dS,$$

and $\Sigma_{\alpha\beta}$ is the specific area of the $\mathcal{S}_{\alpha\beta}$-surface, and $\mathbf{u}$ is the velocity of this surface.

As a useful example, let E be the volume, U, of a phase. Then $G_\alpha \equiv e = 1$, $\overline{e}^\alpha = 1$ and $\overline{e} = \theta$. From (1.2.55), we obtain

$$\frac{\partial \theta}{\partial t} = \frac{1}{U_o} \int_{\mathcal{S}_{\alpha\beta}} \mathbf{u}\cdot\boldsymbol{\nu}\, dS \equiv \overline{u_\nu}^{\alpha\beta}\, \Sigma_{\alpha\beta}, \tag{1.2.56}$$

where u_ν $(\equiv \mathbf{u}\cdot\boldsymbol{\nu})$ is the speed of displacement of the $\mathcal{S}_{\alpha\beta}$-surface.

(d) Average of a spatial derivative

The rule is

$$\overline{\frac{\partial G_{\alpha jkl...}}{\partial x_i}} = \frac{\partial}{\partial x_i}\overline{G_{\alpha jkl...}} + \overline{G_{\alpha jkl...}\nu_{\alpha i}}^{\alpha\beta}\, \Sigma_{\alpha\beta}, \tag{1.2.57}$$

where $G_{\alpha jkl..}$ denotes the $jkl...$ component of G_α, which can be a scalar, a vector, or a tensor of any rank.

Of special interest is the case $E = U$, i.e., $G = e = 1$, $\overline{G} = \theta$. Then (1.2.57) yields

$$\frac{\partial \theta}{\partial x_i} = \overline{-\cos(\boldsymbol{\nu}, 1\mathbf{x}_i)}^{\alpha\beta}\, \Sigma_{\alpha\beta}, \qquad \text{or} \qquad \nabla\theta = -\overline{\boldsymbol{\nu}}^{\alpha\beta}\Sigma_{\alpha\beta}. \tag{1.2.58}$$

Equations (1.2.55) and (1.2.57) may be regarded as extensions of *Leibnitz Rule* which relates the derivative of an integral to the integral of a derivative, taking into account the conditions on the boundaries of integration. We note that the use of (1.2.57), requires information on the detailed shape of $\mathcal{S}_{\alpha\beta}$ and on the value of $G_{\alpha jkl...}$ on this surface (i.e., first kind boundary condition). Obviously, this information is not available. However, sometimes, information is available on the component of the gradient of G_α normal to $\mathcal{S}_{\alpha\beta}$ (i.e., boundary condition of the second kind), as when a boundary is impervious to molecular diffusion and the normal component of the concentration gradient vanishes. This observation motivated Bachmat and Bear (1986) to develop a *modified averaging rule for a spatial derivative* in which the surface integral is on the normal derivative of G_α on $\mathcal{S}_{\alpha\beta}$. However, the modified rule is restricted to cases in which G_α is a scalar that satisfies the condition

$$\nabla^2 G_\alpha = 0 \qquad \text{in} \quad \mathcal{U}_{o\alpha}. \tag{1.2.59}$$

For example, (1.2.59) is valid whenever G_α attains no maximum, or minimum value within $\mathcal{U}_{o\alpha}$.

Under such condition, and with the approximation

$$\int_{S_{\alpha\alpha}} \frac{\partial G_\alpha}{\partial x_i} \mathring{x}_j \nu_i \, dS \;\simeq\; \left(\frac{1}{S_{\alpha\alpha}} \int_{S_{\alpha\alpha}} \frac{\partial G_\alpha}{\partial x_i} \, dS \right) \int_{S_{\alpha\alpha}} \mathring{x}_j \nu_i \, dS$$

$$\simeq\; \frac{\partial \overline{G_\alpha}^\alpha}{\partial x_i} \int_{S_{\alpha\alpha}} \mathring{x}_j \nu_i \, dS, \tag{1.2.60}$$

in which the approximation

$$\frac{1}{S_{\alpha\alpha}} \int_{S_{\alpha\alpha}} \frac{\partial G_\alpha}{\partial x_i} \, dS \simeq \frac{\partial \overline{G_\alpha}^\alpha}{\partial x_i} \tag{1.2.61}$$

has been introduced, the modified averaging rule for a spatial derivative of a scalar, G_α, that satisfies (1.2.59) through (1.2.61), takes the form

$$\overline{\frac{\partial G}{\partial x_j}}^\alpha = \frac{\partial \overline{G}^\alpha}{\partial x_i} T^*_{\alpha i j} + \frac{1}{U_{o\alpha}} \int_{S_{\alpha\beta}} \mathring{x}_j \frac{\partial G}{\partial x_i} \nu_i \, dS, \tag{1.2.62}$$

where

$$T^*_{\alpha i j} = \frac{1}{U_{o\alpha}} \int_{S_{\alpha\alpha}} \nu_i \mathring{x}_j \, dS, \tag{1.2.63}$$

in which $\mathring{x}$ is the distance of points on $S_{\alpha\beta}$ from the centroid of the REV. The subscript α is used in $T^*_{\alpha i j}$, to emphasize that this coefficient depends only on the configuration of the α-phase within $\mathcal{U}_o$.

Thus, this modified rule achieves the goal, albeit as an approximation, of expressing the average of a spatial derivative in terms of the derivative of the average, making use of information in the form of a boundary condition of the second kind on the $S_{\alpha\beta}$-surface.

The macroscopic coefficient, $\mathbf{T}^*_\alpha$, defined by (1.2.63), is a fundamental tensorial property of the configuration of the α-phase within the REV. As such, it depends only on the microscopic configuration of the $S_{\alpha\beta}$-surface that separates the (considered) α-phase from all other phases within the REV. It expresses the total static moment of the oriented elements of the $S_{\alpha\beta}$-surface, with respect to planes passing through the centroid of the REV, per unit volume of the α-phase within $\mathcal{U}_o$.

To obtain an estimate of the magnitude of the $T^*_{\alpha ij}$-components, consider a spherical REV of radius R. Then (Bachmat and Bear, 1986)

$$T^*_{\alpha ij} = \frac{3\theta^S_\alpha}{\theta_\alpha}\overline{\nu_{\alpha i}\nu_{\alpha j}}^{\alpha\alpha}, \tag{1.2.64}$$

where $\theta^S_\alpha (= S_{\alpha\alpha}/S_o)$ denotes the fraction of the $\alpha - -\alpha$-surface in $\mathcal{S}_o$, and

$$\overline{\nu_{\alpha i}\nu_{\alpha j}}^{\alpha\alpha} = \frac{1}{S_{\alpha\alpha}}\int_{S_{\alpha\alpha}} \nu_{\alpha i}\nu_{\alpha j}\, dS. \tag{1.2.65}$$

The term $\nu_{\alpha i}\nu_{\alpha j}$ is a symmetric second rank tensor, and so is its average, $\overline{\nu_{\alpha i}\nu_{\alpha j}}^{\alpha\alpha}$. It can, therefore, be expressed in the form

$$\overline{\nu_{\alpha i}\nu_{\alpha j}}^{\alpha\alpha} = a_1\delta_{1i}\delta_{1j} + a_2\delta_{2i}\delta_{2j} + a_3\delta_{3i}\delta_{3j}, \tag{1.2.66}$$

in which a_1, a_2 and a_3 are the *principal values* of $\overline{\nu_{\alpha i}\nu_{\alpha j}}^{\alpha\alpha}$, i.e., $\overline{\nu^2_{\alpha 1}}^{\alpha\alpha}$, $\overline{\nu^2_{\alpha 2}}^{\alpha\alpha}$, and $\overline{\nu^2_{\alpha 3}}^{\alpha\alpha}$. respectively. Hence, $0 < a_i \leq 1$ for $i = 1, 2, 3$.

For an *isotropic porous medium*

$$\overline{\nu_{\alpha i}\nu_{\alpha j}}^{\alpha\alpha} = a\delta_{ij}, \tag{1.2.67}$$

where $a = \frac{1}{3}$. Thus, for an isotropic porous medium

$$T^*_{\alpha ij} = \frac{\theta^S_\alpha}{\theta_\alpha}\delta_{ij}. \tag{1.2.68}$$

Usually, $\theta^S_\alpha < \theta_\alpha$ (Santalo, 1976). Then

$$0 < T^*_{\alpha ii} < 1 \quad \text{for all } i\text{'s (no summation on } i).$$

In an anisotropic porous medium

$$\frac{\theta^S_\alpha}{\theta_\alpha}\overline{\nu^2_{\alpha i}}^{\alpha\alpha} < \frac{1}{3}. \tag{1.2.69}$$

The ratio $\theta^S_\alpha/\theta_\alpha$ measures the tortuosity of the void space, while the term $\overline{\nu_{\alpha i}\nu_{\alpha j}}^{\alpha\alpha}$ represents the effect of anisotropy on the tortuosity. Altogether, $\mathbf{T}^*$ plays the role of a tortuosity of the α-occupied portion of the void space. We shall see below that, strictly speaking, the term *tortuosity*, as a characteristic feature of the $\mathcal{S}_{\alpha\beta}$-surface only, is justified only for cases in which this surface is 'impervious' to the considered extensive quantity. Otherwise, the tortuosity is also affected by the transport regime in the β-phase(s).

Let us apply (1.2.62) to three cases of special interest.

CASE A. The quantity G_α is such that

$$\nabla G_\alpha \cdot \boldsymbol{\nu}_\alpha = 0 \qquad \text{on} \qquad \mathcal{S}_{\alpha\beta}. \tag{1.2.70}$$

Then, we obtain from (1.2.62)

$$\overline{\frac{\partial G_\alpha}{\partial x_i}}^\alpha = \frac{\partial \overline{G_\alpha}^\alpha}{\partial x_i} T^*_{\alpha ij}. \tag{1.2.71}$$

We have thus related the average of a gradient to the gradient of an average for this particular case.

This case, with G_α representing the concentration of a component, is applicable to the case of molecular diffusion in a fluid that occupies the entire void space, with no diffusion into the solid wall, and without adsorption; α and β denote the fluid and the solid phases, respectively.

CASE B. Only two phases, α and β, occupy the entire space. The quantity G_α in $\mathcal{U}_{o\alpha}$, and G_β in $\mathcal{U}_{o\beta}$, are such that on their common boundary, $\mathcal{S}_{\alpha\beta}$, the conditions

$$-(\lambda_\alpha \nabla G_\alpha \cdot \boldsymbol{\nu}_\alpha)\Big|_{\substack{\alpha\ \text{side} \\ \text{of } \mathcal{S}_{\alpha\beta}}} = (\lambda_\beta \nabla G_\beta \cdot \boldsymbol{\nu}_\beta)\Big|_{\substack{\beta\ \text{side} \\ \text{of } \mathcal{S}_{\alpha\beta}}} , \tag{1.2.72}$$

and

$$G_\alpha\Big|_{\substack{\alpha\ \text{side} \\ \text{of } \mathcal{S}_{\alpha\beta}}} = G_\beta\Big|_{\substack{\beta\ \text{side} \\ \text{of } \mathcal{S}_{\alpha\beta}}} \tag{1.2.73}$$

are satisfied, where λ_α and λ_β are constant coefficients, related to G, that depend on the nature of the α and β phases, respectively, and G_β denotes the value of G in $\mathcal{U}_{o\beta}$. Later we shall see the significance of these conditions.

Applying the modified rule for averaging of a spatial derivative, (1.2.62), to the α-phase, and multiplying the result by λ_α, we obtain

$$\lambda_\alpha \overline{\frac{\partial G_\alpha}{\partial x_j}}^\alpha = \lambda_\alpha \frac{\partial \overline{G_\alpha}^\alpha}{\partial x_i} T^*_{\alpha ij} + \frac{\lambda_\alpha}{U_{o\alpha}} \int_{\mathcal{S}_{\alpha\beta}} \overset{\circ}{x}_j \frac{\partial G_\alpha}{\partial x_i} \nu_{\alpha i}\, dS. \tag{1.2.74}$$

The same procedure, when applied to the β-phase, yields

$$\lambda_\beta \overline{\frac{\partial G_\beta}{\partial x_j}}^\beta = \lambda_\beta \overline{\frac{\partial G_\beta}{\partial x_i}}^\beta T^*_{\beta ij} + \frac{\lambda_\beta}{U_{o\beta}} \int_{S_{\beta\alpha}} \mathring{x}_j \frac{\partial G_\beta}{\partial x_i} \nu_{\beta i}\, dS. \qquad (1.2.75)$$

By adding the two equations, and noting condition (1.2.72), we obtain

$$\lambda_\alpha \theta_\alpha \overline{\frac{\partial G_\alpha}{\partial x_j}}^\alpha + \lambda_\beta \theta_\beta \overline{\frac{\partial G_\beta}{\partial x_j}}^\beta = \lambda_\alpha \theta_\alpha \overline{\frac{\partial G_\alpha}{\partial x_i}}^\alpha T^*_{\alpha ij} + \lambda_\beta \theta_\beta \overline{\frac{\partial G_\beta}{\partial x_i}}^\beta T^*_{\beta ij}.$$

$$(1.2.76)$$

Since

$$\int_{S_o} \mathring{x}_j \nu_{\alpha i}\, dS = \int_{S_{\alpha\alpha}} \mathring{x}_j \nu_{\alpha i}\, dS + \int_{S_{\beta\beta}} \mathring{x}_j \nu_{\beta i}\, dS$$

$$= U_{o\alpha} T^*_{\alpha ji} + U_{o\beta} T^*_{\beta ji} = U_o \delta_{ij},$$

we have

$$\theta_\alpha T^*_{\alpha ij} + \theta_\beta T^*_{\beta ij} = \delta_{ij}. \qquad (1.2.77)$$

By writing the usual averaging rule for a spatial derivative, once for $\partial G_\alpha / \partial x_j$ and then, for $\partial G_\beta / \partial x_j$, and adding the resulting equations, noting condition (1.2.73), we obtain

$$\theta_\alpha \overline{\frac{\partial G_\alpha}{\partial x_j}}^\alpha + \theta_\beta \overline{\frac{\partial G_\beta}{\partial x_j}}^\beta = \frac{\partial \theta_\alpha \overline{G_\alpha}^\alpha}{\partial x_j} + \frac{\partial \theta_\beta \overline{G_\beta}^\beta}{\partial x_j}. \qquad (1.2.78)$$

From these two equations, in the 'unknowns' $\overline{\partial G_\alpha / \partial x_j}^\alpha$, and $\overline{\partial G_\beta / \partial x_j}^\beta$, we obtain

$$\overline{\frac{\partial G_\alpha}{\partial x_j}}^\alpha = \frac{1}{\lambda_\alpha - \lambda_\beta} \left\{ \left(\lambda_\alpha \overline{\frac{\partial G_\alpha}{\partial x_i}}^\alpha - \lambda_\beta \overline{\frac{\partial G_\beta}{\partial x_i}}^\beta \right) T^*_{\alpha ij} - \frac{\lambda_\beta}{\theta_\alpha} \frac{\partial}{\partial x_j} \theta_\alpha (\overline{G_\alpha}^\alpha - \overline{G_\beta}^\beta) \right\},$$

$$(1.2.79)$$

and an analogous equation for $\overline{\partial G_\beta / \partial x_j}^\beta$.

We shall later see the interpretation of (1.2.79) for special cases, e.g., $\lambda_\beta = 0$, and $\overline{G_\alpha}^\alpha = \overline{G_\beta}^\beta$.

CASE C   Here, also, a single fluid phase occupies the entire void space. The condition on $S_{\alpha\beta}$ is

$$\nabla G_\alpha \cdot \boldsymbol{\nu}_\alpha = \frac{G_\alpha - \overline{G_\alpha}^\alpha}{\Delta}, \quad \text{or} \quad \frac{\partial G_\alpha}{\partial x_i} \nu_{\alpha i} = \frac{G_\alpha - \overline{G_\alpha}^\alpha}{\Delta} \quad \text{on} \quad S_{\alpha\beta}.$$

$$(1.2.80)$$

With

$$M_j = \frac{1}{\mathcal{S}_{\alpha\beta}} \int_{\mathcal{S}_{\alpha\beta}} C''_\alpha \mathring{x}_j \, dS, \qquad (1.2.81)$$

and (1.2.80), equation (1.2.62) becomes

$$\overline{\frac{\partial G_\alpha}{\partial x_j}}^\alpha \approx \overline{\frac{\partial G_\alpha}{\partial x_i}}^\alpha T^*_{\alpha ij} + M_j \frac{\widetilde{G_\alpha}^{\alpha\beta} - \overline{G_\alpha}^\alpha}{\Delta_\alpha^2} \qquad (1.2.82)$$

where Δ is a macroscopic length that characterizes the distance between $\mathcal{S}_{\alpha\beta}$ and the interior of $\mathcal{U}_{o\alpha}$. For example, $\Delta = \Delta_\alpha / C''_\alpha$, where C''_α is an empirical coefficient and Δ_α is the hydraulic radius of $\mathcal{U}_{o\alpha}$.

We note that an additional variable, $\widetilde{G_\alpha}^{\alpha\beta}$, which is the average of G_α over $\mathcal{S}_{\alpha\beta}$-surface within the REV, and an additional coefficient, M_j / Δ_α^2 have been introduced.

(e) Average of a material derivative

For a density e of an α-phase, we have

$$\begin{aligned}
\overline{\dot{e}}^\alpha &= \dot{\overline{e}}^\alpha - \overline{e \nabla \cdot \mathbf{V}^E}^\alpha + \overline{\nabla \cdot \mathbf{V}^E}^\alpha \overline{e}^\alpha \\
&= \dot{\overline{e}}^\alpha - \overline{\mathring{e}(\nabla \cdot \mathbf{V}^E)}^\alpha .
\end{aligned} \qquad (1.2.83)$$

For an *isochoric continuum* (at the microscopic level), i.e., when $\nabla \cdot \mathbf{V}^E \equiv 0$, we obtain $\overline{\dot{e}}^\alpha = \dot{\overline{e}}^\alpha$.

We now have

- A general microscopic balance equation for any extensive quantity, E, with examples for mass, mass of a component, momentum and energy, all of a phase within a porous medium domain.

- Averaging rules that transform averages of derivatives to derivatives of average values.

We are ready now to derive the macroscopic balance equations.

1.3   Lecture Three: Macroscopic Balance Equations

In this lecture, macroscopic balance equations for mass, mass of a component, momentum and energy will be derived. We shall also discuss the concept of *effective stress* employed in treating transport phenomena in deformable porous media.

1.3.1   General balance equation

The general microscopic balance equation for an extensive quantity, E (used as a symbol for E of an α-phase, or E^γ of a γ-component of an α-phase), is given by (1.2.11), or (1.2.15). We recall that $e\mathbf{V}^E(\equiv \mathbf{j}^{tE})$ denotes the total flux of E, $e\mathbf{V}$ denotes the advective flux of E, and $\mathbf{j}^{EU}$ denotes the diffusive flux of E (with respect to the volume averaged velocity).

By applying the averaging rules, for the average of a time derivative and for the average of a spatial derivative, to (1.2.15), we obtain the general macroscopic balance equation

$$
\frac{\partial \theta \overline{e}^\alpha}{\partial t} = -\nabla\cdot\theta(\overline{e}^\alpha \overline{\mathbf{V}}^\alpha + \overline{\overset{\circ}{e}\overset{\circ}{\mathbf{V}}}^\alpha + \overline{\mathbf{j}^{EU}}^\alpha)
$$

$$
- \frac{1}{U_o}\int_{\mathcal{S}_{\alpha\beta}} \{e(\mathbf{V}-\mathbf{u})+\mathbf{j}^{EU}\}\cdot\boldsymbol{\nu}\,dS + \theta\overline{\rho\Gamma^{E}}^\alpha, \qquad (1.3.1)
$$

where $\theta(\equiv \theta_\alpha)$ denotes the volumetric fraction of the α-phase, and $\mathbf{u}$ denotes the speed of displacement of the $\mathcal{S}_{\alpha\beta}$-surface, recalling that β denotes all phases other than α in the REV.

By comparing the macroscopic balance equation (1.3.1) with the microscopic one, (1.2.15), we note the following similarities and differences.

- Whereas each term in (1.2.15) expresses added E per unit volume of the considered phase, per unit time, each term in (1.3.1) represents the rate at which E is added *per unit volume of porous medium*.

- In both cases, the first term on the r.h.s. expresses the addition of E due to the total flux of E. However, the composition of the total flux is different.

- Whereas in (1.2.15), the total microscopic flux is made up of advection and diffusion, an additional flux term appears in (1.3.1), called *dispersive flux* of E. In LECTURE 8 we shall explain the reason for calling the term $\overline{\overset{\circ}{e}\overset{\circ}{\mathbf{V}}}^\alpha$ dispersive flux.

- In (1.2.15), sources of E express the production of E at points within the considered phase (recalling that a sink is a negative source). In addition to an analogous term, $\theta\rho\overline{\Gamma^{E}}^{\alpha}$, at the macroscopic level, we have in (1.3.1), an additional source term, expressed by the surface integral over the $\mathcal{S}_{\alpha\beta}$-surface. We note that the two terms in the integrand, express microscopic fluxes at points on the $\mathcal{S}_{\alpha\beta}$-surface. The fluxes are advection with respect to the (possibly) moving surface, and diffusion (of E). By multiplying these fluxes by the outward unit vector, $\boldsymbol{\nu}$, on $\mathcal{S}_{\alpha\beta}$, we obtain their component normal to the surface. Thus, minus the integral expresses the rate at which E is added to the α-phase (from the β-phase) through $\mathcal{S}_{\alpha\beta}$. Division by U_o leads to the quantity added per unit volume of porous medium.

If the $\mathcal{S}_{\alpha\beta}$-surface is a material surface with respect to the volume of the α-phase (i.e., no α-phase crosses it), then

$$(\mathbf{V} - \mathbf{u})\cdot\boldsymbol{\nu} = 0. \qquad (1.3.2)$$

If the surface is 'impervious' to diffusion of E (i.e., no E can diffuse from the interior of the α-phase to and across $\mathcal{S}_{\alpha\beta}$ (or from $\mathcal{S}_{\alpha\beta}$ to the interior of the α-phase), then $\mathbf{j}^{EU}\cdot\boldsymbol{\nu} = 0$.

By averaging (1.2.16), we obtain

$$\frac{\partial\theta\overline{e}^{\alpha}}{\partial t} = -\nabla\cdot\theta(\overline{e}^{\alpha}\overline{\mathbf{V}^{m}}^{\alpha} + \overline{\mathring{e}\mathring{\mathbf{V}}^{m}}^{\alpha} + \overline{\mathbf{j}^{Em}}^{\alpha})$$

$$-\frac{1}{U_o}\int_{\mathcal{S}_{\alpha\beta}} \{e(\mathbf{V}^{m} - \mathbf{u}) + \mathbf{j}^{Em}\}\cdot\boldsymbol{\nu}\, dS + \theta\rho\overline{\Gamma^{E}}^{\alpha}. \qquad (1.3.3)$$

Let us consider a number of particular macroscopic balance equations.

1.3.2 Particular cases

(a) **Mass of a phase.** In this case, $E = m$, $e = \rho$, $\Gamma^{E} = \Gamma^{m} = 0$ (i.e., no total mass is produced) and $\mathbf{j}^{mU}$ is the diffusive flux of the total mass. We obtain from (1.3.1)

$$\frac{\partial\theta\overline{\rho}^{\alpha}}{\partial t} = -\nabla\cdot\theta(\overline{\rho}^{\alpha}\overline{\mathbf{V}}^{\alpha} + \overline{\mathring{\rho}\mathring{\mathbf{V}}}^{\alpha} + \overline{\mathbf{j}^{mU}}^{\alpha})$$

$$-\frac{1}{U_o}\int_{\mathcal{S}_{\alpha\beta}} \{\rho(\mathbf{V} - \mathbf{u}) + \mathbf{j}^{mU}\}\cdot\boldsymbol{\nu}\, dS, \qquad (1.3.4)$$

or, since $\mathbf{j}^{mm} \equiv 0$ and $\mathbf{V}^E = \mathbf{V}^m$, from (1.3.3)

$$\frac{\partial \theta \overline{\rho}^{\alpha}}{\partial t} = - \nabla \cdot \theta (\overline{\rho}^{\alpha} \overline{\mathbf{V}^m}^{\alpha} + \overline{\overset{\circ}{\rho} \overset{\circ}{\mathbf{V}^m}}^{\alpha})$$

$$- \frac{1}{U_o} \int_{\mathcal{S}_{\alpha\beta}} \rho(\mathbf{V}^m - \mathbf{u}) \cdot \boldsymbol{\nu} \, dS. \qquad (1.3.5)$$

When the $\mathcal{S}_{\alpha\beta}$-surface is material with respect to the α-phase, the integral terms in both (1.3.4) and (1.3.5) vanish.

If, also, the advective mass flux is much larger than the dispersive one, i.e.

$$|\overline{\rho}^{\alpha} \overline{\mathbf{V}^m}^{\alpha}| \gg |\overline{\overset{\circ}{\rho} \overset{\circ}{\mathbf{V}^m}}^{\alpha}|,$$

equation (1.3.5) reduces to

$$\frac{\partial \theta \overline{\rho}^{\alpha}}{\partial t} = -\nabla \cdot \theta \overline{\rho}^{\alpha} \overline{\mathbf{V}^m}^{\alpha}. \qquad (1.3.6)$$

This is the basic differential mass balance equation employed in most studies of single phase flow through porous media. By adding subscript α to θ, to ρ and to $\mathbf{V}^m$, this equation is also applicable to a phase in a multiphase system (in the absence of external sources of mass of the α-phase.

Assuming that $\mathcal{S}_{\alpha\beta}$ is material with respect to the total mass of the α-phase (so that the surface integral vanishes), and that

$$|\overline{\rho}^{\alpha} \overline{\mathbf{V}}^{\alpha}| \gg |\overline{\overset{\circ}{\rho} \overset{\circ}{\mathbf{V}}}^{\alpha} + \overline{\mathbf{j}^{mU}}^{\alpha}|,$$

equation (1.3.4) reduces to

$$\frac{\partial \theta \overline{\rho}^{\alpha}}{\partial t} = -\nabla \cdot \theta \overline{\rho}^{\alpha} \overline{\mathbf{V}}^{\alpha}$$

$$= -\nabla \cdot \overline{\rho}^{\alpha} \mathbf{q}, \qquad (1.3.7)$$

where $\mathbf{q} = \theta \overline{\mathbf{V}}^{\alpha}$ denotes the *specific discharge* (= volume of phase passing per unit area *of porous medium* normal to $\overline{\mathbf{V}}^{\alpha}$, per unit time).

Equation (1.3.7) is also commonly used in flow through porous media. Again, it is applicable to a single phase that occupies the entire void space, or to an α-phase in a multiphase flow, by adding subscript α to θ, to ρ and to $\mathbf{V}$.

(b) **Volume of a phase.** Here, $E = U$, $e = 1$, $\mathbf{V}^E = \mathbf{V}$, $\mathbf{j}^E = \mathbf{j}^U = 0$, and the source is given by $\rho \Gamma^U = \nabla \cdot \mathbf{V}$. This follows from the interpretation of $-\nabla\cdot\mathbf{V}$ as the excess of inflow over outflow per unit volume and unit time. Hence, (1.3.1) becomes

$$\frac{\partial\theta}{\partial t} = -\nabla\cdot\theta\overline{\mathbf{V}}^{\alpha} - \frac{1}{U_o}\int_{\mathcal{S}_{\alpha\beta}}(\mathbf{V}-\mathbf{u})\cdot\boldsymbol{\nu}\,dS + \theta\overline{\nabla\cdot\mathbf{V}}^{\alpha},$$

$$(1.3.8)$$

or, in view of (1.2.56)

$$\nabla\cdot\theta\overline{\mathbf{V}}^{\alpha}(\equiv\nabla\cdot\mathbf{q}) = \theta\overline{\nabla\cdot\mathbf{V}}^{\alpha} - \overbrace{\mathbf{V}\cdot\boldsymbol{\nu}}^{\alpha\beta}\Sigma_{\alpha\beta}. \tag{1.3.9}$$

For an $\mathcal{S}_{\alpha\beta}$-surface that is material with respect to the phase, the surface integral term vanishes. For *microscopically isochoric* flow, i.e., a flow in which $\nabla\cdot\mathbf{V} = 0$. Under both conditions, (1.3.8) reduces to

$$\begin{aligned}\frac{\partial\theta}{\partial t} &= -\nabla\cdot\theta\overline{\mathbf{V}}^{\alpha} \\ &= -\nabla\cdot\mathbf{q},\end{aligned} \tag{1.3.10}$$

which is commonly used to describe flow of a homogeneous fluid phase in multiphase flow in a porous medium, with $\theta\equiv\theta_{\alpha}$, $\overline{\mathbf{V}}^{\alpha}\equiv\overline{\mathbf{V}_{\alpha}}^{\alpha}$. In (1.3.10), $\mathbf{q}\,(\equiv\theta\overline{\mathbf{V}}^{\alpha})$ is the specific discharge of the phase. We note that (1.3.10) can be obtained from (1.3.7) when $\overline{\rho}^{\alpha}=$ const., or $D\overline{\rho}^{\alpha}/Dt = 0$.

If, *in addition*, the microscopic boundary of the phase is such that $\partial\theta/\partial t = 0$, equation (1.3.10) reduces to

$$\nabla\cdot\theta\overline{\mathbf{V}}^{\alpha} = 0, \qquad \text{or} \qquad \nabla\cdot\mathbf{q} = 0. \tag{1.3.11}$$

Equation (1.3.11) is commonly used as the continuity equation for the flow of a single component, incompressible, homogeneous fluid phase that occupies the entire void space of a rigid porous medium, viz., $\theta - \phi$, with $\partial\phi/\partial t = 0$. A flow that satisfies (1.3.11) is referred to as *macroscopically isochoric flow*.

(c) **Mass of a γ - component of a phase.** Here $E = m^{\gamma}$, $e = \rho^{\gamma}$ (often also denoted as c^{γ}), $\mathbf{j}^{EU}\equiv\mathbf{j}^{m^{\gamma}U}\equiv\mathbf{j}^{\gamma}$, $\Gamma^{E}=\Gamma^{m^{\gamma}}\,(\equiv\Gamma^{\gamma})$. By averaging (1.2.22), or directly from (1.3.1), we obtain

$$\begin{aligned}\frac{\partial\theta\overline{c^{\gamma}}^{\alpha}}{\partial t} &= -\nabla\cdot\theta(\overline{c^{\gamma}}^{\alpha}\overline{\mathbf{V}}^{\alpha} + \overline{\overset{\circ}{c}^{\gamma}\overset{\circ}{\mathbf{V}}}^{\gamma} + \mathbf{j}^{\gamma^{\alpha}}) \\ &\quad - \frac{1}{U_o}\int_{\mathcal{S}_{\alpha\beta}}\{c^{\gamma}(\mathbf{V}-\mathbf{u})+\mathbf{j}^{\gamma}\}\cdot\boldsymbol{\nu}\,dS + \theta\overline{\rho\Gamma^{\gamma}}^{\alpha}.\end{aligned}$$

$$(1.3.12)$$

48   *MATHEMATICAL MODELLING*

We note the three (macroscopic) fluxes that together comprise the total flux of m^γ, and the two kinds of sources of γ: the transfer of m^γ across $\mathcal{S}_{\alpha\beta}$ into (or onto) adjacent phases, and the rate of production within the phase, both *per unit volume of porous medium*. When $\mathcal{S}_{\alpha\beta}$ is material with respect to the α-phase, $(\mathbf{V} - \mathbf{u})\cdot\boldsymbol{\nu} = 0$, and the surface integral in (1.3.12) reduces to $-\frac{1}{U_o}\int_{\mathcal{S}_{\alpha\beta}} \mathbf{j}^\gamma\cdot\boldsymbol{\nu}\, dS$. If the $\mathcal{S}_{\alpha\beta}$-surface is material to both advective and diffusive fluxes of the mass of the γ-component, the surface integral vanishes.

Equation (1.3.12) is the macroscopic differential balance equation of the mass of a component of a phase. It is often called the *equation of hydrodynamic dispersion*. This name will be understood in LECTURE 8, in which we shall also discuss in detail the dispersive flux of the γ-component, expressed by $\overline{\overset{\circ}{c^\gamma}\overset{\circ}{\mathbf{V}}}^\alpha$. In fact, we still have to express both the macroscopic diffusive flux, $\overline{\mathbf{j}^\gamma}^\alpha$, and the dispersive flux, $\overline{\overset{\circ}{c^\gamma}\overset{\circ}{\mathbf{V}}}^\alpha$, in terms of the macroscopic, or averaged concentration, $\overline{c^\gamma}^\alpha$, which constitutes the state variable of the problem. This topic will be discussed in LECTURE 7.

(d) **Linear momentum of a phase.** In this case, $E = \mathbf{M}$, $e = \rho\mathbf{V}^m$, $\mathbf{j}^{\mathbf{M}m} = \rho\mathbf{V}^m(\mathbf{V}^\mathbf{M} - \mathbf{V}^m) = -\boldsymbol{\sigma}$, where $\boldsymbol{\sigma}$ is the stress in the phase and $\Gamma^\mathbf{M} = \mathbf{F}$ represents the magnitude of the external body force acting on the phase. Note that we introduce here, and henceforth, everywhere in the present lectures, the stress as the *diffusive flux of linear momentum of a phase*.

Making use of the averaged equations (1.3.3), we obtain

$$\frac{\partial\theta\overline{\rho\mathbf{V}^m}^\alpha}{\partial t} = -\nabla\cdot\theta\{\overline{\rho\mathbf{V}^m}^\alpha\overline{\mathbf{V}^m}^\alpha + \overline{(\overset{\circ}{\rho}\mathbf{V}^m)\overset{\circ}{\mathbf{V}^m}}^\alpha - \overline{\boldsymbol{\sigma}}^\alpha\}$$

$$- \frac{1}{U_o}\int_{\mathcal{S}_{\alpha\beta}} \{\rho\mathbf{V}^m(\mathbf{V}^m - \mathbf{u}) - \boldsymbol{\sigma}\}\cdot\boldsymbol{\nu}\, dS + \theta\overline{\rho\mathbf{F}}^\alpha,$$

$$\tag{1.3.13}$$

where

$$\overline{\rho\mathbf{V}^m}^\alpha = \overline{\rho}^\alpha\overline{\mathbf{V}^m}^\alpha + \overline{\overset{\circ}{\rho}\overset{\circ}{\mathbf{V}^m}}^\alpha. \tag{1.3.14}$$

If we assume

$$|\overline{\overset{\circ}{\rho}\overset{\circ}{\mathbf{V}^m}}^\alpha| \ll |\overline{\rho}^\alpha\overline{\mathbf{V}^m}^\alpha|, \tag{1.3.15}$$

and

$$|\overline{(\overset{\circ}{\rho}\mathbf{V}^m)\overset{\circ}{\mathbf{V}^m}}^\alpha| \ll |\overline{\rho\mathbf{V}^m}^\alpha\overline{\mathbf{V}^m}^\alpha| \simeq |\overline{\rho}^\alpha\overline{\mathbf{V}^m}^\alpha\overline{\mathbf{V}^m}^\alpha|, \tag{1.3.16}$$

and when $\mathcal{S}_{\alpha\beta}$ is a material surface with respect to the mass of the phase (as is often the case), i.e., $\rho(\mathbf{V}^m - \mathbf{u})\cdot\boldsymbol{\nu} = 0$, equation (1.3.13) reduces to

$$\frac{\partial \theta \overline{\rho}^\alpha \overline{\mathbf{V}^m}^\alpha}{\partial t} = -\nabla\cdot\theta\left(\overline{\rho}^\alpha \overline{\mathbf{V}^m}^\alpha \overline{\mathbf{V}^m}^\alpha - \overline{\boldsymbol{\sigma}}^\alpha\right)$$
$$+ \frac{1}{U_o}\int_{\mathcal{S}_{\alpha\beta}} \boldsymbol{\sigma}\cdot\boldsymbol{\nu}\,dS + \theta\overline{\rho\mathbf{F}}^\alpha, \qquad (1.3.17)$$

or, in another form, taking into account the mass balance equation, (1.3.6)

$$\theta\overline{\rho}^\alpha \frac{\mathrm{D_m}\overline{\mathbf{V}^m}^\alpha}{\mathrm{D}t} = \nabla\cdot\theta\overline{\boldsymbol{\sigma}}^\alpha + \frac{1}{U_o}\int_{\mathcal{S}_{\alpha\beta}} \boldsymbol{\sigma}\cdot\boldsymbol{\nu}\,dS + \theta\overline{\rho\mathbf{F}}^\alpha. \qquad (1.3.18)$$

(e) **Heat balance of a fluid phase (f) and of a porous medium containing only one fluid phase.** To simplify the discussion, we shall use equation (1.2.40) as a starting point. In this equation, we shall assume that the term $\boldsymbol{\tau} : \nabla\mathbf{V}^m$, that expresses the rate of energy increase (per unit volume) by viscous dissipation, is much smaller than the other terms. Then the microscopic heat balance equation for a fluid phase reduces to the form

$$\frac{\partial \rho_f C_{Vf} T_f}{\partial t} = -\nabla\cdot\left(\rho_f C_{Vf} T_f \mathbf{V}_f^m + \mathbf{j}_f^H\right)$$
$$- T_f \frac{\partial p_f}{\partial T_f}\bigg|_{v_f} \nabla\cdot\mathbf{V}_f^m + T_f\rho_f \frac{\mathrm{D_m}C_{Vf}}{\mathrm{D}t}. \qquad (1.3.19)$$

We have selected this particular form of the energy balance equation, to emphasize the physical interpretation of the term $\rho_f C_{Vf} T_f$. When $C_{Vf} = $ constant, the last term on the r.h.s. of (1.3.19) vanishes.

By averaging (1.3.19), assuming that $C_{Vf} = $ const., and that

$$|\overline{\stackrel{\circ}{\rho}_f \stackrel{\circ}{T}_f}^f| \ll |\overline{\rho_f}^f \overline{T_f}^f|, \quad \text{so that} \quad \overline{\rho_f T_f}^f \simeq \overline{\rho_f}^f \overline{T_f}^f,$$

$$\overline{T_f \frac{\partial p_f}{\partial T_f}\bigg|_{v_f} \nabla\cdot\mathbf{V}_f^m}^f \simeq \overline{T_f}^f \overline{\frac{\partial p_f}{\partial T_f}\bigg|_{v_f}}^f \overline{\nabla\cdot\mathbf{V}_f^m}^f,$$

$$\overline{\rho_f T_f \nabla\cdot\mathbf{V}_f^m}^f \simeq \overline{\rho_f}^f \overline{T_f}^f \overline{\nabla\cdot\mathbf{V}_f^m}^f,$$

and

$$\overline{\mathbf{V}_f^m}^f \simeq \overline{\mathbf{V}_f}^f,$$

i.e., neglecting the diffusive mass flux (as defined by (1.2.6) fro $E = m$), we obtain for a single fluid phase that occupies the entire void space (porosity, ϕ)

$$\frac{\partial \phi \overline{\rho_f}^f C_{vf} \overline{T_f}^f}{\partial t} = -\nabla \cdot \phi \left[\overline{\rho_f}^f C_{vf} \overline{T_f}^f \overline{\mathbf{V}_f}^f + \overline{(\rho_f \mathring{C}_{vf} T_f) \mathring{\mathbf{V}}_f}^f + \overline{\mathbf{j}_f^H}^f \right]$$

$$- \frac{1}{U_o} \int_{\mathcal{S}_{sf}} \mathbf{j}_f^H \cdot \boldsymbol{\nu}_f \, dS - \overline{T_f}^f \frac{\overline{\beta_T}^f}{\beta_p} \bigg|_{\rho_f} \nabla \cdot \phi \overline{\mathbf{V}_f}^f, \qquad (1.3.20)$$

where

$$\beta_p = \frac{1}{\rho_f} \frac{\partial \rho_f}{\partial p_f} \bigg|_{T_f}, \qquad \text{and} \qquad \beta_T = -\frac{1}{\rho_f} \frac{\partial \rho_f}{\partial T_f} \bigg|_{p_f}$$

are the *coefficient of fluid compressibility at constant temperature*, and the *coefficient of thermal (volumetric) expansion at constant pressure*, respectively. The term $\overline{(\rho_f \mathring{C}_{vf} T_f) \mathring{\mathbf{V}}_f}^f$ expresses the *thermal dispersive flux* in the fluid.

For an isotropic *thermoelastic solid*, denoted by subscript s, the microscopic heat balance equation, (1.2.45), for $C_s = \text{const.}$, can also be written as

$$\frac{\partial \rho_s C_s T_s}{\partial t} = -\nabla \cdot (\rho_s C_s T_s \mathbf{V}_s + \mathbf{j}_s^H) - T_s \eta \frac{\partial \varepsilon_s}{\partial t}. \qquad (1.3.21)$$

Assuming that $\eta = \text{const.}$, $C_s = \text{const.}$, and that

$$|\overline{\rho_s}^s \overline{T_s}^s| \gg |\overline{\mathring{\rho}_s \mathring{T}_s}^s|, \qquad \text{hence} \quad \overline{\rho_s T_s}^s \simeq \overline{\rho_s}^s \overline{T_s}^s,$$

$$\overline{T_s \eta \frac{\partial \varepsilon_s}{\partial t}} \simeq (1 - \phi)\overline{T_s}^s \eta \overline{\frac{\partial \varepsilon_s}{\partial t}}^s \simeq \overline{T_s}^s \eta \frac{\partial (1 - \phi)\overline{\varepsilon_s}^s}{\partial t},$$

$$|\overline{(\mathring{\rho}_s \mathring{T}_s) \mathring{\mathbf{V}}_s}^s| \ll |\overline{\rho_s}^s \overline{T_s}^s \overline{\mathbf{V}_s}^s|, \qquad (1 - \phi)\left| \overline{\frac{\partial \varepsilon_s}{\partial t}}^s \right| \simeq \frac{\partial (1 - \phi)\overline{\varepsilon_s}^s}{\partial t},$$

and recalling that $(\mathbf{V}_s - \mathbf{u}) \cdot \boldsymbol{\nu}_s = 0$ on $\mathcal{S}_{\alpha\beta}$ ($=$ a material surface with respect to the solid), we obtain by averaging (1.3.21)

$$\frac{\partial (1 - \phi)\overline{\rho_s}^s C_s \overline{T_s}^s}{\partial t} = -\nabla \cdot (1 - \phi)\left(\overline{\rho_s}^s C_s \overline{T_s}^s \overline{\mathbf{V}_s}^s + \overline{\mathbf{j}_s^H}^s \right)$$

$$- \frac{1}{U_o} \int_{\mathcal{S}_{fs}} \mathbf{j}_s^H \cdot \boldsymbol{\nu}_s \, dS - \overline{T_s}^s \eta \frac{\partial (1 - \phi)\overline{\varepsilon_s}^s}{\partial t}. \qquad (1.3.22)$$

Equations (1.3.20) and (1.3.22) are two macroscopic differential balance equations, one for the fluid (having an averaged temperature $\overline{T_f}^f$) and one for the solid (at an averaged temperature of $\overline{T_s}^s$). By expressing the dispersive and diffusive fluxes also in terms of $\overline{T_f}^f$ and $\overline{T_s}^s$, we obtain two equations that have to be solved simultaneously for $\overline{T_f}^f$ and $\overline{T_s}^s$. In both equations, the surface integrals represent the exchange of heat, *by conduction*, between the two phases. By adding the two equations, the surface integral terms vanish. Obviously, we also need information on $\overline{\rho_f}^f = \overline{\rho_f}^f(\overline{p_f}^f, \overline{T_f}^f)$, $\overline{\mathbf{V}_f^m}^f = \overline{\mathbf{V}_f^m}^f(\overline{\rho_f}^f, \overline{p_f}^f)$ and $\overline{\varepsilon_s}^s = \overline{\varepsilon_s}^s(\overline{\sigma_s}^s(\overline{p_f}^f))$, which means that we have to solve simultaneously for more variables.

Often, because solid grains, in a granular porous medium, are relatively small, the thermal conductivity of the solid relatively high and the fluid velocity relatively small, the *averaged* temperatures of the two phases are maintained equal (or more or less so) at every (macroscopic) point, i.e., $\overline{T_f}^f = \overline{T_s}^s = T$. Under such conditions, by adding the two heat balance equations, we obtain

$$\frac{\partial(\rho C)_{pm}T}{\partial t} =$$
$$-\nabla\cdot\left[(\phi\overline{\rho_f}^f C_{Vf}\overline{\mathbf{V}_f}^f + (1-\phi)\overline{\rho_s}^s C_s \overline{\mathbf{V}_s}^s)T + \overline{\phi(\rho_f \mathring{C}_{Vf}T)\mathring{\mathbf{V}}_f}^f\right]$$
$$-\nabla\cdot\left[\phi\overline{\mathbf{j}_f^H}^f + (1-\phi)\overline{\mathbf{j}_s^H}^s\right] - \left\{\overline{\frac{\beta_T}{\beta_p}}^f \nabla\cdot\phi\overline{\mathbf{V}_f}^f + \eta\frac{\partial(1-\phi)\overline{\varepsilon_s}^s}{\partial t}\right\}T,$$

$$(1.3.23)$$

where

$$(\rho C)_{pm} = \phi\overline{\rho_f}^f C_{Vf} + (1-\phi)\overline{\rho_s}^s C_s$$

is the specific heat of the porous medium.

We shall discuss the total conductive heat flux, $\phi\mathbf{j}_f^H + (1-\phi)\mathbf{j}_s^H$, in LECTURE 7.

1.3.3 Stress in porous media

We consider a solid matrix, denoted by subscript s, that is *deformable*. To simplify the presentation, we shall first limit the discussion to the case of a single fluid, f, that occupies the entire void space.

By writing the (macroscopic) momentum balance equation (1.3.18) for each of the two phases, we obtain:

For the fluid

$$\phi\overline{\rho_f}^f \frac{D_m \overline{V_f^m}^f}{Dt} = \nabla\cdot\phi\overline{\sigma_f}^f + \frac{1}{U_o}\int_{S_{sf}} \sigma_f\cdot\nu_f \, dS + \phi\overline{\rho_f F_f}^f. \tag{1.3.24}$$

For the solid

$$(1-\phi)\overline{\rho_s}^s \frac{D_m \overline{V_s^m}^s}{Dt} = \nabla\cdot(1-\phi)\overline{\sigma_s}^s + \frac{1}{U_o}\int_{S_{sf}} \sigma_s\cdot\nu_s \, dS + (1-\phi)\overline{\rho_s F_s}^s. \tag{1.3.25}$$

To further simplify the discussion, we shall assume that the acceleration of both the fluid and the solid phases can be neglected. Then, by adding (1.3.24) and (1.3.25), we obtain the total momentum balance equation, also called the *equilibrium equation*, for the porous medium as a whole

$$\nabla\cdot\overline{\sigma} + \overline{\rho F} = 0, \tag{1.3.26}$$

where

$$\overline{\sigma} \equiv \overline{\sigma_f} + \overline{\sigma_s} = \phi\overline{\sigma_f}^f + (1-\phi)\overline{\sigma_s}^s$$

is the macroscopic *total stress*, while

$$\overline{\rho F} \equiv \overline{\rho_f F_f} + \overline{\rho_s F_s} = \phi\overline{\rho_f F_f}^f + (1-\phi)\overline{\rho_s F_s}^s$$

is the total macroscopic body force, per unit volume of porous medium, e.g., due to gravity, and we have made use of the condition

$$\sigma_f\cdot\nu_f = -\sigma_s\cdot\nu_s, \qquad \nu_f = -\nu_s \quad \text{on} \quad S_{fs}.$$

Usually we assume that $\overline{F_f}^f = \overline{F_s}^s = -g\nabla z$, i.e., that gravity is the only body force, with g denoting gravity acceleration.

For the fluid phase, we make use of the relation

$$\sigma_f = \tau_f - p_f \mathbf{I},$$

where p_f is the pressure, τ_f is the viscous stress and $\mathbf{I}$ is the unit tensor. Here we consider pressure to be positive for compression, while stresses are taken as positive for tension. When τ_f is assumed negligible, or whenever

$$|\nabla\cdot\tau_f| \ll |\nabla p_f|,$$

equation (1.3.26) reduces to

$$\nabla\cdot(\overline{\sigma_s} - \overline{p_f}\mathbf{I}) + \overline{\rho F} = 0. \tag{1.3.27}$$

We cannot make use of (1.3.27) directly in order to study porous medium deformation, say in response to changes in external stresses imposed on the porous medium, or in response to changes in fluid pressure, $\overline{p_f}^f$, say due to pumping, because the average stress, $\overline{\sigma_s}$, is not the *strain producing stress*. We have to introduce the concept of *effective stress*.

The concept of *effective stress*, was first introduced in soil mechanics by Terzaghi (1925). Essentially, this concept assumes that in a granular material, as the solid comprising the solid matrix (i.e., the granular material) is (*almost*) completely surrounded by an ambient fluid, the latter's pressure (actually stress, unless we neglect τ_f), acting on the solid-fluid interface, produces a stress of equal magnitude in the solid (say, within each individual grain), without contributing to the solid matrix' deformation which is produced mainly by forces that are transmitted (in a granular material) from grain to grain at contact points. Thus, the *strain producing stress*, or *intergranular streee*, is obtained by subtracting the pressure in the fluid from the stress in the solid, where both pressure and stress are average values, and we recall that in our lectures, stress is taken as positive for tension, while pressure is assumed positive for compression.

The concept of effective stress stems from the observation that the deformation of a granular material is much larger than can be explained by the compression of the solid material itself. This suggests that deformation is produced mainly by the rearrangement of grains, with localized slipping and rolling, implying that deformation is governed by the transmission of localized normal and shear forces at contact points. As these forces are not affected by the pressure in the water, a change in fluid pressure, accompanied by an equal change in total stress produces no deformation and, hence, should produce no change in effective stress.

Accordingly, when the shear stress, τ_f, is neglected, we can express the average stress, $\overline{\sigma}$, appearing in (1.3.26), in the form

$$\overline{\sigma} = (1 - \phi)\overline{\sigma_s}^s - \phi\overline{p_f}^f \mathbf{I}$$
$$= (1 - \phi)\{\overline{\sigma_s}^s + \overline{p_f}^f \mathbf{I}\} - (1 - \phi)\overline{p_f}^f \mathbf{I} - \phi\overline{p_f}^f \mathbf{I},$$

or

$$\overline{\sigma} = \sigma'_s - \overline{p_f}^f \mathbf{I}, \tag{1.3.28}$$

where $\overline{p_f}^f$ is the (average) pressure in the fluid, and

$$\overline{\sigma}'_s = (1 - \phi)\{\overline{\sigma_s}^s + \overline{p_f}^f \mathbf{I}\} \tag{1.3.29}$$

is the *effective stress*.

Written in this form, it is evident that the effective stress is made up of two parts: one is an average stress (*positive for tension*) within the solid matrix, and the other is an average pressure (*positive for compression*) in the fluid filling up the void space. The stresses $\overline{\sigma}$ and $\overline{\sigma'_s}$ are forces *per unit area of porous medium cross-section*. In soil mechanics, the minus sign in (1.3.28) is usually replaced by a plus sign, i.e., both $\overline{\sigma}$ and $\overline{\sigma'_s}$ are positive for compression.

Let us extend the concept of effective stress to two (or more) fluid phases (e.g., oil and water) that together fill up the void space. We shall refer to one fluid (subscript w) as the *wetting fluid*, and to the other one (subscript n) as the *nonwetting fluid*. By writing the averaged momentum balance equation, (1.3.18) for each of the three phases, two fluid phases and a solid one, we obtain

$$\theta_n \overline{\rho_n}^n \frac{D_m \overline{\mathbf{V}^{m_n}}^n}{Dt} \;=\; \nabla \cdot \theta_n \overline{\sigma_n}^n + \theta_n \overline{\rho_n \mathbf{F}_n}^n + \frac{1}{U_o} \int_{\mathcal{S}_{ns}+\mathcal{S}_{nw}} \sigma_n \cdot \nu_n \, dS,$$

$$(1.3.30)$$

$$\theta_w \overline{\rho_w}^w \frac{D_m \overline{\mathbf{V}^{m_w}}^w}{Dt} \;=\; \nabla \cdot \theta_w \overline{\sigma_w}^w + \theta_w \overline{\rho_w \mathbf{F}_w}^w + \frac{1}{U_o} \int_{\mathcal{S}_{ws}+\mathcal{S}_{wn}} \sigma_w \cdot \nu_w \, dS,$$

$$(1.3.31)$$

$$\theta_s \overline{\rho_s}^s \frac{D_m \overline{\mathbf{V}^{m_s}}^s}{Dt} \;=\; \nabla \cdot \theta_s \overline{\sigma_s}^s + \theta_s \overline{\rho_s \mathbf{F}_s}^s + \frac{1}{U_o} \int_{\mathcal{S}_{sn}+\mathcal{S}_{sw}} \sigma_s \cdot \nu_s \, dS,$$

$$(1.3.32)$$

where $\mathcal{S}_{wn}$, $\mathcal{S}_{sw}$, and $\mathcal{S}_{ns}$ denote the wetting fluid-nonwetting fluid, solid-wetting fluid and nonwetting fluid-solid contact areas. We note that the outward normals are such that $\nu_n \equiv -\nu_s$ on $\mathcal{S}_{ns}$, $\nu_s \equiv -\nu_w$ on $\mathcal{S}_{sw}$ and $\nu_n \equiv -\nu_w$ on $\mathcal{S}_{nw}$. The symbol $D_m(\;)/Dt$ denotes the material derivative, taken from the point of view of an observer that travels at the average mass weighted velocity of the considered phase.

By adding the three equations, and neglecting the inertial forces, we obtain

$$\nabla \cdot \overline{\sigma} + \overline{\rho \mathbf{F}} \;+\; \frac{1}{U_o} \int_{\mathcal{S}_{nw}} [\sigma]_{n,w} \cdot \nu_n \, dS + \frac{1}{U_o} \int_{\mathcal{S}_{ws}} [\sigma]_{w,s} \cdot \nu_w \, dS$$

$$+ \frac{1}{U_o} \int_{\mathcal{S}_{sn}} [\sigma]_{s,n} \cdot \nu_s \, dS = 0, \qquad (1.3.33)$$

where

$$\overline{\sigma} \;=\; \frac{1}{U_o} \int_{U_o} \sigma \, dU = \frac{1}{U_o} \sum_{(\alpha=s,n,w)} \int_{U_{o\alpha}} \sigma_\alpha \, dU_\alpha$$

$$= \sum_{(\alpha=s,n,w)} \overline{\sigma_\alpha} = \overline{\sigma_s} + \overline{\sigma_n} + \overline{\sigma_w} \qquad (1.3.34)$$

defines the volume averaged stress, or total stress, and

$$\overline{\rho \mathbf{F}} \equiv \overline{\rho_n \mathbf{F}_n} + \overline{\rho_w \mathbf{F}_w} + \overline{\rho_s \mathbf{F}_s}$$

defines the total body force, per unit volume of porous medium at a (macroscopic) point.

We have used the symbol

$$[(\cdot\cdot)]_{1,2} \equiv (\cdot\cdot)|_{\text{side 1}} - (\cdot\cdot)|_{\text{side 2}},$$

to denote the jump from side 1 to side 2 across the microscopic interface. By examining (1.3.33), we note that the interaction between each pair of adjacent phases is accounted for by a surface integral.

The first integral in (1.3.33) describes the interaction across the (microscopic) interface between the nonwetting and wetting fluid phases. We usually assume continuity of traction, i.e., $[\sigma]_{w,s}\cdot\nu_w = 0$, and $[\sigma]_{s,n}\cdot\nu_s = 0$, and neglect *surface tension* phenomena at fluid-solid interfaces. Hence, the last two surface integrals in (1.3.33) vanish.

Back to the first integral, actually, at every point on the microscopic interface between two *immiscible fluids* (here the wetting and the nonwetting ones), regarding them as two continua separated by a sharp interface, the continuity of momentum transfer is expressed by (Landau and Lifschitz, 1960)

$$
\begin{aligned}
\left[\rho V_i^m (V_j^m - u_j) - \sigma_{ij}\right]_{n,w} \nu_j &= \left(\frac{1}{r'} + \frac{1}{r''}\right)\gamma_{wn}\nu_i + \frac{\partial \gamma_{wn}}{\partial r_i} \\
&= \frac{2}{r^*}\gamma_{wn}\nu_i + \frac{\partial \gamma_{wn}}{\partial x_i}, \qquad (1.3.35)
\end{aligned}
$$

where γ_{wn} denotes the magnitude of *surface-tension* between the wetting and the nonwetting fluids. This is a concept that introduces the molecular level effects between the two fluids in the form of a force (per unit length) that is tangent to the interface, and r^* is the mean radius of curvature of the latter, with r' and r'' denoting its principal radii of curvature. The l.h.s. of (1.3.35) expresses a jump in the component in the ith direction of the total momentum flux. The r.h.s. may be interpreted as the rate of production of linear momentum per unit area of the interface. In this equation, γ_{nw} may be nonuniform, e.g., because of impurities and temperature variations.

Note that because γ_{wn} exists only in the interphase surface, the gradient of γ_{wn} in (1.3.35) should be interpreted as

$$\frac{\partial \gamma_{wn}}{\partial x_i} \equiv |\nabla \gamma_{wn}| t_i,$$

in which $t_i = \cos(\nabla \gamma_{wn}, \mathbf{1x}_i)$.

When the $\mathcal{S}_{nw}$-interface is a material surface with respect to fluid mass, the advective flux of momentum vanishes normal to it, i.e.

$$[\rho \mathbf{V}^m (\mathbf{V}^m - \mathbf{u})]_{n,w} \cdot \boldsymbol{\nu}_n = 0.$$

For a stationary fluid, or, as an approximation also for a moving one, when we assume that the viscous stress $[\boldsymbol{\tau}]_{n,w} \simeq 0$, the jump $-[\boldsymbol{\sigma}]_{n,w} \cdot \boldsymbol{\nu}_n$ reduces to $[p]_{n,w} \boldsymbol{\nu}_n$. If also $\nabla \gamma_{wn} = 0$, equation (1.3.35) reduces to

$$p_n - p_w = \left(\frac{1}{r'} + \frac{1}{r''}\right) \gamma_{wn}, \tag{1.3.36}$$

known as the *Laplace formula*. Since $(p_n - p_w) > 0$, the *pressure is greater in the nonwetting fluid for which the surface is convex*. The difference $p_n - p_w$ is called the *capillary pressure*. The difference

$$p_c = \overline{p_n}^n - \overline{p_w}^w = p_c(\theta_w) \tag{1.3.37}$$

is called *macroscopic capillary pressure*. When working at the macroscopic level, we usually refer to it as *capillary pressure*. It is a function of the fluid content, say, of the wetting-fluid, in the void space.

A better understanding of (1.3.35) can be obtained by rewriting it in a local coordinate system at a point on the interface. With $\mathbf{t}_1, \mathbf{t}_2$ denoting the unit vectors in the plane tangent to the interface, and recalling that $\gamma_{wn} = \gamma(t_1, t_2)$, $\partial \gamma_{wn}/\partial s_\nu = 0$, we write a balance along the normal to the plane, in the form

$$\begin{aligned} -\nu_i[\sigma_{ij}]_{n,w}\nu_j &= -\nu_i[\tau_{ij} - p\delta_{ij}]_{n,w}\nu_j \\ &= -[\tau_{ij}]_{n,w}\nu_i\nu_j + [p]_{w,n} = \frac{2}{r^*}\gamma_{n,w}, \end{aligned} \tag{1.3.38}$$

in which $-\tau_{ij}\nu_i\nu_j$ expresses a force component normal to the interface, per unit area of the latter, due to shear. A balance in the tangential plane takes the form

$$\begin{aligned} -t_i[\sigma_{ij}]_{n,w}\nu_j &= -t_i[\tau_{ij}]_{n,w}\nu_j + [p]_{n,w}t_i\nu_i \\ &= \frac{2}{r^*}\gamma_{wn}\nu_i t_i + \frac{\partial \gamma_{wn}}{\partial x_i}t_i. \end{aligned} \tag{1.3.39}$$

Since $t_i \nu_i = 0$, this equation reduces to

$$- t_i [\tau_{ij}]_{n,w} \nu_j = \frac{\partial \gamma_{wn}}{\partial x_i} t_i. \qquad (1.3.40)$$

With (1.3.36), i.e., a stationary fluid and $\nabla \gamma_{wn} = 0$, equation (1.3.33) reduces to

$$\nabla \cdot \overline{\sigma} + \overline{\rho F} + \mathbf{F}_c = 0 \qquad (1.3.41)$$

where

$$\mathbf{F}_c = -\frac{1}{U_o} \int_{S_{nw}} (p_n - p_w) \mathbf{I} \cdot \boldsymbol{\nu}_n \, dS = -\frac{S_{nw}}{U_o} \overline{(p_n - p_w)\nu_n}^{nw}, \qquad (1.3.42)$$

and the total stress is given by

$$\overline{\sigma} = \overline{\sigma_s} + \overline{\tau_n} + \overline{\tau_w} - \overline{p_n}\mathbf{I} - \overline{p_w}\mathbf{I}. \qquad (1.3.43)$$

For the considered *multiphase* case, with negligible shear stress within the fluid, let us define an average pressure of the fluids that together fill the void space (subscript and superscript v), by

$$\overline{p_v}^v = \frac{1}{\phi}\{\theta_w \overline{p_w}^w + \theta_n \overline{p_n}^n\}, \qquad (1.3.44)$$

with $\overline{\sigma} = \overline{\sigma_s'} - \overline{p_v}^v I$.

Another form of (1.3.44) is

$$\overline{p_v}^v = \frac{1}{\phi}\{\phi\overline{p_n}^n - \theta_w(\overline{p_n}^n - \overline{p_w}^w)\} = \frac{1}{\phi}(\phi\overline{p_w}^w + \theta_n p_c). \qquad (1.3.45)$$

We can now rewrite (1.3.28) in the form

$$\begin{aligned}
\overline{\sigma} &= (1-\phi)\overline{\sigma_s}^s - \theta_w \overline{p_w}^w \mathbf{I} - \theta_n \overline{p_n}^n \mathbf{I} \\
&= (1-\phi)\{\overline{\sigma_s}^s + \overline{p_v}^v \mathbf{I}\} - (1-\phi)\overline{p_v}^v \mathbf{I} - \theta_w \overline{p_w}^w \mathbf{I} - \theta_n \overline{p_n}^n \mathbf{I} \\
&= \overline{\sigma_s'} - \frac{\theta_w}{\phi}\overline{p_w}^w \mathbf{I} - \frac{\theta_n}{\phi}\overline{p_n}^n \mathbf{I},
\end{aligned} \qquad (1.3.46)$$

where

$$\overline{\sigma_s'} = (1-\phi)\{\overline{\sigma_s}^s + \overline{p_v}^v \mathbf{I}\} \qquad (1.3.47)$$

is the effective stress for this case.

In *unsaturated flow* (= air - water flow), the nonwetting fluid is air, while the wetting one is water. When we take the air pressure to be atmospheric, i.e., $\overline{p_a}^a \equiv 0$, equation (1.3.46) reduces to

$$\overline{\sigma} = \overline{\sigma'_s} - \frac{\theta_w}{\phi}\overline{p_w}^w \mathbf{I}. \tag{1.3.48}$$

In determining the average void pressure, $\overline{p_v}^v$, we have taken a volume average of the pressure in the various fluids occupying the void space. Other weights in determining $\overline{p_v}^v$ would lead to equations which are different from (1.3.46) and (1.3.48). For example, some authors (e.g., Aitchison and Donald, 1956) use for air–water flow (with $\overline{p_a}^a = 0$)

$$\overline{\sigma} = \overline{\sigma'_s} - \chi(\theta_w)\overline{p_w}^w \mathbf{I}, \tag{1.3.49}$$

where χ is some function of the moisture content, θ_w. Bear *et al.* (1984) used (1.3.49) with $\chi(\theta_w) = \theta_w/\phi$.

In the case of a single fluid phase, or neglecting $\mathbf{F}_c$, as an approximation, in the case of multiphase flow, (1.3.41) reduces to

$$\nabla\cdot\overline{\sigma} + \overline{\rho}\mathbf{F} = 0 \tag{1.3.50}$$

known as the *equilibrium equation*.

To summarize, in this lecture we have

- derived the general macroscopic balance equation, by averaging the microscopic one over an REV,

- obtained macroscopic balance equations for mass of a phase, mass of a component of a phase, momentum of a phase, and energy of a phase and of the porous medium as a whole,

- introduced the macroscopic concepts of total and effective stresses, obtained from the averaged momentum balance equation, and

- extended the concept of effective stress to multiphase flow. The effective stress will be used in LECTURE 6, when we shall consider flow in a deformable porous medium.

We now have all the (macroscopic) balance equations of mass, mass of a component, momentum and energy, all of a phase. However, they contain advective, dispersive and diffusive flux terms which have to be expressed in terms of the macroscopic state variables of the problem.

1.4 Lecture Four: Advective Flux

Equation (1.3.1) is the *general macroscopic differential balance equation* for any extensive quantity, E. In order to solve it for the macroscopic variables $\theta(\mathbf{x},t)$ and $\overline{e}^{\alpha}$, we need additional relationships for the source functions: $\overline{\rho \Gamma^{E}}^{\alpha}$, $\overline{e(\mathbf{V}-\mathbf{u})\cdot\boldsymbol{\nu}}^{\alpha\beta}\ \Sigma_{\alpha\beta}$ and $\overline{\mathbf{j}^{E}\cdot\boldsymbol{\nu}}^{\alpha\beta}\ \Sigma_{\alpha\beta}$, for the fluxes: $\overline{\overset{\circ}{e}\mathbf{V}}^{\alpha}$ and $\overline{\mathbf{j}^{E}}^{\alpha}$, and for the velocity $\overline{\mathbf{V}}^{\alpha}$ (or $\overline{\mathbf{V}^{m}}^{\alpha}$). These additional relations depend on the specific nature of the considered phase and extensive quantity.

In the present lecture, we start by considering the advective flux of a phase occupying the entire void space, or part of it. Then, we'll discuss advective fluxes in multiphase flow. Whenever it is clear that we refer to a single (say, α)-phase only, the subscript α will be omitted, except in the average symbol, $\overline{(..)}^{\alpha}$.

1.4.1 Advective flux of a single fluid that occupies the entire void space.

The general macroscopic balance equation (1.3.1) contains the expression $\theta\overline{e}^{\alpha}\overline{\mathbf{V}}^{\alpha}$, called the (macroscopic) *advective flux* of E, i.e., the quantity of E advected by the average volume weighted velocity, $\overline{\mathbf{V}}^{\alpha}$, of a fluid phase, *per unit area of a porous medium*. Similarly, the expression $\theta\overline{e}^{\alpha}\overline{\mathbf{V}^{m}}^{\alpha}$ is the (macroscopic) advective flux of E, carried by the average mass weighted velocity, $\overline{\mathbf{V}^{m}}^{\alpha}$, of a fluid phase.

For the fluid's volume ($e_{\alpha}=1$), we define two kinds of fluxes. One is

$$\mathbf{q}^{m}=\theta\overline{\mathbf{V}^{m}}^{\alpha},\tag{1.4.1}$$

called the *mass weighted specific discharge*. The other is

$$\mathbf{q}\equiv\theta\overline{\mathbf{V}}^{\alpha},\tag{1.4.2}$$

called *specific discharge*. In both examples, the flux is *per unit area of the considered porous medium*.

In principle, an expression for the advective mass flux, $\theta\overline{\rho}^{\alpha}\overline{\mathbf{V}^{m}}^{\alpha}$, can be obtained by solving a macroscopic momentum balance equations (LECTURE 3), say (1.3.17), for $\overline{\mathbf{V}^{m}}^{\alpha}$, supplemented by constitutive equations that describe the behavior of the considered fluid phase at the macroscopic level. Specifically, we need expressions which relate the stress, $\overline{\sigma}^{\alpha}$, and the various dispersive fluxes, to the average velocity, $\overline{\mathbf{V}^{m}}^{\alpha}$. However, under certain conditions, the momentum balance equation, say (1.3.17), can be reduced to simpler forms. For example, we may assume that certain terms are much smaller than others and may, therefore, be deleted from the equation. In what follows we shall introduce a number of simplifying, *albeit approximate*, assumptions in order to derive simplified expressions for the velocity $\overline{V^m}^{\alpha}$.

As a point of departure for determining the macroscopic velocity of a fluid phase, *that occupies the entire void space*, (i.e., $\theta = \phi$), let us start from the *macroscopic* momentum balance equation (1.3.18), which can be rewritten here in the form

$$\phi\overline{\rho}^f \frac{D_m \overline{V_i^m}^f}{Dt} \equiv \phi\overline{\rho}^f \left(\frac{\partial \overline{V_i^m}^f}{\partial t} + \overline{V_j^m}^f \frac{\partial \overline{V_i^m}^f}{\partial x_j} \right)$$

$$= \phi \frac{\overline{\partial \sigma_{ij}}^f}{\partial x_j} + n\overline{\rho F_i}^f , \tag{1.4.3}$$

where, the symbol f indicates the fluid phase. We recall that underlying (1.4.3) are the assumptions that the fluid-solid interface is a material surface with respect to the fluid's mass, and that all the momentum dispersive fluxes have been neglected.

In terms of the viscous stress, τ_{ij}, and the pressure, p, with $\sigma_{ij} = \tau_{ij} - p\delta_{ij}$, and for a body force that is due to gravity, viz., $F_i = -g\partial z/\partial x_i$, with the z-coordinate directed upward, equation (1.4.3) takes the form

$$\phi\overline{\rho}^f \frac{D_m \overline{V_i^m}^f}{Dt} = -\phi \overline{\frac{\partial p}{\partial x_i}}^f - \phi\rho g \overline{\frac{\partial z}{\partial x_i}}^f + \phi \overline{\frac{\partial \tau_{ij}}{\partial x_j}}^f . \tag{1.4.4}$$

The first term on the r.h.s. represents the total force due to pressure. Since

$$\phi \overline{\frac{\partial p}{\partial x_i}}^f = \frac{1}{U_o} \int_{S_{fs}} p\nu_i \, dS + \frac{1}{U_o} \int_{S_{ff}} p\nu_i \, dS, \tag{1.4.5}$$

this term includes the force of pressure exerted on the fluid across the fluid-solid interface and across the fluid portion of S_o. Note that $\nu \equiv \nu_f \equiv -\nu_s$.

The second term represents the force due to gravity. The third term represents the force due to the viscous resistance to flow.

Let us further develop the terms on the r.h.s. of (1.4.4), that still involve microscopic variables, in order to express them in terms of macroscopic ones.

As shown in LECTURE 3, under certain conditions, the relationship between the average of a gradient and the gradient of an average, can be presented in a modified form. Accordingly, if we assume here that *the fluid pressure within the void space of an* REV *varies monotonously*, i.e., with no maximum or minimum, then

$$\nabla^2 p = 0 \qquad \text{in} \quad \mathcal{U}_{ov}. \tag{1.4.6}$$

Under this condition, (1.2.62) is applicable, and we may use it to write

$$\overline{\frac{\partial p}{\partial x_i}}^f = \frac{\partial \overline{p}^f}{\partial x_j} T_{ji}^* + \frac{1}{\phi U_o} \int_{S_{fs}} \mathring{x}_i \frac{\partial p}{\partial x_j} \nu_j \, dS, \tag{1.4.7}$$

where the coefficient $T_{ji}^*, \equiv T_{fji}^*$, defined by (1.2.63), takes the form

$$T_{ij}^* = \frac{1}{\phi U_o} \int_{S_{ff}} \mathring{x}_i \nu_j \, dS. \tag{1.4.8}$$

In order to express the surface integral in (1.4.7) in terms of macroscopic variables, let us focus our attention on the normal derivative, $(\partial p/\partial x_j)\nu_j$, on the S_{fs}-surface, that appears in that integral. Assuming that in the vicinity of the (possibly moving) fluid-solid surface, the components normal to the latter, of both the inertial force and the viscous resistance to the flow, are negligible, relative to the normal components of pressure gradient and gravity, i.e.

$$\left| \left(\rho \frac{D_m V_j^m}{Dt} - \frac{\partial \tau_{ij}}{\partial x_i} \right) \nu_j \right| \ll \left| \left(\frac{\partial p}{\partial x_j} + \rho g \frac{\partial z}{\partial x_j} \right) \nu_j \right|,$$

we obtain, say from (1.4.4), written at the microscopic level

$$\frac{\partial p}{\partial x_j} \nu_j \cong -\rho g \frac{\partial z}{\partial x_j} \nu_j = -\rho g \delta_{3j} \nu_j \qquad \text{on} \quad S_{fs}. \tag{1.4.9}$$

Hence

$$\frac{1}{\phi U_o} \int_{S_{fs}} \frac{\partial p}{\partial x_j} \mathring{x}_i \nu_j \, dS = -\overline{\rho}^f g \delta_{3j} \frac{1}{\phi U_o} \int_{S_{fs}} \mathring{x}_i \nu_j \, dS$$

$$= -\overline{\rho}^f g (\delta_{3i} - T_{3i}^*). \tag{1.4.10}$$

In writing (1.4.10), we have made the approximation $\mathring{\rho} \ll \overline{\rho}^f$ within $\mathcal{U}_{of}$ and, hence

$$\frac{1}{\phi U_o} \int_{S_{of}} \mathring{x}_i \nu_j \, dS \equiv \frac{1}{\phi U_o} \left\{ \int_{S_{fs}} \mathring{x}_i \nu_j \, dS + \int_{S_{ff}} \mathring{x}_i \nu_j \, dS \right\}$$

$$\equiv \frac{1}{\phi U_o} \int_{S_{fs}} \mathring{x}_i \nu_j \, dS + T_{ij}^* = \delta_{ij}. \tag{1.4.11}$$

The gravity term in (1.4.4) is expressed by

$$\overline{\rho g \frac{\partial z}{\partial x_i}}^f = \overline{\rho}^f g \delta_{3i}.$$ (1.4.12)

By combining (1.4.7), (1.4.10) and (1.4.12), we obtain for the first two terms on the r.h.s. of (1.4.4)

$$-\phi\left(\overline{\frac{\partial p}{\partial x_i}}^f + \overline{\rho g \frac{\partial z}{\partial x_i}}^f\right) = -\phi\left[\frac{\partial \overline{p}^f}{\partial x_j}T_{ji}^* - \overline{\rho}^f g(\delta_{3i} - T_{3i}^*) + \overline{\rho}^f g \delta_{3i}\right]$$

$$= -\phi\left(\frac{\partial \overline{p}^f}{\partial x_j} + \overline{\rho}^f g \frac{\partial z}{\partial x_j}\right)T_{ji}^*,$$ (1.4.13)

noting that $T_{3i}^* = \delta_{3j}T_{ji}^* = \frac{\partial z}{\partial x_j}T_{ji}^*$. The r.h.s. of (1.4.13) represents another form of the resultant force acting on the fluid, per unit volume of porous medium, due tc pressure and gravity.

Our next objective is to express the average viscous resistance force, $\phi(\overline{\partial \tau_{ij}/\partial x_j})^f$, appearing in (1.4.4), in terms of the macroscopic velocity. As we shall see below, this force is made up of two parts: one within the fluid and the other at the fluid-solid interface. Before proceeding with the development of the required expressions,, we recall that (1.4.4) is valid for *any* fluid phase. In order to derive an equation for the macroscopic advective flux of a *particular fluid*, an appropriate constitutive equation, representing the relationship between the components τ_{ij} and the velocity components, V_i^m, for that particular fluid, must be introduced. For this purpose, let us *consider* the particular case of a *Newtonian fluid* (i.e., linearly viscous fluid), for fluid), for which the constitutive equation is given by

$$\tau_{ij} = \mu\left(\frac{\partial V_i^m}{\partial x_j} + \frac{\partial V_j^m}{\partial x_i}\right) + \lambda''\frac{\partial V_k^m}{\partial x_k}\delta_{ij}.$$ (1.4.14)

Hence

$$\frac{\partial \tau_{ij}}{\partial x_j} = \frac{\partial \mu}{\partial x_j}\left(\frac{\partial V_i^m}{\partial x_j} + \frac{\partial V_j^m}{\partial x_i}\right) + \frac{\partial \lambda''}{\partial x_i}\left(\frac{\partial V_j^m}{\partial x_j}\right)$$
$$+ \mu\frac{\partial^2 V_i^m}{\partial x_j \partial x_j} + (\mu + \lambda'')\frac{\partial}{\partial x_i}\left(\frac{\partial V_j^m}{\partial x_j}\right).$$ (1.4.15)

At this point we compare the first and third terms on the r.h.s. of (1.4.15), and assume

$$\left|\frac{\partial \mu}{\partial x_j}\frac{\partial V_i^m}{\partial x_j}\right| \ll \left|\mu\frac{\partial^2 V_i^m}{\partial x_j \partial x_j}\right|.$$ (1.4.16)

Practically, this condition is always satisfied, and, hence, we may delete the first term on the r.h.s. of (1.4.15). Similar considerations will also hold for the second term. Thus, we may replace (1.4.15) by

$$\frac{\partial \tau_{ij}}{\partial x_j} \simeq \mu \frac{\partial^2 V_i^m}{\partial x_j \partial x_j} + (\mu + \lambda'') \frac{\partial}{\partial x_i}\left(\frac{\partial V_j^m}{\partial x_j}\right). \tag{1.4.17}$$

By averaging (1.4.17), and assuming

$$\left|\overline{(\mu + \overset{\circ}{\lambda''})\left\{\frac{\partial}{\partial x_i}\left(\frac{\overset{\circ}{\partial V_j^m}}{\partial x_j}\right)\right\}}^f\right| \ll \left|\overline{(\mu + \lambda'')}^f \overline{\frac{\partial}{\partial x_i}\left(\frac{\partial V_j^m}{\partial x_j}\right)}^f\right|, \tag{1.4.18}$$

and

$$\left|\overline{\overset{\circ}{\mu}\left(\frac{\overset{\circ}{\partial^2 V_i^m}}{\partial x_j \partial x_j}\right)}^f\right| \ll \left|\overline{\mu}^f \overline{\left(\frac{\partial^2 V_i^m}{\partial x_j \partial x_j}\right)}^f\right|, \tag{1.4.19}$$

we obtain

$$\overline{\frac{\partial \tau_{ij}}{\partial x_j}}^f = \overline{\mu}^f \overline{\frac{\partial^2 V_i^m}{\partial x_j \partial x_j}}^f + (\overline{\mu}^f + \overline{\lambda''}^f)\overline{\frac{\partial}{\partial x_i}\left(\frac{\partial V_j^m}{\partial x_j}\right)}^f. \tag{1.4.20}$$

By employing (1.2.57) twice, we obtain

$$\begin{aligned}
\phi \overline{\frac{\partial^2 V_i^m}{\partial x_j \partial x_j}}^f &= \frac{\partial}{\partial x_j}\left(\phi \overline{\frac{\partial V_i^m}{\partial x_j}}^f\right) + \overline{\frac{\partial V_i^m}{\partial x_j}\nu_j}^{fs} \Sigma_{fs} \\
&= \frac{\partial}{\partial x_j}\left[\frac{\partial}{\partial x_j}(\phi \overline{V_i^m}^f) + \overline{V_i^m \nu_j}^{fs}\Sigma_{fs}\right] + \overline{\frac{\partial V_i^m}{\partial s_\nu}}^{fs}\Sigma_{fs} \\
&= \frac{\partial^2 q_i^m}{\partial x_j \partial x_j} + \frac{\partial}{\partial x_j}(\overline{V_i^m \nu_j}^{fs}\Sigma_{fs}) + \overline{\frac{\partial V_i^m}{\partial s_\nu}}^{fs}\Sigma_{fs},
\end{aligned} \tag{1.4.21}$$

and

$$\phi \overline{\frac{\partial}{\partial x_i}\left(\frac{\partial V_j^m}{\partial x_j}\right)}^f = \frac{\partial^2 q_j^m}{\partial x_i \partial x_j} + \frac{\partial}{\partial x_i}(\overline{V_j^m \nu_j}^{fs}\Sigma_{fs}) + \overline{\frac{\partial V_j^m}{\partial x_j}\nu_i}^{fs}\Sigma_{fs}. \tag{1.4.22}$$

Hence

$$\phi\overline{\frac{\partial \tau_{ij}}{\partial x_j}}^{f} = \overline{\mu}^f\left[\frac{\partial^2 q_i^m}{\partial x_j \partial x_j} + \frac{\partial}{\partial x_j}(\overline{V_i^m \nu_j}^{fs}\Sigma_{fs}) + \overline{\frac{\partial V_i^m}{\partial s_\nu}}^{fs}\Sigma_{fs}\right]$$

$$+ \ (\overline{\mu}^f + \overline{\lambda''}^f)\left[\frac{\partial^2 q_j^m}{\partial x_i \partial x_j} + \frac{\partial}{\partial x_i}(\overline{V_j^m \nu_j}^{fs}\Sigma_{fs}) + \overline{\frac{\partial V_j^m}{\partial x_j}\nu_i}^{fs}\Sigma_{fs}\right].$$

$$(1.4.23)$$

Another form of (1.4.23) is obtained by making use of (1.2.56). By taking a derivative of the last equation, we obtain

$$\frac{\partial^2 q_j^m}{\partial x_i \partial x_j} + \frac{\partial}{\partial x_i}(\overline{V_j^m \nu_j}^{fs}\Sigma_{fs}) = \frac{\partial}{\partial x_i}\left[\phi\overline{\left(\frac{\partial V_j^m}{\partial x_j}\right)}^{f}\right]. \qquad (1.4.24)$$

With this equation, (1.4.23) can be written as

$$\phi\overline{\frac{\partial \tau_{ij}}{\partial x_j}}^{f} = \overline{\mu}^f\left[\frac{\partial^2 q_i^m}{\partial x_j \partial x_j} + \frac{\partial}{\partial x_j}(\overline{V_i^m \nu_j}^{fs}\Sigma_{fs}) + \overline{\frac{\partial V_i^m}{\partial s_\nu}}^{fs}\Sigma_{fs}\right]$$

$$+ (\overline{\mu}^f + \overline{\lambda''}^f)\left[\frac{\partial}{\partial x_i}\phi\overline{\left(\frac{\partial V_j^m}{\partial x_j}\right)}^{f} + \overline{\frac{\partial V_j^m}{\partial x_j}\nu_i}^{fs}\Sigma_{fs}\right],$$

$$(1.4.25)$$

where we note that the last term on the r.h.s. vanishes for *microscopically isochoric flow*, i.e., when $\partial V_j^m / \partial x_j \equiv 0$.

Our next objective is to express the surface integrals appearing in (1.4.23) in terms of macroscopic variables.

At this point we introduce the important assumption that the *fluid adheres to the solid walls*, i.e.

$$\mathbf{V}^m|_{S_{fs}} \equiv \mathbf{V}_s|_{S_{fs}},$$

with $\mathbf{V}_s$ denoting the solid's velocity. This assumption is often referred to as the *'no–slip' condition*. Under this condition

$$\overline{V_i^m \nu_j}^{fs}\Sigma_{fs} \equiv \frac{1}{U_o}\int_{S_{fs}} V_i^m \nu_j \, dS = \frac{1}{U_o}\int_{S_{fs}} V_{si}\nu_j \, dS$$

$$\cong \ -\overline{V_{si}}^{s}\frac{\partial \phi}{\partial x_j} = -\frac{\partial \phi \overline{V_{si}}^{s}}{\partial x_j} + \phi\frac{\partial \overline{V_{si}}^{s}}{\partial x_j}, \qquad (1.4.26)$$

where we have introduced the approximation $V_{oi}|_{\mathcal{S}_{fs}} \cong \overline{V_{si}}^s$, and have made use of (1.2.58).

Similarly

$$\overline{V_j^m \nu_j}^{fs} \Sigma_{fs} = -\frac{\partial \phi \overline{V_{sj}}^s}{\partial x_j} + \phi \frac{\partial \overline{V_{sj}}^s}{\partial x_j}. \tag{1.4.27}$$

By making use of (1.3.8), recalling that $\mathcal{S}_{fs}$ is a material surface with respect to the fluid's mass, and that, approximately, $\mathbf{q} \simeq \mathbf{q}^m$, i.e., neglecting the diffusive flux of the fluid's mass, we obtain

$$\overline{\frac{\partial V_j^m}{\partial x_j} \nu_i}^{fs} \Sigma_{fs} \simeq -\overline{\frac{\partial V_j^m}{\partial x_j}}^f \frac{\partial \phi}{\partial x_i} = -\frac{1}{\phi}\left(\frac{\partial \phi}{\partial t} + \frac{\partial q_j^m}{\partial x_j}\right)\frac{\partial \phi}{\partial x_i}. \tag{1.4.28}$$

At any point in the close proximity of the $\mathcal{S}_{fs}$-surface, the fluid's velocity vector, $\mathbf{V}^m$, can be expressed as

$$\begin{aligned} \mathbf{V}^m &= V_\nu^m \boldsymbol{\nu} + V_{t'}^m \mathbf{t}' + V_{t''}^m \mathbf{t}'' \\ &= V_j^m \nu_j \boldsymbol{\nu} + V_j^m t_j' \mathbf{t}' + V_j^m t_j'' \mathbf{t}'', \end{aligned} \tag{1.4.29}$$

where $\boldsymbol{\nu}, \mathbf{t}'$ and $\mathbf{t}''$ are the unit vector normal to $\mathcal{S}_{sf}$ ($=$ the *principal normal*) and the two tangential ones to $\mathcal{S}_{fs}$, in the 'local' Cartesian coordinates at a point of $\mathcal{S}_{fs}$; s_ν, $s_{t'}$, and $s_{t''}$ are lengths along these vectors, respectively. These three mutually orthogonal unit vectors are obtained as the intersections of the *osculating plane*, the *normal plane* and the *rectifying plane*.

A similar expression can be written for $\mathbf{V}_s|_{\mathcal{S}_{fs}}$. By taking the difference $\mathbf{V}^m - \mathbf{V}_s|_{\mathcal{S}_{fs}}$, we obtain

$$V_i^m \quad V_{st}|_{\mathcal{S}_{fs}} = (V_j^m - V_{sj}|_{\mathcal{S}_{fs}})(\nu_j \nu_i + t_j' t_i' + t_j'' t_i''). \tag{1.4.30}$$

Since $\mathcal{S}_{fs}$ is a material surface with respect to the fluid's mass, we have in its immediate neighborhood

$$(V_j^m - V_{sj}|_{\mathcal{S}_{fs}})\nu_j = 0. \tag{1.4.31}$$

Then recalling that

$$\nu_i \nu_j + t_i' t_j' + t_i'' t_j'' = \delta_{ij}, \tag{1.4.32}$$

we obtain for the velocity components in the tangential plane

$$V_i'^m - V_{si}'|_{S_{fs}} = (V_j^m - V_{sj}|_{S_{sf}})(t_i't_j' + t_i''t_j'') = (V_j^m - V_{sj}|_{S_{fs}})(\delta_{ij} - \nu_i\nu_j),$$
$$(1.4.33)$$

where we have introduced the prime $(')$ symbol to emphasize that the components $V_i'^m$ here are in the plane tangential to the solid. At any *internal* point, the transformation $V_j^m = V_i^m \delta_{ij}$ holds. We have thus related the velocity components in the close neighborhood of the S_{fs}-surface to the latter's configuration.

With (1.4.33), we can now express the surface integral $\overline{\partial V_i^m/\partial s_\nu}^{fs} \Sigma_{fs}$, appearing in (1.4.23), in the form

$$\overline{\frac{\partial V_i^m}{\partial s_\nu}}^{fs} \Sigma_{fs} \equiv \frac{1}{U_o} \int_{S_{fs}} \frac{\partial V_i^m}{\partial s_\nu} \, dS$$

$$\simeq \frac{1}{U_o} \int_{S_{fs}} \frac{V_{si}'|_{S_{fs}} - V_i^m|_\Delta}{\Delta} \, dS$$

$$\simeq \Sigma_{sf}(\overline{V_{sj}|_{S_{fs}}}^{fs} - \overline{V_j^m}^f)\frac{1}{\Delta_c}(\delta_{ij} - \overline{\nu_i\nu_j}^{fs})$$

$$\simeq \Sigma_{sf}(\overline{V_{sj}}^s - \overline{V_j^m}^f)\frac{1}{\Delta_c}(\delta_{ij} - \overline{\nu_i\nu_j}^{fs}), \qquad (1.4.34)$$

where $V_i^m|_\Delta$ denotes the velocity at a point *within the fluid*, located at an elementary distance Δ from S_{fs}, and the characteristic value, Δ_c, of Δ, is defined by (1.4.34).

In deriving (1.4.34), we have made use of the 'no - slip' condition on S_{fs}, and the approximation $\overline{V_{sj}}^{fs} \cong \overline{V_{sj}}^s$, with the approximation sign changing into an equality one when the solid (not the solid matrix) is rigid, or approximated as such.

The distance Δ_c appearing in the approximation of the velocity gradient in (1.4.34), is a characteristic distance from the solid walls to the interior of the fluid phase. It is some measure of the pore size. For example, it can be taken as proportional to the *hydraulic radius*, $\Delta_v \equiv \Delta_f$, defined as the ratio of void space volume, U_{ov}, to the interface surface area, S_{sf}, i.e.

$$\Delta_c = \Delta_f/C_f, \quad \Delta_f = U_{of}/S_{fs} = \phi/\Sigma_{fs},$$

where C_f is a macroscopic dimensionless shape factor, and Σ_{fs} $(= S_{fs}/U_o)$ is the specific area of the f–s-interface within U_o, per unit volume of U_o.

With the above approximations, (1.4.34) takes the form

$$\overline{\frac{\partial V_i^m}{\partial s_\nu}}^{fs} \Sigma_{sf} \equiv \frac{1}{U_o}\int_{S_{fs}} \frac{\partial V_i^m}{\partial s_\nu}\, dS \;=\; -C_f\phi\alpha_{ij}\frac{\overline{V_j^m}^f - \overline{V_{sj}}^s}{\Delta_f^2}$$

$$= \; -\frac{C_f}{\Delta_f^2}\alpha_{ij}q_{rj}^m, \qquad (1.4.35)$$

where

$$\alpha_{ij} = \delta_{ij} - \overline{\nu_i\nu_j}^{fs} \qquad (1.4.36)$$

is another coefficient that characterizes the configuration of S_{fs}, and

$$\mathbf{q}_r^m = \phi(\overline{\mathbf{V^m}}^f - \overline{\mathbf{V}_s}^s) \qquad (1.4.37)$$

is the *relative mass weighted specific discharge* (i.e., relative to the solid).

By inserting (1.4.26), (1.4.27), (1.4.28), and (1.4.35) into (1.4.23), we obtain

$$\phi\overline{\frac{\partial \tau_{ij}}{\partial x_j}}^f \;=\; \overline{\mu}^f\left[\frac{\partial^2 q_{ri}^m}{\partial x_j\partial x_j} + \frac{\partial}{\partial x_j}\left(\phi\frac{\partial \overline{V_{si}}^s}{\partial x_j}\right)\right]$$

$$+(\overline{\mu}^f + \overline{\lambda''}^f)\left[\frac{\partial^2 q_{rj}^m}{\partial x_i\partial x_j} - \frac{1}{\phi}\left(\frac{\partial\phi}{\partial t} + \frac{\partial q_j^m}{\partial x_j}\right)\frac{\partial\phi}{\partial x_i} + \frac{\partial}{\partial x_i}\left(\phi\frac{\partial \overline{V_{sj}}^s}{\partial x_j}\right)\right]$$

$$-\overline{\mu}^f\frac{C_f}{\Delta_f^2}\alpha_{ij}q_{rj}^m. \qquad (1.4.38)$$

When we assume that the flow *at the microscopic level is isochoric*, i.e., $\nabla\cdot\mathbf{V}^m = 0$, in view of (1.4.35) and (1.4.27), the total viscous resistance reduces to

$$\phi\overline{\frac{\partial \tau_{ij}}{\partial x_j}}^f \;=\; \mu^f\left[\frac{\partial^2 q_{ri}^m}{\partial x_j\partial x_j} + \frac{\partial}{\partial x_j}\phi\left(\frac{\partial \overline{V_{si}}^s}{\partial x_j}\right)\right] - \mu^f\frac{C_f}{\Delta_f^2}\alpha_{ij}q_{rj}^m$$

$$= \; \overline{\mu}^f\frac{\partial}{\partial x_j}\left(\frac{\partial q_i^m}{\partial x_j} - \overline{V_{si}}^s\frac{\partial\phi}{\partial x_j}\right) - \overline{\mu}^f\frac{C_f}{\Delta_f^2}\alpha_{ij}q_{rj}^m, \qquad (1.4.39)$$

where we note the effect of solid deformation.

Finally, by assuming, *still for microscopically isochoric fluid motion*, that

$$\left|\frac{\partial q_{ri}^m}{\partial x_j}\right| \gg \left|\phi\frac{\partial \overline{V_{si}}^s}{\partial x_j}\right|,$$

equation (1.4.39) simplifies to

$$\phi\overline{\frac{\partial\tau_{ij}}{\partial x_j}}^f = \overline{\mu}^f\left[\frac{\partial^2 q_{ri}^m}{\partial x_j\partial x_j} - \frac{C_f}{\Delta_f^2}\alpha_{ij}q_{rj}^m\right]. \tag{1.4.40}$$

Henceforth, this equation will be employed as the expression for the *total* resistance to the flow.

It should be interesting to examine the term $\phi\overline{\partial\tau_{ij}/\partial x_j}^f$ that appears in (1.4.4). As stated above, this term expresses the total viscous resistance to the flow. By making use of (1.2.57), we can express this resistance as

$$\phi\overline{\frac{\partial\tau_{ij}}{\partial x_j}}^f = \frac{\partial\overline{\phi\tau_{ij}}^f}{\partial x_j} + \frac{1}{U_o}\int_{S_{fs}}\tau_{ij}\nu_j\,dS, \tag{1.4.41}$$

where all terms express resistance forces per unit volume of porous medium. We note here that the total viscous resistance is made up of two parts. The first is a resistance resulting from the internal friction in the fluid, while the second results from friction at the fluid-solid interface.

By comparing (1.4.41) with (1.4.40), we may conclude that only the last term on the r.h.s. of the latter expresses the drag at the fluid-solid interface, i.e.

$$\frac{1}{U_o}\int_{S_{fs}}\tau_{ij}\nu_j\,dS = -\overline{\mu}^f\frac{C_f}{\Delta_f^2}\alpha_{ij}q_{rj}^m. \tag{1.4.42}$$

The remaining term, $\overline{\mu}^f\partial^2 q_{ri}^m/\partial x_j\partial x_j$, represents the resistance to flow due to the internal friction *inside* the fluid.

Since the fluid-solid interface is a material surface, (1.3.10) holds when the flow is assumed to be microscopically isochoric. Under such conditions, the l.h.s. of (1.4.3) can be rewritten in the form

$$\phi\overline{\rho}^f\frac{\mathrm{D_m}\overline{V_i^m}^f}{\mathrm{Dt}} \equiv \overline{\rho}^f\left[\frac{\partial q_i^m}{\partial t} + \frac{\partial}{\partial x_j}\left(\frac{q_i^m q_j^m}{\phi}\right)\right]. \tag{1.4.43}$$

We have thus achieved our goal of expressing all the terms that appear in the averaged balance equation (1.4.4) in terms of macroscopic variables.

By inserting (1.4.13), (1.4.40) and (1.4.43) into the averaged momentum balance equation (1.4.4), we obtain

$$\overline{\rho}^f\left[\frac{\partial q_i^m}{\partial t} + \frac{\partial}{\partial x_j}\left(\frac{q_i^m q_j^m}{\phi}\right)\right] = -\phi\left(\frac{\partial\overline{p}^f}{\partial x_j} + \overline{\rho}^f g\frac{\partial z}{\partial x_j}\right)T_{ji}^*$$

$$+ \overline{\mu}^f\frac{\partial^2 q_{ri}^m}{\partial x_j\partial x_j} - \overline{\mu}^f\alpha_{ij}\frac{C_f}{\Delta_f^2}q_{rj}^m, \tag{1.4.44}$$

where T_{ij}^* and $\alpha_{ij}(\equiv \delta_{ij} - \overline{\nu_i\nu_j}^{sf})$ are two tensorial properties of the config-uration of the $\mathcal{S}_{fs}$-surface in saturated, single phase flow. Both coefficients are second rank tensors that constitute macroscopic representations of the microscopic configuration of the fluid - solid interface within the REV. The first, T_{ji}^*, transforms the local body force into a macroscopic one. The second, α_{ij}, introduces the effect of the configuration of the solid-fluid surface in the term that transforms part of the force resisting the flow at a point to an averaged resistance force at the fluid-solid interface.

If, in addition, $\mathbf{V}_s = 0$, then, by (1.3.9), written for $\beta \equiv s$, we have $\nabla\cdot\mathbf{q} = 0$, and $\mathbf{q} \equiv \mathbf{q}_r$, so that (1.4.44) further reduces to

$$\overline{\rho}^f\left(\frac{\partial q_i^m}{\partial t} + q_j\frac{\partial(q_i/\phi)}{\partial x_j}\right) = -\phi\left(\frac{\partial\overline{p}^f}{\partial x_j} + \overline{\rho}^f g\frac{\partial z}{\partial x_j}\right)T_{ji}^*$$
$$+ \overline{\mu}^f\frac{\partial^2 q_i^m}{\partial x_j\partial x_j} - \overline{\mu}^f\alpha_{ij}\frac{C_f}{\Delta_f^2}q_j^m. \tag{1.4.45}$$

Equation (1.4.44) represents an approximation of the macroscopic momentum balance equation for a fluid phase that fully occupies the void space of a porous medium domain. All the variables appearing in it, viz. q_{ri}^m, q_i^m, $\overline{p}^f$ and ϕ, are macroscopic ones.

We note that the effect of the bulk viscosity, λ'', has been eliminated, because we have assumed that the flow is microscopically isochoric. Henceforth, in this series of lectures, we shall continue to adopt this approximation.

We have presented here different levels of approximation of the macroscopic momentum balance equation for a fluid phase that fully occupies the void space of a porous medium domain. All the variables appearing in each of these equations are macroscopic ones.

Each term in (1.4.45) represents a force per unit volume of porous medium. The term on the l.h.s. represents the inertial force acting on the fluid, per unit volume of porous medium.

The first term on the r.h.s. of (1.4.45), represents the resultant force acting on the fluid, due to gravity and to pressure gradient, per unit volume of porous medium.

The second term on the r.h.s. of (1.4.45) represents the force acting on the fluid, due to the viscous resistance to its flow inside the fluid, per unit volume of porous medium. Actually, this force is exerted on the fluid within the REV, across the fluid-fluid portion of the surface bounding the REV.

The last term on the r.h.s. of (1.4.45), expresses the viscous resistance, or viscous drag force, exerted by the solid phase on the flowing fluid, at their contact surfaces within the REV, per unit volume of porous medium.

Often, in a given problem, one of these forces is much smaller with respect to the remaining ones and, therefore, may be deleted from the momentum balance equation. This may be so for the entire flow domain, or only for a part of it, on which our attention is focused. Hence, we may now proceed to consider simplified cases of (1.4.45).

1.4.2 Particular cases

CASE A. When the flow in a given domain is such that the viscous resistance force, due to the momentum transfer *at the solid-fluid interface*, is much larger than both the inertial force and the viscous resistance to the flow *inside the fluid*, i.e.

$$\left| \overline{\mu}^f \alpha_{ij} \frac{C_f}{\Delta_f^2} q_{rj}^m \right| \gg \left| \phi \overline{\rho}^f \frac{\mathrm{D}_m \overline{V_i^m}^f}{\mathrm{D}t} \right|, \tag{1.4.46}$$

and

$$\left| \overline{\mu}^f \alpha_{ij} \frac{C_f}{\Delta_f^2} q_{rj}^m \right| \gg \left| \overline{\mu}^f \frac{\partial^2 q_{ri}^m}{\partial x_j \partial x_j} \right|, \tag{1.4.47}$$

the momentum balance (or motion) equation (1.4.44) reduces to

$$q_{rj}^m = \phi(\overline{V_j^m}^f - \overline{V_{sj}}^s) = -\frac{k_{jl}}{\overline{\mu}^f}\left(\frac{\partial \overline{p}^f}{\partial x_l} + \overline{\rho}^f g \frac{\partial z}{\partial x_l} \right), \tag{1.4.48}$$

known as *Darcy's generalized law*. In this equation

$$k_{jl} = \frac{\phi \Delta_f^2}{C_f}(\alpha_{ji})^{-1} T_{il} = \frac{\phi^3}{C_f (\Sigma_{sf})^2}(\alpha_{ji})^{-1} T_{il}^* \tag{1.4.49}$$

is a coefficient related only to macroscopic parameters that describe the geometrical configuration of the void space. The coefficient k_{jl}–a second rank symmetrical tensor–is called the *permeability* of the porous medium (since we have assumed here that the fluid occupies the entire void space).

Although we present the general expression for permeability in the form of (1.4.49), the *actual value of k_{jl}-components of the tensor* **k**, *for any particular porous medium of interest, must be determined experimentally.*

By employing the methodology of deletion of terms that represent non-dominant effects (see, for example, Bear and Bachmat, 1990), it can be shown that conditions (1.4.46) and (1.4.47) prevail when the *Reynolds number*, Re, the *Darcy number*, Da, and the *Strouhal number*, St, defined by

$$\mathrm{Re} = \frac{V_c(k_c/\phi T_c^*)^{\frac{1}{2}}}{\nu_c}, \qquad \mathrm{Da} = \frac{k_c/\phi_c T_c^*}{L_c^2}, \qquad \text{and} \quad \mathrm{St} = \frac{L_c}{V_c t_c},$$

in which subscript c denotes characteristic values, are such that

$$\mathrm{St} \le 1, \qquad \text{and} \qquad \mathrm{ReDa}^{\frac{1}{2}} \ll 1. \tag{1.4.50}$$

In most regional groundwater problems, $\mathrm{Da}^{\frac{1}{2}} \ll 1$ and $\mathrm{St} \le 1$, and, therefore, (1.4.48) is valid even for Re equal to several tens.

It may be of interest to compare the above expressions for the permeability, as given by (1.4.49), with any of the forms of *Kozeny's equation* (see, for example, Bear, 1972, p. 166), recalling that $C_f = \phi/\Sigma_{sf}\Delta$. For example, one such form is $k = c_o T\phi^3/M^2$, where c_o is a dimensionless coefficient, $M \equiv \Sigma_{sf}$ and T is a coefficient called *tortuosity*. Also, in (1.4.49) we note the dependence of the permeability on Δ_f^2 and on a tensorial factor that represents the geometry of the void space.

Equation (1.4.48) is the more common form of the *motion equation*, for saturated flow in an anisotropic porous medium, when conditions (1.4.50) are satisfied.

Two particular cases of equation (1.4.48) are of practical interest.

• For the flow of a fluid of constant density, i.e., $\overline{\rho}^f = \text{const.}$, we may introduce the *piezometric head*

$$\overline{\varphi}^f = z + \frac{\overline{p}^f}{\overline{\rho}^f g}, \tag{1.4.51}$$

that expresses the mechanical energy (due to elevation and pressure) per unit weight of fluid. Then, $\mathbf{q}_r^m = \mathbf{q}_r$, and (1.4.48) reduces to

$$\mathbf{q}_{rj} = -K_{ji}\frac{\partial \overline{\varphi}^f}{\partial x_i}, \qquad \mathbf{q}_r = -\mathbf{K} \cdot \nabla \overline{\varphi}^f, \tag{1.4.52}$$

where the second rank symmetrical tensor

$$K_{ji} = k_{ji}\frac{\overline{\rho}^f g}{\overline{\mu}^f} \tag{1.4.53}$$

is a coefficient *called hydraulic conductivity*, and $-\nabla \overline{\varphi}^f$ is called the *hydraulic gradient*. We note that $\mathbf{K}$ depends on properties of both the fluid phase (in the form of $\overline{\rho}^f/\overline{\mu_f}^f$, often referred to as *fluidity*, equal to the reciprocal of the fluid's *kinematic viscosity*), and the solid matrix (through the permeability tensor, $\mathbf{k}$). The motion equation (1.4.52) is usually referred to as *Darcy's law*, as it was proposed, *on the basis of experiments* for one-dimensional flow in a column of homogeneous, nondeformable sand, by the French engineer Henry Darcy in 1856. Here, it was developed from first principles, as an *approximate macroscopic momentum balance equation.*

- For the flow of a *compressible fluid*, $\overline{\rho}^f = \overline{\rho}^f(\overline{p}^f)$, we may use the potential, $\overline{\varphi^*}^f$, defined by Hubbert (1940) in the form

$$\overline{\varphi^*}^f = z + \int_{p_o}^{\overline{p}^f} \frac{d\overline{p}^f}{g\overline{\rho}^f(\overline{p}^f)}, \tag{1.4.54}$$

where p_o is a reference fluid pressure, to rewrite (1.4.48) in the form

$$\mathbf{q}_r^m = -\mathbf{K}\cdot\nabla\overline{\varphi^*}^f. \tag{1.4.55}$$

The potential $\overline{\varphi^*}^f$ is often referred to as *Hubbert's potential.*

CASE B. When (1.4.46) is valid, i.e., when the inertial effects are negligible, but (1.4.47) is not, i.e., we do not neglect the effects of internal friction, expressed by the second term on the r.h.s. of (1.4.44), we obtain

$$-\frac{k_{jp}(T_{pk}^*)^{-1}}{\phi}\frac{\partial^2 q_{rj}^m}{\partial x_i\partial x_i} + \frac{k_{kj}}{\overline{\mu}^f}\left(\frac{\partial\overline{p}^f}{\partial x_j} + \overline{\rho}^f g\frac{\partial z}{\partial x_j}\right) + q_{rk}^m = 0.$$

$$\tag{1.4.56}$$

This equation is a good approximation of (1.4.44) when

$$\mathrm{Re}\,\mathrm{Da}^{\frac{1}{2}} \ll 1. \tag{1.4.57}$$

For an isotropic porous medium, and macroscopically isochoric flow, i.e., $\partial q_{ri}^m/\partial x_i = 0$, equation (1.4.56) reduces to

$$\frac{\overline{\mu}^f}{\phi T^*}\nabla\cdot\nabla\mathbf{q}_r^m - (\nabla\overline{p}^f + \overline{\rho}^f g\nabla z) - \frac{\overline{\mu}^f}{k}\mathbf{q}_r^m = 0 \tag{1.4.58}$$

where the scalars T^* and k are the tortuosity and permeability of the isotropic porous medium, respectively.

Equation, (1.4.58), with $\mathbf{q}^m \cong \mathbf{q}$, but without the coefficient $1/\phi T^*$ appearing in the first term, and without the gravity term, was proposed by

Brinkman (1948) and is known as *Brinkman's equation*. The coefficient $1/\phi T^*$ introduces at the macroscopic level the effect of the geometry of the microscopic fluid-solid interface.

CASE C. We may encounter situations, especially at the onset of flow and in oscillatory flows, in which the Strouhal number, St, may be large, such that the local acceleration, $\partial \overline{V_i^m}^f / \partial t$, may not be neglected in (1.4.44). Then, (1.4.44) reduces to

$$\frac{\bar{\rho}^f}{\bar{\mu}^f}(T_{ik}^*)^{-1}k_{kj}\frac{\partial \overline{V_j^m}^f}{\partial t} + \frac{k_{ij}}{\bar{\mu}^f}\left(\frac{\partial \bar{p}^f}{\partial x_j} + \bar{\rho}^f g\frac{\partial z}{\partial x_j}\right) + q_{ri}^m = 0. \qquad (1.4.59)$$

CASE D. Finally, let us consider the situation during a period, for example, immediately following the onset of flow from rest, in which the inertial effects are much larger than the viscous ones, as manifested by the second and third terms on the r.h.s. of (1.4.44). Employing the methodology of comparing pairs of terms appearing in (1.4.44), with each term representing a certain force, it follows that when

$$\text{Re Da}^{\frac{1}{2}} \gg 1 \qquad (1.4.60)$$

the contribution of the convective acceleration to the inertial force is much larger than the force of viscous resistance exerted by the solid matrix on the fluid.

Similarly, we compare the force resulting from the acceleration, with that acting on the fluid, due to the viscous resistance to the flow inside the fluid. We find that the inertial term will dominate when

$$\frac{t_c \nu_c}{L_c^2} \ll 1, \qquad \text{or} \qquad \text{Fo}^\nu \ll 1, \qquad (1.4.61)$$

and

$$\text{Fo}^\nu = \frac{t_c}{L_c^2/\nu_c} \qquad (1.4.62)$$

is the *Fourier number* associated with the fluid's viscosity, that expresses the ratio between the time interval during which a significant change in velocity (= momentum per unit mass of fluid), at a point, occurs, and the time required for significantly smoothing out spatial velocity differences by molecular transfer of momentum.

Under conditions (1.4.60) and (1.4.61), the effects of viscosity, both in the form of drag on the solid and as internal friction within the fluid, are eliminated. The motion equation (1.4.44) reduces to

$$\frac{1}{g}\frac{D_m\overline{V_i^m}^f}{Dt} = -\left(\frac{1}{\overline{\rho}^f g}\frac{\partial \overline{p}^f}{\partial x_j} + \frac{\partial z}{\partial x_j}\right)T_{ji}^*. \qquad (1.4.63)$$

1.4.3   Multiphase flow

So far, we have assumed that a single fluid occupies the entire void space. When two *immiscible fluids* occupy (disjoint) parts of the void space, because of *surface tension phenomena*, one of the fluids, called the *wetting fluid*, (subscript w), tends to adhere to the solid wall of the void space, while the other, called the *nonwetting fluid* (subscript n), tends to stay away from the solid wall. This means that a wetting fluid always coats the *entire* solid surface. This phenomenon takes place in pores originally occupied by a wetting fluid, where a small amount of the latter will *always* remain on the solid wall in the form of a very thin *film*, with a thickness of some tens of molecules, that adheres to the wall by strong molecular forces and cannot be displaced. Even if initially a pore is not occupied by a wetting fluid, the latter, when invading a pore, will tend to spread on the solid wall by *imbibition*, gradually displacing the nonwetting fluid. This film behaves practically like a " solid in that it can transmit momentum (practically) without changing it.

According to this picture, the total surface area surrounding the wetting fluid is made up of a wetting fluid-solid part (not including the thin film part) and a wetting fluid-nonwetting fluid part. In addition, there exist a wetting fluid-wetting fluid, and nonwetting fluid-nonwetting fluid, interfaces on the external boundary of an REV. Similarly, the nonwetting fluid is in contact with both the solid (across the film) and the wetting fluid, through portions of the total surface surrounding it. This means that in the derivation of an averaged momentum balance equation for the wetting fluid, the momentum transfer expressed by the integral over $\mathcal{S}_{fs}$-surface, say in (1.4.41), should be replaced by the sum of two integrals of the same integrand: one over the $\mathcal{S}_{w-s}$-surface, between the wetting fluid and the solid, and the other, over the $\mathcal{S}_{w-n}$-surface, between the wetting fluid and the nonwetting one. Analogously, when developing an averaged momentum balance equation for the non wetting phase, we shall have integrals of $\tau_{nw}\cdot\nu_n$ over the $\mathcal{S}_{n-s}$-surface and over the $\mathcal{S}_{n-w}$-surface. In this way, two expressions will be derived, one for each surface integral, in terms of averaged velocities and coefficients that represent the configuration of the phase within the REV. This conceptual model can serve as a basis for the derivation of flux expressions that *in principle* will exhibit *coupling* between immiscible phases, due to momentum transfer across the $\mathcal{S}_{w-n}$-surface. For example, a as a result of coupling, pressure gradient in one fluid will also cause movement in the other fluid (e.g., Bensabat, 1986).

We recall that the condition of continuity of momentum transfer across the boundary between two immiscible fluids is given by (1.3.35), repeated here for convenience in the form

$$
\begin{aligned}
[\rho V_i^m (V_j^m - u_j) - \sigma_{ij}]_{n,w} \nu_j &= \left(\frac{1}{r'} + \frac{1}{r''} \right) \gamma_{wn} \nu_i + \frac{\partial \gamma_{wn}}{\partial x_i} \\
&= \frac{2}{r^*} \gamma_{wn} \nu_i + \frac{\partial \gamma_{wn}}{\partial x_i},
\end{aligned} \tag{1.4.64}
$$

where we note the effect of γ_{wn}, which denotes the magnitude of the *surface-tension* between the wetting and the nonwetting fluids.

With this condition on the w-n-interface, we start from (1.4.25), replacing in it the integrals over the $\mathcal{S}_{sf}$ surface, by the sum of integrals over the $\mathcal{S}_{ws}$- and $\mathcal{S}_{ns}$-surfaces. Their development leads to the two flux equations:

Motion equation for the wetting phase

$$
\begin{aligned}
q_{rw\ell}^m = &- \kappa_{\ell j}^{ww} \left(\frac{\partial \overline{p_w}^w}{\partial x_j} + \overline{\rho_w}^w g \frac{\partial z}{\partial x_j} \right) \\
&- \kappa_{\ell j}^{wn} \left(\frac{\partial \overline{p_n}^n}{\partial \varpi_j} + \overline{\rho_n}^n g \frac{\partial z}{\partial \varpi_j} \right) - \kappa_{\ell j}^{\Sigma_{wn}} \overline{\frac{\partial \gamma_{wn}}{\partial \varpi_j}}^{nw}.
\end{aligned} \tag{1.4.65}
$$

motion equation for the nonwetting phase

$$
\begin{aligned}
q_{rn\ell}^m = &- \kappa_{\ell j}^{nn} \left(\frac{\partial \overline{p_n}^n}{\partial x_j} + \overline{\rho_n}^n g \frac{\partial z}{\partial x_j} \right) \\
&- \kappa_{\ell j}^{(nw)} \left(\frac{\partial \overline{p_w}^w}{\partial x_j} + \overline{\rho_w}^w g \frac{\partial z}{\partial x_j} \right) - \kappa_{\ell j}^{\Sigma_{nw}} \overline{\frac{\partial \gamma_{wn}}{\partial x_j}}^{nw},
\end{aligned} \tag{1.4.66}
$$

in which $\kappa_{\ell j}^{ww}$, $\kappa_{\ell j}^{wn}$, etc., are coefficients that involve the viscosities of the fluids, the saturations and various parameters that represent the geometry of the fluids within the void space.

In the motion equations (1.4.66) and (1.4.65), we note two main features:

- *Coupling* between the two phases. The forces due to pressure gradient and to gravity in one fluid, cause motion in the other one (due to the momentum exchange across their common (microscopic) boundary).

- Inhomogeneity in surface tension acts as an additional driving force. Surface tension, in turn, is a function of composition and temperature in the two fluids, e.g., in the schematic form

$$\gamma_{wn} = \gamma_{wn}(c_w, c_n, T),$$

where c_w and c_n represent the concentrations of components in the wetting fluid and in the nonwetting one, respectively, and T represents the temperature, assumed the same in both fluids. Then, we may replace $\overline{\partial \gamma_{wn}/\partial x_i}^{wn}$, as a further approximation, by terms that involve the gradients of the averaged concentrations and temperature.

The resulting flux equations indicate that, in principle, the mass transport problem for each phase is also coupled to those of component and energy transport in the two phases.

Contrary to the distribution of phases within an REV, as dictated by the concept of wettability, observations seem to support the notion that in multiphase flow, *each fluid tends to establish its own flow paths* through the void space. Accordingly, the wetting fluid tends to *completely* fill (and move through) the smaller pores, while the nonwetting fluid occupies the remaining, larger, pores, except for the thin film described above. The size of the largest pore occupied by the wetting fluid is determined, in the vicinity of a point, by the *wetting fluid saturation*, S_w, defined by

$$S_w = \theta_w/\phi, \qquad S_n(\equiv 1 - S_w) = \theta_n/\phi,$$

where S_n denotes the nonwetting fluid saturation.

If we accept this picture of phase distribution within the void space, the total surface area surrounding each fluid phase is made up of two parts: a fluid-fluid surface and a fluid-solid one. However, *as an approximation*, we often assume that

$$\mathcal{S}_{nw} \ll \mathcal{S}_{ns}, \qquad \mathcal{S}_{nw} \ll \mathcal{S}_{ws},$$

and that, therefore, the fluid-fluid momentum transfer is much smaller than the fluid-solid one. Then, the entire development presented above for saturated flow ($\theta_\alpha = \phi$), can be repeated, *separately*, once for the wetting fluid with $\mathcal{S}_{fs}$ replaced by $\mathcal{S}_{ws}$, and once for the nonwetting fluid, with

$\mathcal{S}_{fs}$ replaced by $\mathcal{S}_{ns}$. We shall then introduce the characteristic lengths $\Delta_w = U_{ow}/S_{ws}C_w$ and $\Delta_n = U_{on}/S_{ns}C_n$ for the wetting phase and for the nonwetting one, respectively. The areas S_{ws} and S_{ns} depend on the saturation, S_w, of the wetting fluid.

As a consequence, we shall obtain two *uncoupled* motion equations, one for each of the phases. **For the wetting phase (subscript w), the motion equation** will be

$$q^m_{rwj} = \theta_w(\overline{V^m_{wj}}^w - \overline{V_{sj}}^s) = -\frac{k_{wjm}}{\overline{\mu_w}^w}\left(\frac{\partial \overline{p_w}^w}{\partial x_m} + \overline{\rho_w}^w g\frac{\partial z}{\partial x_m}\right) \qquad (1.4.67)$$

where

$$k_{wjm} = \frac{\theta_w \Delta_w^2}{C_w}(\alpha_{wji})^{-1}T^*_{wim}. \qquad (1.4.68)$$

A similar expression can be written for the nonwetting phase. Because the partial area between the wetting fluid and the solid depends on the saturation, S_w, and so does T^*_{wim}, the permeability of each phase is a second rank tensor that varies with the saturation of that phase, viz.

$$k_{wij} = k_{wij}(S_w), \qquad k_{nij} = K_{nij}(S_n). \qquad (1.4.69)$$

We refer to these permeabilities as *effective permeabilities* to the wetting phase and to the nonwetting one, respectively. For an *isotropic* porous medium, the effective permeabilities reduce to scalar functions

$$k_w = k_w(S_w), \quad \text{and} \quad k_n = k_n(S_n).$$

For such a medium, a *relative permeability*, defined as the ratio between effective permeability and the permeability at full saturation, is often introduced.

For an *anisotropic* porous medium, each of the effective permeability components, of the second rank tensor of permeability, depends on the geometrical configuration of the void space and its characteristics (e.g., porosity) and on the saturation (that represents the fluid configuration within the void space). In general, the dependence on saturation may be different for the different components. Hence, the use of the relative permeability concept is not permitted.

To summarize this lecture, we now have an expression for the macroscopic advective flux of a fluid phase in single or multiphase flow. In the latter case, we have a flux equation for each fluid phase. Each of the flux expressions is nothing but an averaged momentum balance equation of the considered fluid phase, to which we have added certain simplifying assumptions.

1.5   Lecture Five: Complete Transport Model

In earlier lectures, we have presented the partial differential equations that describe the balance of extensive quantities of phases and components, at the macroscopic (continuum) level. These equations, one for every extensive quantity of every phase and/or component that is relevant to a given problem, constitute the core of the model that describes the transport of these quantities. We recall that the momentum balance equation for a fluid has been reduced to the simplified form of a motion equation. However, as we shall see shortly, additional information is required before a complete, or *well posed* mathematical model can be stated for a specific problem. This information includes the geometry of the boundary of the domain in which the transport occurs, and conditions that have to be satisfied on it. In addition, information is required on *constitutive relations* that describe the behavior, e.g., stress-strain relations, of the *specific* materials involved.

Our objectives in this lecture are to discuss, first, the various boundary conditions that are encountered in problems of transport of extensive quantities through porous media, and then to outline the standard content of a complete mathematical model of such problems.

1.5.1   Boundary conditions

In earlier lectures, we have seen that the partial differential balance equations that describe the transport of extensive quantities, have to be supplemented by constitutive equations that provide information on the behavior of the materials involved in any particular case of interest. We also need information on the numerical values of all parameters pertinent to the domain within which the transport problem takes place, and to the phases contained in it, as well as on all rates of production of all relevant extensive quantities. When all this information is put together, we obtain a closed set of equations that can be solved for the values of the state variables that describe the future behavior of the relevant extensive quantities in the considered domain.

However, the above set of equations has an infinite number of solutions. To obtain a unique solution that corresponds to a particular case of interest, it is necessary to provide supplementary information, namely:

- Geometry of the boundaries of the domain within which the transport phenomena under consideration take place.

- Description of the initial state in the considered domain (= *initial conditions*).

- Description of the interaction of the domain under consideration with its environment (referred to as *boundary conditions*).

Different boundary conditions lead to different solutions. Hence, the importance of stating them in a way that reflects the actual conditions of the problem on hand.

As we shall see soon, and more specifically in LECTURE 8, where we shall consider mass transport of a fluid phase, the various boundary conditions take the form of equalities between either the values of the dependent variables, or of fluxes, on 'both sides' of a considered boundary. In such equalities, *the information related to the external side must be known*. It is obtained from the conditions that are known to exist, or that are expected to be maintained in the environment. Sometimes, we are forced to *assume*, on the basis of past experience (or even to guess, subject to a-posteriori verification) the situation that will prevail on the external side of the boundaries in the future.

It is usually implicitly assumed that the interaction between the external world and the investigated domain is such that the former imposes known conditions on the latter, but is not affected by any processes that take place in it. However, sometimes the interaction between the two domains across the common boundary is such that the behavior in one domain is affected by what happens in the other, so that the conditions that will prevail on the common boundary are unknown a-priori. Under such conditions, the lack of information related to the external side of a considered domain, requires a simultaneous solution for both the considered domain and the one that is external to it.

Since we are interested in the statement of problems at the *macroscopic level*, the boundary conditions must also be stated at that level.

The boundary

Let us start by defining the boundary of a domain in which transport phenomena take place. Such a domain is bounded by a closed surface, possibly with segments at infinity. A considered domain may, for example, be a subdomain of a larger porous medium domain, separated from the former by an arbitrarily chosen (mathematical) surface, possibly coinciding with a *surface of discontinuity* in any macroscopic parameter of the solid matrix.

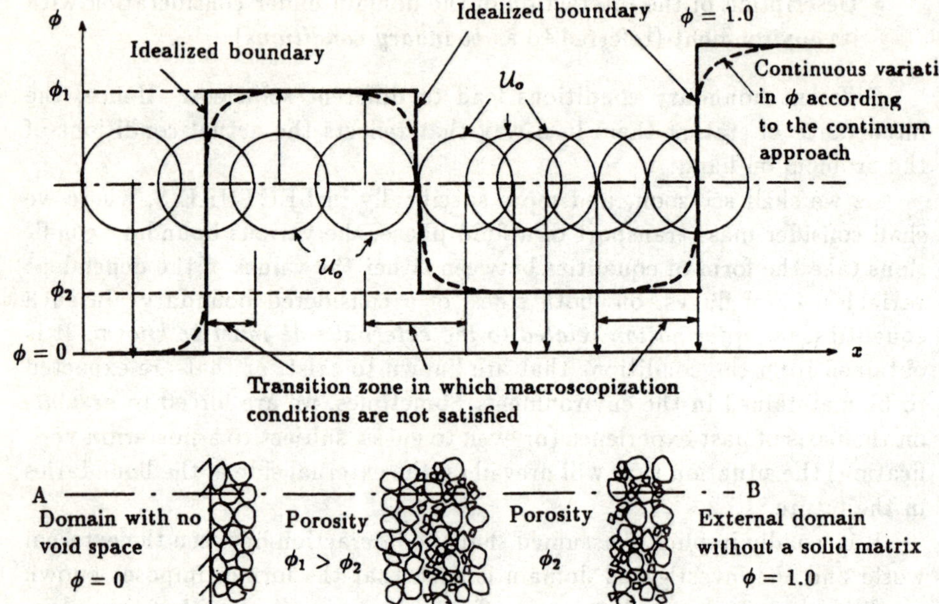

Figure 1.5.1: Macroscopic boundaries as an approximation.

Another example is the surface that separates a porous medium domain from its environment, when the latter may be devoid of solid matrix (i.e., $\phi = 1$), or it may be devoid of any pore space (i.e., impervious, with $\phi = 0$).

Figure 1.5.1 shows four regions of different media: a region with no void space, two regions containing solid matrices of different porosities and a region with no solid matrix. By employing the methodology of averaging over an REV in order to determine the porosity, we note that the latter varies *gradually* as we move along the x-axis. Although we observe rather steep changes in porosity (shown by the S-shaped dashed lines), no abrupt change occurs. Thus, in the strict sense of the continuum approach, sharp boundaries, that delineate the different media at the macroscopic level, do not exist anywhere along the x-axis. However, we recall that in defining a porous medium in the first lecture, we required that the variation of any macroscopic quantity (here porosity) over the REV be *linear*. If this condition *is not satisfied* in the region of transition from one porosity to the other, as shown, for example in 1.5.1, the actual variation in porosity in every region of transition must be replaced by an *idealized boundary in the form of a surface across which an abrupt change in porosity takes place.*

The boundary surfaces introduced in this way, divide the entire domain into subdomains separated from each other by sharp boundary surfaces. The continuum approach is applicable to each such subdomain. Across the boundaries, we *assume* the existence of a jump in porosity and in other macroscopic matrix properties. On the two sides of each such boundary, the values of these properties are obtained by extrapolating the spatial trend in property values, as the boundary is approached from within each subdomain. For convenience, we usually locate this boundary at the point corresponding to the mean porosity between the two adjacent regions In this way we have regular continuum domains for all phases present in the system *up to the boundary surface* on both its sides, with discontinuities along such boundary surfaces. Under such conditions, a given transport problem must be formulated separately for each subdomain, with appropriate conditions specified on the common boundaries. These conditions describe the interactions between the adjacent subdomains across their common boundaries.

A jump in volumetric porosity implies also a jump in the areal porosity. Figure 1.5.2 illustrates this jump, at the macroscopic level, in the case of a single fluid that occupies the entire void space. We note the presence of fluid-fluid, fluid-solid, solid-fluid and solid-solid segments along the boundary. Similar considerations apply to a number of fluid phases that occupy the void space. Using the nomenclature of Fig. 1.5.2, this means that

$$A_o^{sf}\big|_2 = A_o^{fs}\big|_1 \neq A_o^{sf}\big|_1 = A_o^{fs}\big|_2,$$

and

$$A_o^{fs}\big|_1 + A_o^{ff}\big|_1 \neq A_o^{ff}\big|_1 + A_o^{fs}\big|_2.$$

One may argue that because the probability of a fluid-solid interface coinciding with the boundary is negligibly small, the schematic diagram shown in Fig. 1.5.2 is unrealistic. Indeed, in the real (microscopic) multiphase porous medium, no jump can occur in any of the phases across a (macroscopic) boundary. Also, in the strict sense of a continuum model, we cannot

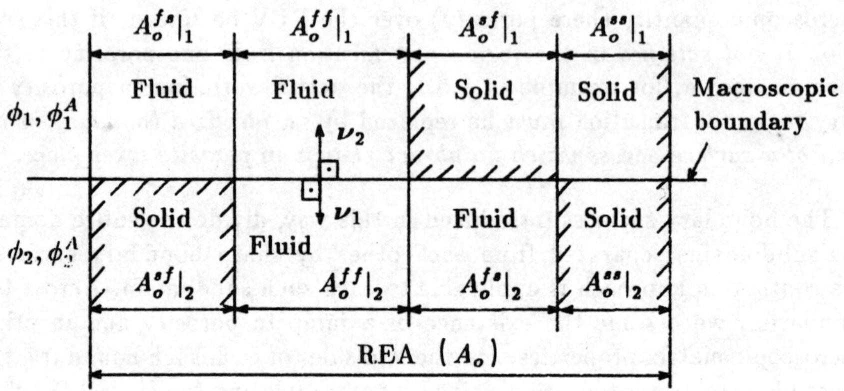

Figure 1.5.2: Schematic representation of a boundary with a jump in areal porosity.

have two areal porosities (or two values of any other solid matrix property) at the same point. However, as illustrated earlier, in the discussion that led to the introduction of abrupt changes in porosity, the proposed *conceptual model* does not attempt to describe the *real* medium exactly. Instead, our conceptual model *assumes* discontinuities in continuum properties across the boundary surface.

So far we have considered boundaries which are either some arbitrary surfaces, or surfaces of (hypothetical) discontinuities in solid matrix properties. However, a sharp boundary may be introduced *as an approximation* in three additional cases:

(a) **A fictitious abrupt boundary between two miscible fluids.** Here, the 'two miscible fluids' may, actually, be the same fluid, but with different concentrations of certain components. Due to phenomena to be discussed in LECTURE 8, a *transition zone* is usually created between two adjacent domains containing different concentrations of a considered component of the fluid occupying the void space. The concentration varies gradually across this transition zone. When the latter is narrow, relative to the two domains of interest, we may *approximate* it as a sharp boundary between two fluids. Across such a boundary, the concentration changes abruptly from that of one fluid to that of the other. The *interface* between fresh water and salt water in a coastal aquifer may serve as an example of such a fictitious sharp boundary.

(b)   **A boundary between two immiscible fluids.** Here, due to capillary effects, the saturation of each fluid varies gradually across a transition zone between the two fluids. Again, if this zone is narrow, relative to the domains of interest on both its sides, it may be approximated as a sharp boundary across which *we stipulate* a jump in the saturation of the considered fluids. The phreatic surface (LECTURE 6) may serve as an example; the two fluids are air and water.

(c)   **A boundary between different states of aggregation of the same material.** Under certain conditions, the material in the void space undergoes a change of phase. Evaporation, condensation, freezing, thawing and melting, may serve as examples. When this phenomenon takes place within a relatively narrow zone in a porous medium domain, we may introduce, *as an approximation*, an abrupt boundary surface across which the phase change is assumed to take place. We assume that the void space on each side of such a boundary is completely occupied by a different state of aggregation of the same material. This interface boundary may move as a result of the change of phase. Transport problems having such a moving boundary are called *Stefan problems*.

Because of the approximation involved in the introduction of sharp boundaries to replace the transition zones that exist in reality in all the cases mentioned above, measurements *within* the transition zones should not be expected to compare with predictions obtained by solving the mathematical models which include such (hypothetical, sharp) boundaries.

In general, a boundary surface may be stationary or moving. It may also be material or non-material with respect to any considered extensive quantity.

Let a surface within a continuum be described by the equation $F(\mathbf{x}, t) = C_1 = \text{const}$. As the surface moves, its shape may change, yet its equation remains unchanged. When the surface consists always of the same E-particles, it is called a *material surface* of the E-continuum.

With $\mathbf{u}(\mathbf{x}, t)$ denoting the velocity vector of a point, $\mathbf{x}$, on the surface $F = C_1$, its material derivative is given by

$$\frac{\mathrm{D}_F F}{\mathrm{D}t} \equiv \frac{\partial F}{\partial t} + \mathbf{u} \cdot \nabla F = 0, \tag{1.5.1}$$

where the subscript F is introduced in the material derivative to indicate that as the points belonging to the surface, F, move, they are observed by an observer moving with F. Hence

$$\mathbf{u}\cdot\nabla F \equiv u_\nu \frac{\partial F}{\partial \nu} = -\frac{\partial F}{\partial t}, \quad \frac{\partial F}{\partial \nu} = |\nabla F|, \tag{1.5.2}$$

where ν is the unit vector normal to the surface (always on the same side of the latter). Hence, u_ν is given by

$$u_\nu = -\frac{\partial F/\partial t}{\partial F/\partial \nu}, \quad \nu = \frac{\nabla F}{|\nabla F|}. \tag{1.5.3}$$

General boundary condition

The general boundary condition for any extensive quantity of a phase in a multiphase continuum, arises from the balance of that quantity as it is transported across the boundary. Intuitively, we could state that *in the absence of sources and sinks of a considered extensive quantity on the boundary*, the total amount of that quantity, transported by all phases present in the porous medium domain, must be conserved as it is being transported across the boundary. However, in general, we need information regarding the transport of the various extensive quantities by each phase separately. Can we assume that an extensive quantity is conserved separately for each phase? Had we assumed that this is the case, then the quantity transported across the boundary in any phase should enter *only that phase* on the other side of the boundary. However, this assumption violates our conceptualization of the boundary as shown in Fig. 1.5.2, where the jump in areal porosity requires that the phases comprising the medium cannot be completely continuous across the hypothesized macroscopic boundary. This implies that surfaces of contact must exist on the boundary between any given phase and all other phases. Across such surfaces of contact, extensive quantities may be transported from one phase to another.

To obtain the balance of an extensive quantity across a boundary for an individual phase in a multiphase system, let us consider a domain with only two phases, α and β. Figure 1.5.3 shows a domain, $\mathcal{U}(t)$, that is composed of two parts: $\mathcal{U}_1(t)$ and $\mathcal{U}_2(t)$, separated from each other by a common macroscopic surface, $\mathcal{S}^*(t)$. The domain $\mathcal{U}(t)$ is bounded by the surface $\mathcal{S}(t)$ $(= \mathcal{S}_1(t) + \mathcal{S}_2(t))$. The outward normal unit vector on $\mathcal{S}^*$, to $\mathcal{U}_\ell$, $\ell = 1, 2$, is denoted by ν_ℓ, with $\nu_1 = -\nu_2 \equiv \nu$. The outward normal unit vector on $\mathcal{S}$ is denoted by $\mathbf{N}$. Each surface may be moving at a velocity $\mathbf{u} \neq \mathbf{V}_E$.

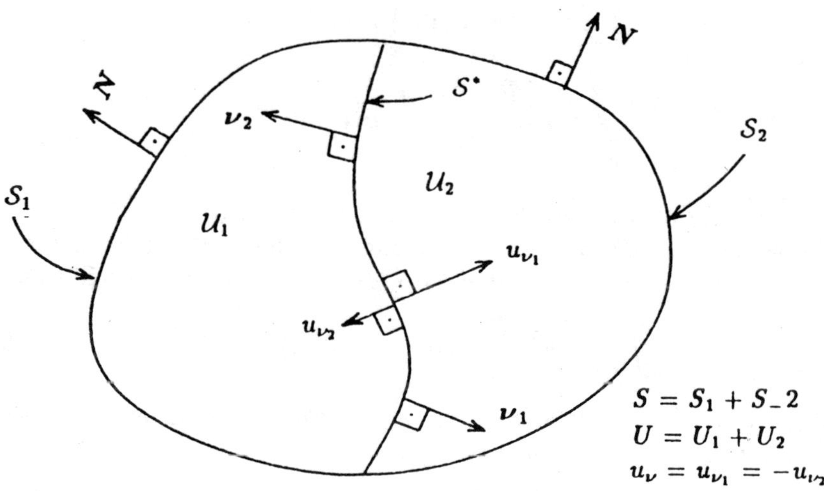

Figure 1.5.3: Definition sketch for developing the general boundary condition.

The macroscopic global balance of an extensive quantity, E, having a density, e, within $\mathcal{U}_\ell$, $\ell = 1, 2$, is given by (Bear and Bachmat, 1990)

$$\int_{\mathcal{U}_\ell(t)} \overline{\left(\frac{\partial e}{\partial t} + \nabla \cdot e\mathbf{V}^E - \rho\Gamma^E \right)} \, dU$$
$$- \int_{\mathcal{S}_\ell^*(t)} \overline{e(\mathbf{V}^E - \mathbf{u}) \cdot \boldsymbol{\nu}} \, dS + \int_{\mathcal{S}_\ell(t)} \overline{e(\mathbf{V}^E - \mathbf{u}) \cdot \mathbf{N}} \, dS,$$

$$(1.5.4)$$

where the overbar indicates the volume average defined by (1.2.50). On the l.h.s. of this equation, we note the terms that appear in the microscopic balance equation, averaged and then integrated over $\mathcal{U}_\ell(t)$. On the r.h.s., we note the possibility that the surfaces $\mathcal{S}_\ell(t)$ and $\mathcal{S}_\ell^*(t)$ are not material with respect to E.

The macroscopic global balance equation of E over $\mathcal{U}$ $(= \mathcal{U}_1 + \mathcal{U}_2)$, bounded by $\mathcal{S}\ (=\mathcal{S}_1+\mathcal{S}_2)$ is given by

$$\int_{\mathcal{U}(t)}\overline{\left(\frac{\partial e}{\partial t}+\nabla e\cdot\mathbf{V}^E-\rho\Gamma^E\right)}\,dU$$

$$=-\int_{\mathcal{S}^*(t)}\overline{\Gamma^{SE}}\,dS+\int_{\mathcal{S}_1(t)+\mathcal{S}_2(t)}\overline{e(\mathbf{V}^E-\mathbf{u})}\cdot\mathbf{N}\,dS,$$

$$(1.5.5)$$

where $\overline{\Gamma^{SE}}$ denotes the sources of E on $\mathcal{S}^*$, if such sources exist.

By writing (1.5.4), once for $\ell=1$ and then for $\ell=2$, adding the two equations, and comparing the sum with (1.5.5), we obtain the condition

$$\int_{\mathcal{S}^*(t)}\{[\overline{e(\mathbf{V}^E-\mathbf{u})}]_{1,2}\cdot\boldsymbol{\nu}+\overline{\Gamma^{SE}}\}dS=0. \qquad (1.5.6)$$

Since (1.5.6) is valid for any size of $\mathcal{S}^*$, the integrand must vanish at every point of $\mathcal{S}^*$. Hence, the condition

$$[\overline{e(\mathbf{V}^E-\mathbf{u})}]_{1,2}\cdot\boldsymbol{\nu}+\overline{\Gamma^{SE}}=0 \qquad (1.5.7)$$

must be satisfied at *every* point of the boundary $\mathcal{S}^*$.

By making use of (1.2.50), we rewrite (1.5.7) in the form

$$\sum_{(\alpha)}[\theta_\alpha\overline{e_\alpha(\mathbf{V}_\alpha^E-\mathbf{u})}^\alpha]_{1,2}\cdot\boldsymbol{\nu}+\sum_{(\alpha)}\theta_\alpha\overline{\Gamma_\alpha^{SE}}^\alpha=0 \qquad (1.5.8)$$

or

$$\sum_{(\alpha)}[\theta_\alpha\{\overline{e_\alpha}^\alpha(\overline{\mathbf{V}_\alpha}^\alpha-\mathbf{u})+\mathbf{J}_\alpha^{*E}+\overline{\mathbf{j}_\alpha^{EU}}^\alpha\}]_{1,2}\cdot\boldsymbol{\nu}+\sum_{(\alpha)}\theta_\alpha\overline{\Gamma_\alpha^{SE}}^\alpha=0, \qquad (1.5.9)$$

where $\sum_{(\alpha)}\theta_\alpha=1$, while $\mathbf{J}_\alpha^{*E}(\equiv\overline{\overset{\circ}{e}_\alpha\overset{\circ}{\mathbf{V}}_\alpha}^\alpha)$ denotes the dispersive flux of E and $\overline{\mathbf{j}_\alpha^{EU}}^\alpha$ denotes the averaged diffusive flux of E.

Equation (1.5.9) represents the *general macroscopic boundary condition for any extensive* quantity, E, in a porous medium domain. We note that it expresses the notion that E does not accumulate on the boundary. The difference in flux between the two sides is balanced by the (possible) production of E on the boundary. In the absence of sources on the boundary, (i.e., $\Gamma^{SE}\equiv 0$), equation (1.5.9) reduces to the continuity of the total flux across the boundary

$$\sum_{(\alpha)}[\theta_\alpha\{\overline{e_\alpha}^\alpha(\overline{\mathbf{V}_\alpha}^\alpha-\mathbf{u})+\mathbf{J}^{*E}+\overline{\mathbf{j}^{EU}}^\alpha\}]_{1,2}\cdot\boldsymbol{\nu}=0, \qquad (1.5.10)$$

as suggested from the beginning. We note that the total flux is made up of advection relative to the (possibly moving) boundary, dispersion and diffusion. We also note that this condition is written as a single one for the total flux *through all phases together.* We have no separate condition for each individual phase, unless the interphase boundary is material with respect to the extensive quantity.

In practice, usually, the boundary conditions stated in models describing transport phenomena in porous media, do not take the no-jump forms developed above. Instead, they take the form of specification of the values of state variables, or of their derivatives, on the boundary. However, since these conditions are supposed to represent the true physical aspects of the transport phenomena, they must satisfy the no-jump condition developed above, as the latter expresses nothing but a statement of continuity of the total flux of the considered extensive quantities across the boundary. One may think of side 1 of a boundary as being the considered porous medium domain, and side 2 as the external environment, for which *all* the information on values of state variables and fluxes are *known*.

Let us demonstrate the application of the general boundary condition to a number of particular cases. In LECTURES 6 and 8, the general boundary condition developed here will be applied to cases of transport of fluid mass and of mass of a component. We shall see there how the general condition–expressing flux continuity–is reduced to the more familiar and simple forms of boundary conditions, specified in terms of variables and/or their derivatives.

In presenting boundary conditions for specific extensive quantities, the intrinsic phase average symbol is omitted, as the entire discussion will be at the macroscopic level.

Boundary between two porous media in single phase flow

We consider a (macroscopic) boundary across which there exists a discontinuity in the (volumetric) porosity. The entire void space is occupied by a single fluid phase. No sources of any of the considered extensive quantities exist on the boundary.

The general boundary condition for any extensive quantity, (1.5.10), in which α represents either a solid, (s), or a fluid, (f), will always be used as a starting point. With this equation, conditions will refer to the combined transport through all the phases together.

However, whenever the considered extensive quantity is such that any portion of the boundary that is a solid-fluid, or fluid-fluid interface, is material with respect to that quantity, the no-jump boundary condition can be written separately for each phase.

(a) Conditions for the mass of a phase

In the absence of phase change (see below), the fluid-solid portion of the boundary is a material surface with respect to (advective, dispersive and diffusive) mass (transport). Hence

[A5.1] mass is not transported across the solid-fluid portion of the boundary.

Under this assumption, with e_α replaced by the mass density, ρ_α, the general (no-jump) boundary conditions for mass of a phase, written *separately* for each of the two phases, takes the form

$$[\theta_\alpha \rho_\alpha (\mathbf{V}_\alpha - \mathbf{u}) + \theta_\alpha (\mathbf{J}_\alpha^m + \mathbf{J}_\alpha^{*m})]_{1,2} \cdot \boldsymbol{\nu} = 0, \qquad \alpha = f, s, \qquad (1.5.11)$$

where $\mathbf{J}_\alpha^m$ and $\mathbf{J}_\alpha^{*m}$ denote the diffusive and the dispersive mass fluxes, respectively, of the α-phase. For a single fluid that completely occupies the void space, $\theta_f \equiv \phi$, and $\theta_s \equiv 1 - \phi$. No diffusive mass flux of the solid exists. We recall that all symbols in (1.5.11), and in what follows, denote intrinsic phase averages.

Let us consider a number of particular cases of (1.5.11).
We shall assume that

[A5.2] the dispersive flux of any extensive quantity in the solid phase may
be neglected.

Therefore, *for the solid phase*, (1.5.11) reduces to

$$[\rho_s (1 - \phi)(\mathbf{V}_s - \mathbf{u})]_{1,2} \cdot \boldsymbol{\nu} = 0. \qquad (1.5.12)$$

Let us further limit the discussion, here and henceforth in this lecture, to cases in which we assume that

[A5.3] the boundary is a material surface with respect to the mass of the
solid phase,

i.e., despite deformation and movement of the boundary, no solid mass is transported across it.

Hence

$$(\mathbf{V}_s - \mathbf{u})|_1 \cdot \boldsymbol{\nu} \;=\; (\mathbf{V}_s - \mathbf{u})|_2 \cdot \boldsymbol{\nu} = 0,$$

$$(1.5.13)$$

$$\mathbf{V}_s|_1 \cdot \boldsymbol{\nu} \;=\; \mathbf{V}_s|_2 \cdot \boldsymbol{\nu} = \mathbf{u} \cdot \boldsymbol{\nu}.$$

We note that assumptions [A5.2] and [A5.3] thus reduce (1.5.12) to an expression that provides no information on $[\rho_s]_{1,2}$. The mass density of the solid phase on each side of the boundary, may take on a different value. Assumption [A5.3], or (1.5.14), also implies that there is no jump in the normal component of the solid phase displacement, $\mathbf{w}$, i.e.

$$[\mathbf{w}]_{1,2} \cdot \boldsymbol{\nu} = 0.$$

$$(1.5.14)$$

The velocity of the boundary, defined by $\mathbf{u} \cdot \boldsymbol{\nu} = d(\mathbf{w} \cdot \boldsymbol{\nu})/dt$, should be used in all the expressions in which $\mathbf{u}$ appears.

For the fluid phase, (1.5.11) takes the form

$$[\phi\rho_f(\mathbf{V}_f - \mathbf{u}) + \phi(\mathbf{J}_f^m + \mathbf{J}_f^{*m})]_{1,2} \cdot \boldsymbol{\nu} = 0,$$

$$(1.5.15)$$

or, in view of (1.5.14)

$$[\rho_f \mathbf{q}_{rf} + \phi(\mathbf{J}_f^m + \mathbf{J}_f^{*m})]_{1,2} \cdot \boldsymbol{\nu} = 0,$$

$$(1.5.16)$$

where $\mathbf{q}_{rf}(= \phi(\mathbf{V}_f - \mathbf{V}_s))$ is the fluid's specific discharge relative to the solid. It can be expressed by an appropriate motion equation.

This is the most general boundary condition that can be stated for the fluid's mass. However, it may be further simplified by the introduction of additional assumptions. To begin with, we recall that in the continuum model employed here, a discontinuity in solid and fluid properties (including, for example, density) is permissible. However, we may raise the question as to whether such discontinuity is indeed possible when we consider *nonequilibrium thermodynamic processes*, such as mass diffusion, heat conduction, etc. Accordingly, although in the fictitious continuum model employed here, a discontinuity in any state variable is acceptable, we shall assume that

[A5.4] at every point on a macroscopic boundary, there exists no discontinuity in the (intrinsic phase averages of) scalar intensive quantities. *including* points on the boundary.

Under this assumption, we must have $[\rho_f]_{1,2} = 0$, as otherwise, the infinite density gradient associated, with the discontinuity in ρ_f, will instantaneously equalize the densities on both sides of the boundary, through the process of diffusion.

With this assumption, (1.5.16), reduces to

$$\rho_f[\mathbf{q}_{rf}]_{1,2}\cdot\boldsymbol{\nu} - [\phi\mathbf{D}^m_{fh}\cdot\nabla\rho_f]_{1,2}\cdot\boldsymbol{\nu} = 0, \tag{1.5.17}$$

where, as we shall see in LECTURES 7 and 8, we can express the sum of the diffusive and dispersive mass fluxes by

$$\mathbf{J}^m_f + \mathbf{J}^{*m}_f = -\boldsymbol{\mathcal{D}}^{*m}_f\cdot\nabla\rho_f - \mathbf{D}^m_f\cdot\nabla\rho_f = -\mathbf{D}^m_{fh}\cdot\nabla\rho_f. \tag{1.5.18}$$

Equation (1.5.17) is the most general boundary condition for mass flux on the boundary between two porous media, in the case of a single fluid phase that occupies the entire void space. Let us further assume that (see assumption [A5.1])

[A5.5] the sum of the fluid's diffusive and dispersive mass fluxes is negligible, with respect to the advective one.

Then, (1.5.17), combined with $[\rho_f]_{1,2} = 0$, reduces to

$$[\mathbf{q}_{rf}]_{1,2}\cdot\boldsymbol{\nu} = 0, \qquad [\phi\mathbf{V}_f]_{1,2}\cdot\boldsymbol{\nu} \equiv [\mathbf{q}_f]_{1,2}\cdot\boldsymbol{\nu} = [\phi]_{1,2}\mathbf{u}\cdot\boldsymbol{\nu}. \tag{1.5.19}$$

Since $[\phi]_{1,2} \neq 0$, equation (1.5.19) indicates that the normal components of the volume weighted velocity of the fluid, on both sides of the boundary, are not the same.

With [A5.5], but without [A5.4], the general boundary condition (1.5.17) reduces to

$$[\phi\rho_f(\mathbf{V}_f - \mathbf{u})]_{1,2}\cdot\boldsymbol{\nu} = 0, \quad \text{or} \quad [\rho_f\mathbf{q}_{rf}]_{1,2}\cdot\boldsymbol{\nu} = 0. \tag{1.5.20}$$

(b)   Condition for the mass of a component of a phase

For the mass of a component of a fluid phase, we also invoke assumption [A5.4], which in this case means

$$[c^\gamma]_{1,2} = 0, \tag{1.5.21}$$

where c^γ is the concentration of the γ-component in the fluid phase.

Actually, we have to be careful in writing the no-jump condition (1.5.21), as this condition may be valid only for the *chemical potential* of the component. Nevertheless, we shall continue to use here the notation c^γ, with the understanding that whenever necessary, it will be replaced by an appropriate measure of the concentration, or of the chemical potential of the considered component.

When also, the fluid-solid portion of the boundary is material with respect to the mass of the γ-component. Then, similar to the considerations leading to (1.5.17), we obtain here, for a component of a fluid

$$c^\gamma[\mathbf{q}_{rf}]_{1,2}\cdot\boldsymbol{\nu} - [\phi\mathbf{D}_{fh}^\gamma\cdot\nabla c^\gamma]_{1,2}\cdot\boldsymbol{\nu} = 0, \tag{1.5.22}$$

where we have expressed the sum of the diffusive and dispersive fluxes of the γ-component, in terms of ∇c^γ, as we did above for the mass of the phase.

By summing (1.5.22) for all components, with $\sum_{(\gamma)} c^\gamma = \rho_f$, we obtain (1.5.17). Equation (1.5.22) is the most general condition for the mass flux of a γ-component across a boundary, in terms of c^γ, provided we accept [A5.4], i.e., that $[\rho_f]_{1,2} = 0$.

If we also invoke assumption [A5.5], which leads to (1.5.19), then (1.5.22) reduces to

$$[\phi\mathbf{D}_{fh}^\gamma\cdot\nabla c^\gamma]_{1,2}\cdot\boldsymbol{\nu} = 0. \tag{1.5.23}$$

(c) Condition for the momentum of a phase

Unlike the two mass of a phase, momentum can cross (by a diffusive process) from one phase to another through a portion of the boundary that is common to both. Hence, we can only write a boundary condition for the porous medium as a whole.

As can be seen in the first term on the r.h.s. of (1.3.13), the total momentum flux of a phase is composed of advective, dispersive and diffusive fluxes. Accordingly, the general no-jump condition, (1.5.10), is written for the momentum flux of the *porous medium as a whole*, in the form

$$\sum_{(\alpha=f,s)} [\theta_\alpha\{\rho_\alpha\mathbf{V}_\alpha^m(\mathbf{V}_\alpha^m - \mathbf{u}) + \mathbf{J}_\alpha^{*M} - \sigma_\alpha\}]_{1,2}\cdot\boldsymbol{\nu} = 0, \tag{1.5.24}$$

where $\mathbf{J}_\alpha^{*M}$ denotes the dispersive momentum flux, actually, only in the fluid.

Making use of [A5.2] and [A5.3], equation (1.5.24) reduces to

$$[\phi\{\rho_f\mathbf{V}_f^m(\mathbf{V}_f^m - \mathbf{u}) + \mathbf{J}_f^{*M} - \sigma_f\} - (1 - \phi)\sigma_s]_{1,2}\cdot\boldsymbol{\nu} = 0, \tag{1.5.25}$$

which states a condition of no-jump, in the component of the total momentum flux normal to the boundary.

If we now assume that

[A5.6] the sum of the advective and dispersive momentum fluxes of the fluid across the boundary is much smaller than the diffusive one, expressed by $-\sigma_f \cdot \boldsymbol{\nu}$,

then (1.5.25) reduces to

$$[\sigma]_{1,2} \cdot \boldsymbol{\nu} \equiv [\phi\sigma_f + (1 - \phi)\sigma_s]_{1,2} \cdot \boldsymbol{\nu} = 0, \tag{1.5.26}$$

where $\sigma(\equiv \overline{\sigma})$ denotes the total stress in the porous medium.

As we have seen in LECTURE 3, deformation in a porous medium is not produced by the phase average stress in the solid, $\overline{\sigma}_s$, alone, but by the effective stress, $\overline{\sigma}'_s$, that takes into account the effect of the pressure in the fluid enveloping the solid phase. Accordingly, another way of writing (1.5.26) is

$$[\sigma'_s + \sigma_f]_{1,2} \cdot \boldsymbol{\nu} \equiv [\sigma'_s + \tau - p\mathbf{I}]_{1,2} \cdot \boldsymbol{\nu} = 0, \tag{1.5.27}$$

where $\sigma'_s = (1 - \phi)(\sigma_s - \sigma_f)$ is the *effective stress* that produces solid matrix deformation in a saturated porous medium, (with $\sigma'_s \equiv \overline{\sigma}'_s$, $\sigma_s \equiv \overline{\sigma}_s{}^s$, $\sigma_f \equiv \overline{\sigma}_f{}^f$).

Often, the effect of τ is neglected in the last equation, assuming that

$$|\tau \cdot \boldsymbol{\nu}| \ll |p\mathbf{I} \cdot \boldsymbol{\nu}|.$$

From [A5.4], we conclude that

$$[p_f]_{1,2} = 0. \tag{1.5.28}$$

When τ is neglected, it follows from (1.5.27) and (1.5.28) that

$$[\sigma'_s]_{1,2} \cdot \boldsymbol{\nu} = 0. \tag{1.5.29}$$

We wish to emphasize that (1.5.28) is obtained from [A5.4] and not from the boundary condition (1.5.24), which expresses the continuity of momentum flux.

(d) **Condition for energy**

Like momentum, energy can also be transported (as a diffusive flux) across portions of a boundary that are common to two phases (e.g., solid and fluid). We can, therefore, write a no-jump condition only for the porous medium as a whole, and not for the individual phases, separately.

The total energy flux in any phase, per unit area of porous medium, is given by

$$\phi \left[\rho(I + \tfrac{1}{2}(V^m)^2)\mathbf{V}^m - \sigma \cdot \mathbf{V}^m + \mathbf{J}^{*H} + \mathbf{J}^H \right].$$

Employing [A5.3] and (1.5.14), and neglecting the dispersive heat flux in the solid, $\mathbf{J}_s^{*H}$, the no-jump condition, (1.5.10), for the energy of the *porous medium as a whole*, takes the form

$$[\phi\rho_f(I_f + \tfrac{1}{2}(V_f^m)^2)(\mathbf{V}_f^m - \mathbf{u}) - \phi\sigma_f \cdot \mathbf{V}_f^m + \phi(\mathbf{J}_f^H + \mathbf{J}_f^{*H})]_{1,2} \cdot \boldsymbol{\nu}$$
$$+[(1 - \phi)\{-\sigma_s \cdot \mathbf{V}_s^m + \mathbf{J}_s^H\}]_{1,2} \cdot \boldsymbol{\nu} = 0. \tag{1.5.30}$$

Let us consider the expression

$$[\phi\sigma_f \cdot \mathbf{V}_f^m + (1 - \phi)\sigma_s \cdot \mathbf{V}_s^m]_{1,2} \cdot \boldsymbol{\nu}$$

appearing in (1.5.30). We can rewrite this expression in the form

$$[\phi(\mathbf{V}_f^m - \mathbf{u}) \cdot \sigma_f + (1 - \phi)(\mathbf{V}_s^m - \mathbf{u}) \cdot \sigma_s]_{1,2} \cdot \boldsymbol{\nu} + [\phi\sigma_f + (1 - \phi)\sigma_s]_{1,2} \mathbf{u} \cdot \boldsymbol{\nu}.$$

In view of (1.5.19), the first term reduces to $\phi(\mathbf{V}_f^m - \mathbf{u})[\sigma_f]_{1,2} \cdot \boldsymbol{\nu}$. However, in view of [A5.4], $[p]_{1,2} = 0$, and $[\tau]_{1,2} = 0$, and hence $[\sigma_f]_{1,2} = 0$. The second term vanishes in view of (1.5.14). The last term vanishes in view of (1.5.26). Hence, (1.5.30) reduces to

$$[\phi\rho_f \left(I_f + \tfrac{1}{2}(V_f^m)^2\right)(\mathbf{V}_f^m - \mathbf{u}) + \phi(\mathbf{J}_f^H + \mathbf{J}_f^{*H})]_{1,2} \cdot \boldsymbol{\nu}$$
$$+[(1 - \phi)\mathbf{J}_s^H]_{1,2} \cdot \boldsymbol{\nu} = 0. \tag{1.5.31}$$

As stated above, the fluid-solid portion of the boundary is not a material surface with respect to (the conductive part of the) energy transport. Hence, energy may be exchanged between the fluid phase and the solid one across their common portion of the boundary, so that energy is not conserved within any of the phases alone. Equation (1.5.31) includes the possible exchange between the two phases. It is impossible to separate this equation into two equations, one for each phase.

Let us add the assumption that

[A5.7] the fluid and solid phases are everywhere, locally, in thermal equilibrium,

i.e., $T_s = T_f \equiv T$. Then only one energy balance equation is required in
order to describe the thermodynamic state of the fluid-solid system as a
whole.

Furthermore, we assume that

[A5.8] the flux of kinetic energy is negligible, with respect to the thermal
one.

Then, equation (1.5.31) reduces to

$$\{\rho_f \mathbf{q}_{rf}[I_f]_{1,2} + [\sum_{(\alpha=f,s)} (\theta_\alpha \mathbf{J}_\alpha^H) + \theta_f \mathbf{J}_f^{*H}]_{1,2}\}\cdot\boldsymbol{\nu} = 0. \qquad (1.5.32)$$

Note that even if we choose to construct a model for $T_s \neq T_f$, we have a
condition only on the combined fluxes in the two phases.

At this point, we have to introduce specific (to the considered phases)
expressions for the internal energy, I_f, and for the diffusive and dispersive
heat fluxes appearing in (1.5.32). For example,

$$I_f = C_{Vf}T, \qquad \text{with} \qquad [C_{Vf}]_{1,2} = 0, \qquad (1.5.33)$$

and

$$\sum_{(\alpha=f,s)} \theta_\alpha(\mathbf{J}_\alpha^H + \mathbf{J}_\alpha^{*H}) = -\Lambda^H\cdot\nabla T,. \qquad (1.5.34)$$

with C_{Vf} denoting the fluid's specific heat at constant volume and Λ^H denoting the sum of the thermal conductivity of the combined fluid-solid continuum and the coefficient of thermal dispersion in the fluid (recalling that
all along we have neglected the dispersive heat flux in the solid).

Then, (1.5.32) can be rewritten in the form

$$\rho_f C_{Vf} \mathbf{q}_{rf}[T]_{1,2}\cdot\boldsymbol{\nu} - [\Lambda^H\cdot\nabla T]_{1,2}\cdot\boldsymbol{\nu} = 0. \qquad (1.5.35)$$

Since, by [A5.4]

$$[T]_{1,2} = 0, \qquad (1.5.36)$$

equation (1.5.32) reduces to the condition

$$[\Lambda^H \cdot \nabla T]_{1,2} \cdot \boldsymbol{\nu} = 0. \tag{1.5.37}$$

Boundary between two fluids

Here we consider the condition on an (assumed) abrupt boundary (= interface) between two fluids that occupy adjacent domains in a porous medium. Earlier in this lecture, this interface was introduced as an *approximation* of a transition zone from one fluid to another, whether these fluids are miscible, or immiscible. The solid matrix itself is assumed here to undergoes no change in porosity (nor in any other property) across this kind of boundary, so that $[\phi]_{1,2} = 0$.

Unlike the situation in the discussion thus far in the present lecture, here (1.5.14) is not valid. Again we wish to consider a number of examples.

(a) Mass of a phase

Since in this case, the boundary contains no fluid-solid part, we may write a no-jump condition separately for the mass of the fluid and for that of the solid. Hence, for the fluid, we obtain from (1.5.10)

$$[\phi \rho_f (\mathbf{V}_f - \mathbf{u}) + \phi(\mathbf{J}_f^m + \mathbf{J}_f^{*m})]_{1,2} \cdot \boldsymbol{\nu} = 0. \tag{1.5.38}$$

No condition is required for the mass of the solid. Note that in this case, the interface is not a material surface with respect to the solid.

Since

[A5.9] the (possibly moving) fluid-fluid interface is a material surface with respect to the mass of each fluid,

and recalling that we have a different fluid on each side of the boundary, equation (1.5.38) reduces to a separate condition for each fluid, viz.

$$\{\rho_f (\mathbf{V}_f - \mathbf{u}) - \mathbf{D}_{fh} \cdot \nabla \rho_f\}|_\ell \cdot \boldsymbol{\nu} = 0, \qquad \ell = 1,2, \tag{1.5.39}$$

where we have expressed the sum of diffusive and dispersive mass fluxes $\mathbf{J}_f^m$ and $\mathbf{J}_f^{*m}$, respectively, in terms of $\nabla \rho_f$. If conditions permit the application of [A5.5] to the considered problem, then for the fluid on each side of the interface, (1.5.39) reduces to

$$(\mathbf{V}_f - \mathbf{u})|_\ell \cdot \boldsymbol{\nu} = 0, \qquad \ell = 1,2, \tag{1.5.40}$$

and

$$V_f|_1 \cdot \boldsymbol{\nu} \equiv \mathbf{V}_f|_2 \cdot \boldsymbol{\nu} = \mathbf{u} \cdot \boldsymbol{\nu}. \tag{1.5.41}$$

Otherwise, (1.5.39) remains the boundary condition for each of the fluids. We note that no information on $[\rho_f]_{1,2}$ can be derived from (1.5.38), and, in general, we may have $[\rho_f]_{1,2} \neq 0$.

(b)   Mass of a component

Although we consider here a hypothetical abrupt interface between either miscible or immiscible fluids, let us discuss the case where

[A5.10] each fluid is composed of a number of components, some, but not all of which are capable of diffusing through the interface, into an adjacent fluid. However, the diffusing components constitute only a very small fraction of the fluid's density.

In this way, the interface remains a material surface with respect to the total mass of each of the fluids, but not necessarily with respect to the mass of a considered γ-component. The no-jump condition for the normal mass flux of a γ-component is expressed by

$$[\phi c_f^\gamma (\mathbf{V}_f - \mathbf{u}) + \phi (\mathbf{J}_f^\gamma + \mathbf{J}_f^{*\gamma})]_{1,2} \cdot \boldsymbol{\nu} = 0, \tag{1.5.42}$$

where c_f^γ denotes the concentration of the γ-component in the fluid phase. We note that in (1.5.42), a different fluid is present on each side of the boundary i.e., the subscript f has a different meaning for $\ell = 1$ and $\ell = 2$. In view of (1.5.40), equation (1.5.42) reduces to

$$[\mathbf{D}_{fh}^\gamma \cdot \nabla c_f^\gamma]_{1,2} \cdot \boldsymbol{\nu} = 0. \tag{1.5.43}$$

Equation (1.5.43) implies nothing with respect to $[c_f^\gamma]_{1,2}$, unless we invoke [A5.4] to yield (1.5.21), i.e., $[c_f^\gamma]_{1,2} = 0$. If [A5.4] is applicable, but [A5.5] is not, then (1.5.42) becomes

$$c_f^\gamma [\mathbf{V}_f]_{1,2} \cdot \boldsymbol{\nu} - [\mathbf{D}_{fh}^\gamma \cdot \nabla c_f^\gamma]_{1,2} \cdot \boldsymbol{\nu} = 0. \tag{1.5.44}$$

We should comment that the actual condition of equilibrium between adjacent phases, is one of no jump in chemical potential. For a component that does not diffuse across the boundary, when neither [A5.4], nor [A5.5] is applicable, the condition is

$$\{c_f^\gamma(\mathbf{V}_f - \mathbf{u}) - \mathbf{D}_{fh}^\gamma \cdot \nabla c_f^\gamma\}|_\ell \cdot \boldsymbol{\nu} = 0, \qquad \ell = 1, 2. \qquad (1.5.45)$$

When [A5.5] is applicable, the condition is

$$(\mathbf{D}_{fh}^\gamma \cdot \nabla c_f^\gamma)|_\ell \cdot \boldsymbol{\nu} = 0, \qquad \ell = 1, 2. \qquad (1.5.46)$$

(c) Momentum

In view of (1.5.40), condition (1.5.10), applied to momentum, becomes

$$[\sigma]_{1,2} \cdot \boldsymbol{\nu} \equiv \phi[\sigma_f]_{1,2} \cdot \boldsymbol{\nu} + (1 - \phi)[\sigma_s]_{1,2} \cdot \boldsymbol{\nu} = 0. \qquad (1.5.47)$$

Because there exists no solid-fluid portion of the boundary, we may write the no-jump condition separately for each phase. Hence, we write

$$\begin{aligned}
[\sigma_s]_{1,2} \cdot \boldsymbol{\nu} &= 0, & (1.5.48) \\
[\sigma_f]_{1,2} \cdot \boldsymbol{\nu} &= -[p_f]_{12}\mathbf{I} \cdot \boldsymbol{\nu} + [\tau_f]_{1,2} \cdot \boldsymbol{\nu}, & (1.5.49) \\
[\sigma_f]_{1,2} &= 0. & (1.5.50)
\end{aligned}$$

At this point, we invoke [A5.4], and neglect the effect of τ. This leads to (1.5.28). Then, (1.5.50) reduces to

$$[\sigma_s']_{1,2} \cdot \boldsymbol{\nu} = 0. \qquad (1.5.51)$$

(d) Energy

From $I_f = C_{Vf}T_f$, equation (1.5.34), $[\rho_s]_{1,2} = 0$, and $[T]_{1,2} = 0$, based on assumption [A5.4], and (1.5.36), the no-jump condition for the total energy flux (through both the solid and the fluid phases), takes the form

$$T\phi[\rho_f C_{Vf}(\mathbf{V}_f - \mathbf{u})]_{1,2} \cdot \boldsymbol{\nu} - [\Lambda^H \cdot \nabla T]_{1,2} \cdot \boldsymbol{\nu} = 0, \qquad (1.5.52)$$

where

$$\Lambda^H = \phi\lambda_f + (1 - \phi)\lambda_s + \phi\mathbf{D}_f^H,$$

defined by (1.5.34), is different on each side of the boundary..

If [A5.5] is also applicable, (1.5.52) is further reduced to

$$[\Lambda^H \cdot \nabla T]_{1,2} = 0. \qquad (1.5.53)$$

Shape of the interface

The interface velocity, $\mathbf{u}$, or its normal component, $\mathbf{u} \cdot \boldsymbol{\nu}$, appears in some of the equations developed above. Since $\mathbf{u}$ is a-priori unknown, these equations, in their present form, cannot be used as boundary conditions. In fact, in many cases, the shape and position of a moving interface, as represented by the equation, $F(\mathbf{x}, t) = 0$, may be regarded as an additional unknown variable for which a solution is sought. It is, therefore, necessary to express $F(\mathbf{x}, t)$ in terms of the state variables of the problem. Once $F(\mathbf{x}, t) = 0$ is known, (1.5.3) may be used to determine $\mathbf{u} \cdot \boldsymbol{\nu}$.

In principle, the explicit functional form of $F(\mathbf{x}, t)$ can be derived from any of the no-jump conditions of a considered problem. For example, if condition (1.5.53) is valid, we may use (1.5.3) to write

$$F(\mathbf{x}, t) \equiv (\Lambda^H \cdot \nabla T)|_1 \cdot \nabla F - (\Lambda^H \cdot T)|_2 \cdot \nabla F = 0. \qquad (1.5.54)$$

However, this equation expresses $F(\mathbf{x}, t)$ in terms of ∇F. It is much more convenient to derive $F(\mathbf{x}, t)$ from a condition of no-jump in the value of the state variable of the problem. Denoting the state variable, e.g., pressure, concentration, or temperature, by $\Omega(\mathbf{x}, t)$, the no-jump condition $[\Omega]_{1,2} = 0$ on the boundary, can be used to define the shape of the interface in the form

$$F(\mathbf{x}, t) \equiv \Omega(\mathbf{x}, t)|_1 - \Omega(\mathbf{x}, t)|_2 = 0. \qquad (1.5.55)$$

Actually, the problem is *non-linear*, as the values of $\Omega(\mathbf{x}, t)$ in both domains are not known a-priori. However, once these values have been determined (say, by some iterative technique), (1.5.55) can be used to obtain an explicit expression for the shape of the interface.

Boundary with fluid phase change

The assumed sharp boundary is between two states of aggregation of the *same* substance within the void space. Across such a (possibly moving) boundary, a change of state (e.g., freezing, thawing, evaporation, condensation) of the fluid phase within the void space may take place. We thus consider a substance, gas or liquid, that, under a considered range of pressure and temperature, may be either in a solid, a liquid, or a gaseous state. The solid matrix remains unchanged, although it is possible to consider cases in which the solid matrix melts, or dissolves. Such cases are not included here. We assume that

[A5.11] only a single state, rather than a mixture of states, of the fluid, is present in the void space on each side of the boundary.

Problems with such a boundary are often referred to as *Stefan type* problems.

Here, as in the previous examples, all boundary conditions are developed from the condition of no-jump in the total flux of a considered extensive quantity in crossing the boundary. Since in this case, $[1 - \phi]_{1,2} = 0$, the sum in (1.5.10) does not include the solid phase.

Accordingly, we shall consider here only conditions related to the fluid (in its different states) that occupies the void space on both sides of the boundary. No subscript will be used to indicate this phase.

(a)   Conditions for mass

From (1.5.10), we obtain

$$[\phi\rho(\mathbf{V} - \mathbf{u}) - \phi\mathbf{D}_h\cdot\nabla\rho]_{1,2}\cdot\boldsymbol{\nu} = 0, \qquad (1.5.56)$$

recalling that a different state of aggregation, gas or liquid, of the considered substance occupies the entire void space on each side of the boundary. Since $[\phi]_{1,2} = 0$, and $[\mathbf{u}]_{1,2}\cdot\boldsymbol{\nu} = 0$, equation (1.5.56) takes the form

$$[\rho\mathbf{V}]_{1,2}\cdot\boldsymbol{\nu} - [\rho]_{1,2}\mathbf{u}\cdot\boldsymbol{\nu} - [\mathbf{D}_h\cdot\nabla\rho]_{1,2}\cdot\boldsymbol{\nu} = 0. \qquad (1.5.57)$$

We are unable to further reduce (1.5.57) because, in general, $[\rho]_{1,2} \neq 0$, and $[\mathbf{V}]_{1,2} \neq 0$.

The fact that $[\rho]_{1,2} \neq 0$, stems from the nature of the phase change, except at the *critical point*, where the densities of the two states of the considered substance are identical. The jump in the normal component of the velocity arises from the change in the density, or specific volume, of the substance, upon the change of state. For example, if on one side of the boundary, the substance is at rest, the change in specific volume that occurs as a result of phase change, induces a velocity on the other side. For a change from solid to liquid, or vice versa, this effect may be negligible. However, changes from liquid to vapor, and vice versa, are associated with significant changes in the specific volume of the considered substances.

The boundary of phase change considered here *is not a material surface*, and

$$(\mathbf{V} - \mathbf{u})|_\ell\cdot\boldsymbol{\nu} \neq 0, \qquad \ell = 1,2,$$

since mass (of the considered substance) does cross the boundary. However, we recognize that in crossing the boundary, the mass assumes a different state of the same substance.

(b)   Condition for energy

Using assumption [A5.8], together with $[\mathbf{u}]_{1,2}\cdot\boldsymbol{\nu} = 0$, and solid phase properties, with $[\phi]_{1,2} = 0$, the condition of no-jump in the total energy flux in the direction normal to the boundary, reduces to

$$\phi[\rho I \mathbf{V}]_{1,2}\cdot\boldsymbol{\nu} - \phi[\rho I]_{1,2}\cdot\mathbf{u}\cdot\boldsymbol{\nu} - [\Lambda_h\cdot\nabla T]_{1,2}\cdot\boldsymbol{\nu} = 0. \tag{1.5.58}$$

Although we have assumed $[T]_{1,2} = 0$, the jump in internal energy, $[I]_{1,2} \equiv [\rho C_V T]_{1,2}$ ($\neq 0$) in (1.5.58), expresses the jump in the energetic state of the substance in the void space, on both sides of the boundary.

When we consider a change of phase from solid to liquid, or from liquid to vapor, the jump in energy represents the additional energy required to produce a more disordered state of the molecular structure, i.e., the energy required to further separate the molecules from each other. The jump in the energetic state of the substances is manifested by the fact that for the thermal flux, we have $[-\Lambda^H\cdot T]_{1,2} \neq 0$. This means that part of the sum of dispersive and diffusive heat fluxes entering, or leaving, the boundary, is compensating for the energy consumed by the phase change. This is an example of a sink, Γ^{SE}, on the boundary.

Alternative forms of (1.5.58), i.e., without phase change, are obtained by expressing the internal energy in terms of the *enthalpy*, h, viz.

$$[(\rho h - p)\mathbf{V}]_{1,2}\cdot\boldsymbol{\nu} - [\rho h - p]_{1,2}\mathbf{u}\cdot\boldsymbol{\nu} - [\Lambda^H\cdot\nabla T]_{1,2}\cdot\boldsymbol{\nu} = 0,$$

$$\tag{1.5.59}$$

where $h = I + p/\rho$, with the quantity p/ρ expressing the energy associated with the change in volume per unit mass. It is often assumed that

[A5.12] only a small part of the energy required to produce a change of phase is derived from the change in volume (Denbigh, 1955).

Then, (1.5.59) is reduced to

$$[\rho h \mathbf{V}]_{1,2}\cdot\boldsymbol{\nu} - L_{1,2}\mathbf{u}\cdot\boldsymbol{\nu} - [\Lambda^H\cdot\nabla T]_{1,2}\cdot\boldsymbol{\nu} = 0, \tag{1.5.60}$$

where $L_{1,2}(= [\rho h]_{1,2})$ is the *latent heat of phase change*, defined per unit volume. It represents the energy required to produce a change in the state of a unit volume of substance. The latent heat may also be defined with respect to the density of one of the states of a considered fluid phase, e.g., in the form

$$L_{1,2} = [\rho h]_{1,2} = \rho_1 L_1 - \rho_2 L_2,$$

where L_1 and L_2 are the latent heat per unit mass of states 1 and 2, present on sides 1 and 2, of a boundary, respectively.

In (1.5.60), the quantity $[\rho h \mathbf{V}]_{1,2}\cdot\boldsymbol{\nu}$, is associated with the change in volume of the fluid in the void space, due to phase change, and the resulting advective energy flux that is induced across the boundary. Some authors neglect this effect.

(c)   Boundary shape

In the case of a solid-liquid (i.e., a liquified solid), or a liquid-vapor (i.e., a gas containing the liquid's vapor), boundary, the shape of the boundary, $F(\mathbf{x},t) = 0$, can be derived from the condition $[T]_{1,2} = 0$. Hence

$$F(\mathbf{x},t) = T(\mathbf{x},t)|_1 - T(\mathbf{x},t)|_2 = 0. \tag{1.5.61}$$

The comment that follows (1.5.55) is also valid here. The relationship expressed by (1.5.61) is valid also for changes from a solid state to a liquid one, and vice versa (of the material that occupies the void space).

1.5.2   Content of a complete model

We now have all the elements needed in order to formulate the complete model of a problem, with the objective of forecasting the distribution of state variables within a porous medium domain at the macroscopic level.

Prior to the construction of a mathematical model for a given transport problem, we should analyze the given problem in order to

- define the geometry of the boundaries of the domain of interest,

- identify the various solid and fluid phases involved, and their components, and

- define those aspects of their behavior which are relevant to the problem
  on hand,

- define the relevant extensive quantities and the state variables of interest,

- identify the relevant processes of transport and transformation that take place,

- identify relevant sources and sinks of the considered extensive quantity (or quantities),

- identify the environment of the considered domain and the interactions between this environment and the considered domain, and

- state various, usually simplifying, assumptions aimed at obtaining a more manageable model of the problem.

The results of this analysis are summarized in the form of an explicitly stated list of statements and assumptions that constitute the *conceptual model* of the problem. The latter then serves as a base for the construction of the *mathematical model*. In fact, the mathematical model represents a quantified form of the conceptual one. While the latter constitutes a *verbal* statement of the problem, the former represents the given problem in a *quantified* form.

Actually, the construction of the conceptual model is an *iterative process*. Results obtained at any stage of the investigations, may lead to revisions of earlier assumptions.

The standard content of a mathematical model, consists of the following items:

- Formulation, at the macroscopic level, of the geometry of the surface that bounds the problem domain.

- A list of the macroscopic variables that will be used to describe the state of the system, e.g., $\overline{\rho_\alpha}^\alpha(\mathbf{x},t)$, $\overline{c_\alpha^\gamma}^\alpha$, $\overline{T_\alpha}^\alpha$ for an α-phase.

- Formulation of the macroscopic differential balance equations for the relevant extensive quantities, in terms of the variables specified above.

- Formulation of the relevant constitutive equations, i.e., the flux equation for each relevant extensive quantity and the equation of state for each relevant phase. The various statements should also include the numerical values of all the coefficients that appear in them.

- Formulations of the various source and sink functions for the relevant extensive quantities, in terms of the considered state variables.

  By inserting the flux equations, the equations of state and the expressions for the various source and sink functions into the appropriate differential balance equations, we obtain a set of partial differential equations, expressed in terms of the state variables selected to describe the state of the problem. This set of equations, to be satisfied at *all (macroscopic) points within the considered domain*, must be a closed one, i.e., it should contain a sufficient number of equations to enable a simultaneous solution for all the dependent variables of interest.

- Formulation of macroscopic *initial conditions*. These take the form of equations that define the values of the relevant dependent variables at all points within the considered domain, at some initial time, usually taken as $t = 0$.

- Formulation of boundary conditions. These take the form of equations that have to be satisfied by the relevant dependent variables at *all points of the domain's boundary.*

As emphasized earlier in this lecture, the type of boundary conditions to be specified in each particular case is motivated by the physical reality of the considered problem and by the interaction that takes place between the considered domain and its environment. The set of selected conditions should be consistent from the mathematical point of view.

A mathematical model is said to be *well posed* if it satisfies the following fundamental requirements:

- The solution must exist *(existence).*

- The solution must be uniquely determined *(uniqueness).*

- The solution should depend continuously on the data *(stability).*

The first requirement simply states that a solution does in fact exist. The second requirement stipulates completeness of the problem statement — redundancy, or ambiguity, should be avoided. The third requirement means that a variation of the given data (e.g., boundary and initial conditions and/or values of the coefficients) in a sufficiently small range should lead to a correspondingly small change in the solution. This requirement is also valid for approximate solutions. We require that a small error in satisfying the equation be reflected in only a small deviation of the approximate solution from the true one. If small errors in the data do not result in correspondingly small errors in the solution, we should conclude that the mathematical model is ill-posed.

In these lectures we have implicitly assumed that if, based on a thorough analysis of the physical reality, a description of this reality, albeit with certain simplifying assumptions, is formulated in the form of a mathematical model, the latter will be a well-posed one.

1.6 Lecture Six: Modelling Mass Transport of a Single Fluid Phase Under Isothermal Conditions

In this lecture, we consider the flow, or mass transport, *at the macroscopic level*, of a single component Newtonian fluid that occupies the entire void space, under isothermal conditions. We often refer to such flow as *saturated flow* under isothermal conditions. Unless otherwise specified, no special symbols will be used to indicate that dependent variables and parameters are macroscopic ones. Subscripts f and s will be used to indicate fluid and solid variables and parameters, respectively. The symbol $\mathbf{V}$ will be used to denote the fluid's velocity, $\mathbf{V}_f$.

Our objective is to construct the complete model for the flow of a compressible, or an incompressible fluid in a rigid, or a deformable solid matrix. In the discussion on boundary conditions, we shall focus on conditions that are required only for the mass transport part of a model. If the model involves the simultaneous transport of additional extensive quantities, additional boundary conditions will be required.

1.6.1 Basic mass balance equations

Let us start from the averaged mass balance equation (1.3.4), written for a single fluid phase that occupies the entire void space, i.e., $\theta_\alpha = \phi$ =porosity. We shall make two assumptions:

[A6.1] The (microscopic) fluid-solid interface is a material surface with respect to the fluid's mass, i.e., no mass of the considered fluid phase crosses it. Hence, the surface integral in (1.3.4) vanishes.

[A6.2] The dispersive and diffusive mass fluxes (to be discussed in detail in subsequent lectures) are much smaller than the advective one, and may, therefore, be neglected.

Since the fluid's advective mass flux is expressed by $\rho_f \mathbf{V}$, while $-\mathbf{D} \cdot \nabla \rho_f$ and $-\mathcal{D}^* \cdot \nabla \rho_f$ express the dispersive and diffusive mass fluxes, respectively (LECTURES 7 and 8), the last assumption implies that

$$|\rho_f \mathbf{V}| \gg |\mathbf{D}\cdot\nabla\rho_f + \boldsymbol{\mathcal{T}}^*\cdot\nabla\rho_f|. \qquad (1.6.1)$$

Expressing the diffusive and dispersive fluxes by (1.7.14), and (1.8.1), condition (1.6.1) can be represented in the form

$$|\rho_f V_i| \gg \left\{ \left| a_{ijkl}\frac{V_k V_l}{V}\frac{\partial\rho_f}{\partial x_j} \right| + \left| \mathcal{D}^*_{ij}\frac{\partial\rho_f}{\partial x_j} \right| \right\}. \qquad (1.6.2)$$

Let us use this opportunity to present in detail the method of deleting nondominant terms (which represent nondominant effects) from a balance equation (see, for example, Bear and Bachmat, 1990). The method has been mentioned several times in earlier lectures.

According to this method, we use characteristic values of state variables and parameters, denoted by subscript c, to rewrite equations in dimensionless forms.

Denoting dimensionless quantities by an asterisk, we write

$$(\Delta x_i)^* = \frac{(\Delta x_i)}{L_c}, \qquad a^*_{ijkl} = \frac{a_{ijkl}}{a_c}, \qquad \rho^*_f = \frac{\rho_f}{\rho_c},$$

$$\left(\frac{\partial\rho_f}{\partial x_j}\right)^* = \left(\frac{\partial\rho_f}{\partial x_j}\right) \bigg/ \left|\frac{\partial\rho_f}{\partial x_j}\right|_c, \qquad V_i = \frac{V_i}{V_c}, \qquad (1.6.3)$$

The characteristic values are selected such that

$$V_c = \left|V_i\right|_{\max}, \qquad \left|\frac{\partial\rho_f}{\partial x_j}\right|_c = \left|\frac{\partial\rho_f}{\partial x_j}\right|_{\max}, \qquad \mathcal{D}^*_c = \mathrm{Max}|\mathcal{D}^*_{ij}| \quad \text{for all } i,j,$$

$$\rho_c = (\rho_f)|_{\max}, \qquad L^{(\rho)}_c = (\rho_f)|_{\max}/|\partial\rho_f/\partial x_j|_{\max},$$

and a_c is the largest dispersivity component. For example, in an isotropic porous medium, we usually select $a_c = a_L$.

With these dimensionless quantities, we rewrite (1.6.1) in the form

$$\frac{\left| a_{ijkl}\frac{V_k V_l}{V}\frac{\partial\rho_f}{\partial x_j} + \mathcal{D}^*_{ij}\frac{\partial\rho_f}{\partial x_j} \right|}{|\rho_f V_i|}$$

$$\equiv \frac{a_c}{L^{(\rho)}_c}\frac{\left| a^*_{ijkl}\frac{V^*_k V^*_l}{V^*}(\frac{\partial\rho_f}{\partial x_j})^* \right|}{|\rho^*_f V^*_i|} + \frac{\mathcal{D}^*_c}{L^{(\rho)}_c V_c}\left| \mathcal{D}^*_{ij}\frac{\partial\rho_f}{\partial x_j} \right|^* \ll 1.$$

$$(1.6.4)$$

Since we have selected the characteristic values such that the ratios of dimensionless quantities are of the *order of magnitude of unity*, the condition expressed by (1.6.1) holds whenever

$$\frac{a_c}{L_c} + \frac{\mathcal{D}_c^*}{L_c^{(\rho)} V_c} = \frac{a_c}{L_c^{(\rho)}} \left(1 + \frac{1}{\mathrm{Pe}^{(a)}} \right) \ll 1, \tag{1.6.5}$$

The dimensionless group

$$\mathrm{Pe}^{(a)} = \frac{a_c V_c}{\mathcal{D}_c^*}$$

is called *Peclet number*. It expresses the magnitude of the dispersive flux relative to the diffusive one. Other kinds of Peclet numbers, each with a different characteristic length, may also be defined. For example

$$\mathrm{Pe}^{(E)} = \frac{L_c^{(E)} V_c^E}{\mathcal{D}_c^E}.$$

We note the possibility of a different characteristic length scale for different variables.

The dimensionless ratio $a_c/L_c^{(\rho)} \equiv (a_c V_c/L_c^{(\rho)} V_c)$ expresses the magnitude of the dispersive flux relative to the advective one.

Since, usually, $L_c \gg a_c$, the above condition always holds for $Pe \geq 1$. Assumption [A6.2] holds as long as $Pe \gg a_c/L_c$.

With these two assumptions, (1.3.4), written for the fluid's mass, reduces to

$$\frac{\partial \phi \rho_f}{\partial t} = -\nabla \cdot \rho_f \mathbf{q}, \tag{1.6.6}$$

where $\rho_f, \mathbf{V}_f$ and $\mathbf{q}\ (= \phi \mathbf{V}_f)$, denote the density, the velocity and the specific discharge of the fluid, respectively. Equation (1.6.6) is identical to (1.3.7), with $\theta = \phi$.

With similar considerations, we could be led to (1.3.6), written in terms of the mass weighted velocity, $\mathbf{V}^m$.

In what follows, we shall apply the mass balance equation (1.6.6) to various specific cases of fluids and solid matrices. Note that (1.6.6), and all subsequent fluid mass balance equations, do not include a fluid source term. Such a term can always be added when necessary. Comments to that effect will be made as we continue.

The mass balance for the solid phase can be expressed by an equation similar to (1.6.6), viz.

$$\frac{\partial(1-\phi)\rho_s}{\partial t} = -\nabla\cdot\{(1-\phi)\rho_s\mathbf{V}_s\}, \qquad (1.6.7)$$

where ρ_s and $\mathbf{V}_s$ denote the solid's density and velocity, respectively.

We recall that Darcy's law gives the fluid's velocity, or specific discharge, relative to the solid. In (1.6.6), for example, we have

$$\mathbf{q} = \phi\mathbf{V}_f = \mathbf{q}_r + \phi\mathbf{V}_s, \qquad (1.6.8)$$

where $\mathbf{q}$, $\mathbf{V}_f$ and $\mathbf{V}_s$ are with respect to a fixed coordinate system, while $\mathbf{q}_r$ is the specific discharge of the fluid relative to the (possibly moving) solid.

The fluid's mass balance, (1.6.6), can now be rewritten in the form

$$\phi\frac{\partial\rho_f}{\partial t} + \rho_f\frac{\partial\phi}{\partial t} + \phi\mathbf{V}_f\cdot\nabla\rho_f + \rho_f\nabla\cdot(\mathbf{q}_r + \phi\mathbf{V}_s) = 0, \qquad (1.6.9)$$

or

$$\rho_f\nabla\cdot\mathbf{q}_r + \phi\frac{D_f\rho_f}{Dt} + \rho_f\frac{D_s\phi}{Dt} + \rho_f\phi\nabla\cdot\mathbf{V}_s = 0, \qquad (1.6.10)$$

where

$$\frac{D_s(..)}{Dt} = \frac{\partial(..)}{\partial t} + \mathbf{V}_s\cdot\nabla(..)$$

is the *material derivative* of $(..)$ with respect to the (possibly moving) solid.

Equation (1.6.7) can be rewritten in the form

$$\frac{1}{1-\phi}\frac{D_s(1-\phi)}{Dt} + \frac{1}{\rho_s}\frac{D_s\rho_s}{Dt} + \nabla\cdot\mathbf{V}_s = 0. \qquad (1.6.11)$$

By eliminating $\nabla\cdot\mathbf{V}_s$ from (1.6.10) and (1.6.11), we obtain

$$\nabla\cdot\mathbf{q}_r + \phi\frac{1}{\rho_f}\frac{D_f\rho_f}{Dt} + \frac{1}{1-\phi}\frac{D_s\phi}{Dt} - \phi\frac{1}{\rho_s}\frac{D_s\rho_s}{Dt} = 0. \qquad (1.6.12)$$

The following assumption is usually introduced already at this stage.

[A6.3] The solid phase (at the microscopic level!) preserves its volume, i.e.,

$$(\dot{dU_s}) = 0.$$

This condition means that at the microscopic level, the solid's rate of dilatation vanishes, i.e.

$$\varepsilon_s \equiv \sum_{(i)} \varepsilon_{ii} = 0, \qquad \text{and} \qquad \frac{D_s \varepsilon_s}{Dt} \equiv \nabla \cdot \mathbf{V}_s = 0.$$

Hence, macroscopically, by (1.2.83), applied to $e = \rho_s$, $\mathbf{V}^E \equiv \mathbf{V}_s$, we have $\overline{\dot{\rho}_s}^s = \overline{\rho_s}^{\dot{s}}$. This means that while assumption [A6.3] implies that microscopically $\dot{\rho}_s = 0$, macroscopically, we also have $\overline{\dot{\rho}_s}^s = 0$, i.e., the solid's macroscopic density remains unchanged, while the porous medium as a whole may undergo deformation. We recall that, as explained in LECTURE 3, with a constant solid density, the deformation of a porous medium, i.e., the deformation of its solid skeleton, is manifested only by changes in porosity. In a granular porous medium, the deformation of the solid skeleton is attributed to the rolling and slipping of grains with respect to each other. This results in the rearrangement of the grains at a different porosity. Obviously, it is possible to take into account also changes in ρ_s.

With assumption [A6.3], equation (1.6.12) reduces to

$$\nabla \cdot \mathbf{q}_r + \psi \frac{1}{\rho_f} \frac{D_f \rho_f}{Dt} + \frac{1}{1 - \phi} \frac{D_s \phi}{Dt} = 0. \tag{1.6.13}$$

Another possible form of this equation is

$$\nabla \cdot \rho_f \mathbf{q}_r + \phi \frac{D_s \rho_f}{Dt} + \rho_f \frac{1}{1 - \phi} \frac{D_s \phi}{Dt} = 0. \tag{1.6.14}$$

Let us consider a number of particular cases of this equation.

1.6.2 Stationary nondeformable solid skeleton

In this subsection we assume that

[A6.4] the solid skeleton is rigid (= nondeformable) and fixed in space.

In such a medium, there is no change in porosity, i.e., $\partial \phi / \partial t = 0$. If the entire porous medium is also stationary in space, i.e., $\mathbf{V}_s = 0$, we have the relationship

$$\frac{D_s \phi}{Dt} = 0. \tag{1.6.15}$$

We could regard (1.6.15) itself as an assumption that is equivalent to [A6.4].

With assumption [A6.4], which also leads to $\mathbf{q} \equiv \mathbf{q}_r$, equation (1.6.6) reduces to

$$\nabla \cdot \rho_f \mathbf{q} + \phi \frac{\partial \rho_f}{\partial t} = 0. \tag{1.6.16}$$

CASE A. The fluid is compressible, i.e., $\rho_f = \rho_f(p)$. By inserting the motion equation (1.4.55), valid for a compressible Newtonian fluid, into (1.6.16), we obtain

$$\nabla \cdot \{\rho_f \mathbf{K} \cdot \nabla \varphi^*\} = \phi \beta_p \rho_f^2 g \frac{\partial \varphi^*}{\partial t}, \tag{1.6.17}$$

where β_p is the fluid's coefficient of compressibility, defined by

$$\beta_p = \frac{1}{\rho_f} \frac{\partial \rho_f}{\partial p}\bigg|_{T,\rho^\gamma},$$

with ρ^γ denoting the concentrations of components, and

$$\mathbf{K} = \frac{\mathbf{k}\rho_f g}{\mu}, \tag{1.6.18}$$

a second rank symmetric tensor, is the *hydraulic conductivity*, and $\mathbf{k}$, also a second rank symmetric tensor, denotes the *permeability* of the porous medium, defined by (1.4.49). In (1.6.17), since $\rho = \rho(p)$, we have made use of the relations

$$\frac{\partial \rho_f}{\partial t} = \frac{d\rho_f}{dp} \frac{\partial p}{\partial t} = \beta_p \rho_f \frac{dp}{dt} = \beta_p \rho_f^2 g \frac{\partial \varphi^*}{\partial t},$$

where φ^* is *Hubbert's potential*, defined by (1.4.54).

In terms of pressure, (1.6.17) can be rewritten in the form

$$\nabla \cdot \left\{ \rho_f \frac{\mathbf{k}}{\mu} \cdot (\nabla p + \rho_f g \nabla z) \right\} = \phi \beta_p \rho_f \frac{\partial p}{\partial t}, \tag{1.6.19}$$

which has to be solved for p, with a known constitutive relation, $\rho_f = \rho_f(p)$, for the considered fluid.

For a gas at moderate to high pressures, we employ the relationship

$$\rho_f = \rho_f(p, T) = \frac{pM}{ZRT}, \tag{1.6.20}$$

where $Z = Z(p,T)$ is an empirical correction factor called *compressibility factor*, commonly employed in the petroleum industry, T is the temperature and M is the molecular weight of the gas. We obtain

$$\phi\frac{\partial}{\partial t}\frac{pM}{Z(p,T)RT} = \nabla\cdot\left\{\frac{pM}{Z(p,T)RT}\frac{\mathbf{k}}{\mu}\cdot\left(\nabla p + \frac{pM}{Z(p,T)RT}g\nabla z\right)\right\}. \quad (1.6.21)$$

For $M/ZRT = \text{const.}$, e.g., $Z = 1$ for an *ideal gas*, $\mu = \text{const.}$, and a homogeneous isotropic porous medium, (1.6.21) reduces to

$$\phi\frac{\partial p}{\partial t} = \frac{k}{2\mu}\nabla^2 p^2 + \frac{kMg}{\mu ZRT}\frac{\partial p^2}{\partial z}. \quad (1.6.22)$$

The last two equations are *nonlinear* with respect to pressure. They are often used to describe gas flow in reservoir engineering.

Let us add the assumption that

[A6.5] $$\left|\frac{\partial\rho_f}{\partial t}\right| \gg |\mathbf{V}_f\cdot\nabla\rho_f|,$$

which, with $\rho_f = \rho_f(p_f)$, is equivalent to

$$\left|\frac{\partial p}{\partial t}\right| \gg |\mathbf{V}_f\cdot\nabla p|.$$

Following the methodology mentioned above, for deleting nondominant effects (for example, Bear and Bachmat, 1990), assumption [A6.5] may be expressed in the form

$$\frac{\partial\rho_f}{\partial t}\bigg|_c\left|\left(\frac{\partial\rho_f}{\partial t}\right)^*\right| \gg V_c(\nabla\rho_f)_c|\mathbf{V}^*\cdot\nabla(\rho_f)^*|, \quad (1.6.23)$$

where the asterisk denotes a dimensionless quantity, or in the form

$$\mathrm{St}^{(\rho)}\left|\left(\frac{\partial\rho_f}{\partial t}\right)^*\right| \gg |\mathbf{V}^*\cdot(\nabla\rho_f)^*|,$$

where the *Strouhal number*, $\mathrm{St}^{(\rho)}$, with respect to density changes, is defined by

$$\mathrm{St}^{(\rho)} = \frac{L_c^{(\rho)}}{t_c^{(\rho)}V_c},$$

in which

$$t_c^{(\rho)} = \frac{(\rho_f)_c}{(\partial \rho_f/\partial t)_c} = \frac{\rho_{\max}}{|\partial \rho_f/\partial t|_{\max}}, \qquad L_c^{(\rho)} = \frac{\rho_c}{|\nabla \rho_f|_c} = \frac{\rho_{\max}}{|\nabla \rho_f|_{\max}}.$$

The Strouhal number, $\text{St}^{(\rho)}$, expresses the ratio between the time required for a moving particle to observe a significant spatial change in ρ and the time required for the same change to occur at a given point.

Similarly, the assumption with respect to pressure in [A6.5] can be written as

$$\text{St}^{(p)}\left|\left(\frac{\partial p_f}{\partial t}\right)^*\right| \gg |\mathbf{V}^* \cdot (\nabla p_f)^*|, \qquad \text{St}^{(p)} = \frac{L_c^{(p)}}{t_c^{(p)} V_c}. \tag{1.6.24}$$

Then, assumption [A6.5] is equivalent to both

$$\text{St}^{(\rho)} \gg 1, \qquad \text{and} \quad \text{St}^{(p)} \gg 1,$$

and (1.6.16) reduces to

$$\phi \beta_p \frac{\partial p}{\partial t} = -\nabla \cdot \mathbf{q}, \tag{1.6.25}$$

in which we recall that $\mathbf{q} \equiv \mathbf{q}_r$.

By inserting (1.4.55) into (1.6.25), we obtain

$$\phi \beta_p \rho_f g \frac{\partial \varphi^*}{\partial t} = \nabla \cdot (\mathbf{K} \cdot \nabla \varphi^*). \tag{1.6.26}$$

CASE B.   When $\rho = \text{const.}$, or $\mathrm{D}_f \rho_f/\mathrm{D}t = 0$, equation (1.6.16) reduces to

$$\nabla \cdot \mathbf{q} = 0, \qquad \mathbf{q} \equiv \mathbf{q}_r, \tag{1.6.27}$$

which describes *macroscopically isochoric flow.*

Equation (1.6.27), is thus an *approximation*, known as the balance equation, (1.6.13), when $\phi = \text{const.}$, and $\mathrm{D}_f \rho_f/\mathrm{D}t = 0$.

When combined with (1.4.52), equation (1.6.27) becomes

$$\nabla \cdot (\mathbf{K} \cdot \nabla \varphi) = 0. \tag{1.6.28}$$

CASE C.   For a homogeneous anisotropic porous medium, (1.6.28) reduces to

$$K_{ij} \frac{\partial^2 \varphi}{\partial x_i \partial x_j} = 0. \tag{1.6.29}$$

If, in a Cartesian coordinate system, the directions of the x, y, z-axes are also *principal directions* of the permeability, (1.6.29) reduces to

$$K_{xx}\frac{\partial^2 \varphi}{\partial x^2} + K_{yy}\frac{\partial^2 \varphi}{\partial y^2} + K_{zz}\frac{\partial^2 \varphi}{\partial z^2} = 0. \qquad (1.6.30)$$

CASE D. If the porous medium is homogeneous and isotropic throughout the considered domain, (1.6.30) reduces to

$$\nabla^2 \varphi = 0, \qquad \frac{\partial^2 \varphi}{\partial x^2} + \frac{\partial^2 \varphi}{\partial y^2} + \frac{\partial^2 \varphi}{\partial z^2} = 0, \qquad (1.6.31)$$

known as the *Laplace equation.*

It is important to note that in (1.6.28) through (1.6.31), φ may still be time dependent, as a result of time dependent boundary conditions.

The entire discussion in the present section has been limited to a domain in which no sources and sinks are present. Indeed, there cannot be a distributed mass source, or sink, in single phase flow in a three-dimensional domain. However, we may consider a case in which sources, or sinks, of fluid are located, *perhaps as an approximation*, at points, $\mathbf{x} = \mathbf{x}^{(m)}$, within a considered domain. At each such point, mass is introduced at a rate $\rho_f^{(m)}Q^{(m)}$, where $Q^{(m)}$ denotes the volume of fluid supplied (say, by injection) to the porous domain, per unit volume of porous medium and per unit time, and $\rho_f^{(m)}$ denotes the (known) density of the injected fluid. When such point sources and sinks are present in a domain, we may surround each of them by a small sphere which is then excluded from the considered domain. The latter remains without sources, or sinks. The surfaces of the spheres become part of the boundary of the domain. The strength of each such source, or sink, specifies the (known) total flux through its respective boundary segment.

Another way of treating point sources and sinks, is to introduce them in the balance equations by making use of the *Dirac delta function*, or the *Dirac distribution*, defined by

$$\delta(\mathbf{x} - \mathbf{x}^{(m)}) = \lim_{a \to 0}\begin{cases} 1/a^3 & |\mathbf{x} - \mathbf{x}^{(m)}| < a, \\ 0 & \text{elsewhere.} \end{cases} \qquad (1.6.32)$$

We may then rewrite (1.6.6) for a domain with point sources in the form

$$\frac{\partial \phi \rho_f}{\partial t} = -\nabla \rho_f \mathbf{q} + \sum_{(m)} \rho_f^{(m)}Q^{(m)}\delta(\mathbf{x} - \mathbf{x}^{(m)}). \qquad (1.6.33)$$

When the $Q^{(m)}$'s denote sinks (e.g., due to pumping), equation (1.6.6) should be changed to the form

$$\frac{\partial \phi \rho_f}{\partial t} = -\nabla \cdot \rho_f \mathbf{q} - \sum_{(m)} \rho_f Q^{(m)} \delta(\mathbf{x} - \mathbf{x}^{(m)}), \qquad (1.6.34)$$

where ρ_f is the fluid's density at $\mathbf{x} = \mathbf{x}^{(m)}$.

These changes can be introduced in all the other forms of the basic mass balance equation.

1.6.3 Deformable porous medium

In the discussion on flow in deformable porous media to be presented below, we shall assume that assumption [A6.3], introduced above, is still valid, but replace assumption [A6.4] by the assumption that

[A6.6] $\phi = \phi(\sigma'_s)$

where σ'_s is Terzaghi's *effective stress* defined in LECTURE 3. Also, here $V_s \neq 0$, and $\mathbf{q}_r \neq \mathbf{q}$.

Let $\overline{\mathbf{V}_s}^s$ and $\overline{\mathbf{w}}^s$ denote the solid's macroscopic velocity and displacement vectors (actually the velocity and displacement of the solid skeleton), respectively, and ε_{sk} denotes the (macroscopic) *volumetric strain*, (or *dilatation*), of) the solid skeleton. We start by averaging the condition $\nabla \cdot V_s = 0$, which follows from assumption [A6.3]. By employing (1.2.57), with

$$\nabla \theta = -\widetilde{\nu}^{\alpha\beta} \Sigma_{\alpha\beta},$$

that follows from it, we obtain

$$\begin{aligned}
\overline{\nabla \cdot V_s}^s &= \frac{1}{\theta_s} \nabla \cdot \theta_s \overline{V_s}^s + \frac{1}{U_{os}} \int_{S_{sv}} V_s \cdot \nu \, dS \\
&= \frac{1}{\theta_s} \nabla \cdot \theta_s \overline{V_s}^s + \frac{1}{U_{os}} \int_{S_{sv}} (V_s - \mathbf{u}) \cdot \nu \, dS + \frac{1}{U_{os}} \int_{S_{sv}} \mathbf{u} \cdot \nu \, dS \\
&= \nabla \cdot \overline{V_s}^s + \frac{1}{\theta_s}(\overline{V_s}^s \cdot \nabla \theta_s + \frac{1}{U_o} \int_{S_{sv}} \mathbf{u} \cdot \nu \, dS) \\
&= \nabla \cdot \overline{V_s}^s + \frac{1}{\theta_s} \frac{D_s \theta_s}{Dt}, \qquad (1.6.35)
\end{aligned}$$

where $\theta_s \equiv 1 - \phi$, and $(\mathbf{V}_s - \mathbf{u})\cdot\boldsymbol{\nu} = 0$, because $\mathcal{S}_{sv}$ is a material surface. Since $\nabla\cdot\mathbf{V}_s = 0$, and hence $\overline{\nabla\cdot\mathbf{V}_s}^s = 0$, we obtain

$$\nabla\cdot\overline{\mathbf{V}_s}^s = -\frac{1}{\theta_s}\frac{\mathrm{D}_s\theta_s}{\mathrm{D}t} = -\frac{1}{1-\phi}\frac{\mathrm{D}_s(1-\phi)}{\mathrm{D}t}. \tag{1.6.36}$$

This equation is identical to (1.6.11) in which, following assumption [A6.3], we insert $\dot{\rho}_s$ $(\equiv \overline{\dot{\rho}_s}^s) = 0$. Note that $\mathbf{V}_s$ in (1.6.11), is identical to $\overline{\mathbf{V}_s}^s$ in (1.6.36).

From the microscopic relationship

$$\mathbf{V}_s \equiv \dot{\mathbf{w}},$$

and applying (1.2.83) to $e_\alpha = \mathbf{w}$, with $\nabla\cdot\mathbf{V}_s = 0$, we obtain at the macroscopic level

$$\overline{\mathbf{V}_s}^s \equiv \overline{\dot{\mathbf{w}}}^s = \dot{\overline{\mathbf{w}}}^s - \overset{\circ}{\mathbf{w}}\overline{(\nabla\cdot\mathbf{V}_s)}^s = \dot{\overline{\mathbf{w}}}^s. \tag{1.6.37}$$

Since,

$$(\dot{dx_i}) = \frac{\partial \overline{V_{si}}^s}{\partial x_i}dx_i, \qquad i = 1, 2, 3,$$

we obtain

$$\nabla\cdot\overline{\mathbf{V}_s}^s = \frac{(\dot{dU_{sk}})}{dU_{sk}} = \dot{\varepsilon}_{sk} = -\frac{1}{1-\phi}\frac{\mathrm{D}_s(1-\phi)}{\mathrm{D}t}, \tag{1.6.38}$$

where ε_{sk} denotes the (macroscopic) volumetric strain of the porous medium as a whole (actually, of the solid skeleton), and the extreme r.h.s. term is based on (1.6.36).

From (1.6.37) and (1.6.38), we obtain for the macroscopic strain of the solid skeleton

$$\varepsilon_{sk} = \nabla\cdot\overline{\mathbf{w}}^s \tag{1.6.39}$$

We can obtain (1.6.39) by averaging the microscopic relationship $\varepsilon_s = \nabla\cdot\mathbf{w}$. The solid skeleton, or solid matrix, should be regarded as a material continuum that undergoes (macroscopic) deformation. The latter manifests itself by macroscopic displacements, $\overline{\mathbf{w}}^s$, and by changes in θ_s, due to the movement of the solid phase, for example, of grains, with respect to each other, resulting in changes in solid phase configuration, and in θ_s. All this occurs while we assume that the solid phase itself is incompressible. In fact, since $\varepsilon_s = 0$ at the microscopic level, we have $\overline{\varepsilon_s}^s = 0$.

With these relationships, the basic mass balance equation (1.6.14) becomes

$$\nabla\cdot\rho_f\mathbf{q}_r + \phi\frac{\mathrm{D}_s\rho_f}{\mathrm{D}t} + \rho_f\frac{\mathrm{D}_s\varepsilon_{sk}}{\mathrm{D}t} = 0. \tag{1.6.40}$$

The dilatation, ε_{sk}, is the *first invariant* of the strain tensor, ε_{sk}, i.e., $\varepsilon_{sk} = \sum_{(i)}(\varepsilon_{sk})_{ii}$.

We often assume that

[A6.7]   the solid skeleton's deformation is such that

$$\left|\frac{\partial\rho_f}{\partial t}\right| \gg |\mathbf{V}_s\cdot\nabla\rho_f|,$$

which, in view of the discussion presented above, can be rewritten as

$$\mathrm{St}^{(\rho_f,\mathbf{V}_s)}\left|\left(\frac{\partial\rho_f}{\partial t}\right)^*\right| \gg |\mathbf{V}_s^*\cdot\nabla^*\rho_f^*|.$$

This condition is satisfied when

$$\mathrm{St}^{(\rho_f,\mathbf{V}_s)}\left(\equiv \frac{L_c^{(\rho_f)}}{t_c^{(\rho_f)}V_{sc}}\right) \gg 1,$$

where

$$t_c^{(\rho_f)} = \frac{|\rho_f|_{\max}}{|\partial\rho_f/\partial t|_{\max}}.$$

Another assumption is that

[A6.8] $|\partial\varepsilon_{sk}/\partial t| \gg |\mathbf{V}_s\cdot\nabla\varepsilon_{sk}|,$

which, in view of the discussion on nondimensionalization presented above, can be rewritten as

$$\mathrm{St}^{(\varepsilon_{sk},\mathbf{V}_s)}\left|\left(\frac{\partial\varepsilon_{sk}}{\partial t}\right)^*\right| \gg |\mathbf{V}_s^*\cdot\nabla^*\varepsilon_{sk}^*|.$$

This condition is satisfied when

$$\mathrm{St}^{(\varepsilon_{sk},\mathbf{V}_s)}\left(\equiv \frac{L_c^{(\varepsilon_{sk})}}{t)_c^{(\varepsilon_{sk})}V_{sc}}\right) \gg 1,$$

where $t_c^{(\varepsilon_{sk})} = |\varepsilon_{sk}|_{\max}/|\partial\varepsilon_{sk}/\partial t|_{\max}.$

With these two assumptions, the mass balance equation (1.6.40) reduces to

$$\nabla\cdot\rho_f\mathbf{q}_r + \phi\frac{\partial\rho_f}{\partial t} + \rho_f\frac{\partial\varepsilon_{sk}}{\partial t} = 0. \tag{1.6.41}$$

For a compressible fluid, $\rho = \rho(p)$, we can now rewrite (1.6.41) in the form

$$\nabla\cdot\rho_f\mathbf{q}_r + \phi\rho_f\beta_p\frac{\partial p}{\partial t} + \rho_f\frac{\partial\varepsilon_{sk}}{\partial t} = 0. \tag{1.6.42}$$

As a further approximation, we often stipulate that assumption [A6.5] is also valid. Then (1.6.42) reduces to

$$\nabla\cdot\mathbf{q}_r + \phi\beta_p\frac{\partial p}{\partial t} + \frac{\partial\varepsilon_{sk}}{\partial t} = 0. \tag{1.6.43}$$

In (1.6.42), $\rho_f = \rho_f(p), \mathbf{q}_r$ can be expressed in terms of φ^*, or in terms of p, and $\phi = \phi(\sigma'_s)$, with the effective stress, σ'_s, related to p and to σ by (1.3.28). Thus (1.6.42) involves three variables: σ, p (or φ^*) and ε_{sk}. To solve for these variables, we need additional equations.

Following the discussion in LECTURE 3, and under the simplifying assumptions introduced there, e.g., no acceleration, the total stress, σ, at a point within a porous medium domain satisfies the macroscopic equilibrium relationship (1.3.26), rewritten here for convenience in the form

$$\nabla\cdot\sigma + \mathbf{f} = 0, \tag{1.6.44}$$

where $\sigma \equiv \overline{\sigma}$, and $\mathbf{f}$ represents the total body force acting on the porous medium, per unit volume of the latter. In the case of gravity only

$$\mathbf{f} \equiv \overline{\rho\mathbf{F}} = [\phi\overline{\rho_f}^f + (1 - \phi)\overline{\rho_s}^s]\mathbf{g}.$$

Using (1.3.28), rewritten in the form

$$\sigma - \sigma'_s - p\mathbf{I} \tag{1.6.45}$$

to express the total stress, σ, in terms of Terzaghi's effective stress, σ'_s ($\equiv \overline{\sigma'_s}$), and the pressure, p ($\equiv \overline{p_f}^f > 0$ for compression), in the fluid filling the void space, let us follow Verruijt (1969) and separate the stresses, σ and σ'_s, the pressure p and the body force $\mathbf{f}$, into initial steady state values, σ^o, σ'^o_s, p^o and $\mathbf{f}^o$, and deformation-producing increments, σ^e, σ'^e_s, p^e and $\mathbf{f}^e$, with

$$\sigma^o = \sigma'^o_s - p^o\mathbf{I}, \quad \text{and} \quad \sigma^e = \sigma'^e_s - p^e\mathbf{I}. \tag{1.6.46}$$

Assuming, as a good approximation, that

[A6.9]   the body force, $\mathbf{f}$, remains unchanged, although ϕ varies, i.e.,
$$\mathbf{f}^e = 0,$$

the equilibrium equation for the initial steady state, is

$$\nabla \cdot \boldsymbol{\sigma}_s'^o + \mathbf{f}^o - \nabla p^o = 0. \qquad (1.6.47)$$

For the incremental (deformation producing) effective stress and pressure, we have

$$\nabla \cdot \boldsymbol{\sigma}_s'^e - \nabla p^e = 0. \qquad (1.6.48)$$

We now make the assumptions that

[A6.10]   the solid matrix is isotropic and, for the relatively small excess
effective stresses considered here, is made of a linear, perfectly elastic
material,

and

[A6.11]   the stress–strain relationship for the solid matrix at the macroscopic level, relating the (average) excess effective stress, $\overline{\boldsymbol{\sigma}_s'}$ (denoted
here by $\overline{\boldsymbol{\sigma}_s'^e}$), to the average displacement $\overline{\mathbf{w}}^s$ (denoted here by $\mathbf{w}$), has
the same form, as at the microscopic one, viz.

$$
\begin{aligned}
(\boldsymbol{\sigma}_s'^e)_{ij} &= \overline{\mu_s'}\left(\frac{\partial w_i}{\partial x_j} + \frac{\partial w_j}{\partial x_i}\right) + \overline{\lambda_s''}\left(\frac{\partial w_k}{\partial x_k}\right)\delta_{ij} \\
&= 2\overline{\mu_s'}\varepsilon_{ij} + \overline{\lambda_s''}\varepsilon\delta_{ij}. \qquad (1.6.49)
\end{aligned}
$$

In (1.6.49), ε_{ij} denotes the ijth component of the strain tensor, ε, $\varepsilon \equiv \varepsilon_{sk}$, and $\overline{\mu_s'}$ and $\overline{\lambda_s''}$ are macroscopic constant coefficients analogous to the *Lamé* coefficients. We shall refer to them as the *Lamé) coefficients of the solid skeleton*. These are not the microscopic Lamé constants, used in the constitutive relations of the *solid material* comprising the solid matrix, but macroscopic coefficients describing the elastic behavior of the *solid matrix*, taking into account the geometry of the pore space. *Their values must be determined experimentally.*

We assume here that only the incremental effective stress causes displacement. We recall that $\mathbf{w} \equiv \overline{\mathbf{w}}^s$ and $\boldsymbol{\sigma}_s' \equiv \overline{\boldsymbol{\sigma}_s'}$.

In principle, there should be no difficulty in replacing the stress-strain relationship expressed by (1.6.49) by any other one, that corresponds to a considered material (e.g., plastic). However, here the discussion is limited to elastic solid matrices.

The mass balance equation (1.6.42) may also be rewritten as two balance equations, one representing the initial steady state (with variables denoted by superscript o), and the other, involving increments of pressure (with variables denoted by superscript e), that produce displacements. Thus, the second equation may be written in the form

$$\nabla \cdot \rho_f \mathbf{q}_r^e + \phi \rho_f \beta_p \frac{\partial p^e}{\partial t} + \rho_f \frac{\partial \varepsilon_{sk}}{\partial t} = 0, \tag{1.6.50}$$

where $\varepsilon_{sk}^e \equiv \varepsilon_{sk}$ since $\varepsilon_{sk}^o \equiv 0$, and

$$\mathbf{q}_r^e = -\frac{k}{\mu}(\nabla p^e + \rho_f g \nabla z). \tag{1.6.51}$$

Note that the porous medium is assumed here to be isotropic.

In writing (1.6.50) and (1.6.51), we have introduced the approximations:

[A6.12] $\rho_f = \rho_f^o + \rho_f^e \simeq \rho_f^o, \qquad \rho_f^e \ll \rho_f^o, \quad \mu^e \ll \mu^o.$

[A6.13] $\phi = \phi^o + \phi^e \simeq \phi^o, \qquad \phi^e \ll \phi^o.$

[A6.14]   The permeability, $\mathbf{k}$, remains unchanged, in spite of the deformation that takes place.

By inserting the expression for $\mathbf{q}_r^e$ into (1.6.50), we obtain the mass balance equation

$$-\nabla \cdot \left[\rho_f \frac{k}{\mu}(\nabla p^e + \rho_f g \nabla z) \right] + \phi \rho_f \beta_p \frac{\partial p^e}{\partial t} + \rho_f \frac{\partial \varepsilon_{sk}}{\partial t} = 0, \tag{1.6.52}$$

which is a single equation in two variables p^e and ε_{sk}.

The complete model describing the flow of a single compressible Newtonian fluid in a deformable porous medium, consists now of the equations and relationships summarized in Table 1.6.1. In making use of this mathematical model, one should always recall the various assumptions ($\equiv$ conceptual model) underlying its development.

By inserting (1.6.49) into (1.6.48), we obtain

$$\overline{\mu'_s}\nabla^2\mathbf{w} + (\overline{\mu'_s} + \overline{\lambda''_s})\nabla(\nabla\cdot\mathbf{w}) - \nabla p^e = 0, \qquad (1.6.53)$$

to be used for determining the displacement, $\mathbf{w}$.

From Table 1.6.1, it follows that we have a sufficient number of equations to solve for the various dependent variables involved. In principle, this is the model introduced by Biot (1941). We note that this model also yields the displacement vector, $\mathbf{w}$. It can be used, for example, for determining soil compaction and land subsidence.

By inserting (1.6.39) and (1.6.51) into (1.6.50), we obtain

$$-\frac{\partial}{\partial x_j}\left[\rho_f\frac{k}{\mu}\left(\frac{\partial p^e}{\partial x_j} + \rho_f g\frac{\partial z}{\partial x_j}\right)\right] + \phi\frac{\partial \rho_f}{\partial t} + \rho_f\frac{\partial}{\partial t}\left(\frac{\partial w_i}{\partial x_i}\right) = 0. \qquad (1.6.54)$$

Consider a number of simplified versions of (1.6.52).

CASE A. We assume that [A6.5] is valid. Then, (1.6.52) reduces to

$$-\nabla\cdot\left[\frac{k}{\mu}(\nabla p^e + \rho_f g\nabla z)\right] + \phi\beta_p\frac{\partial p^e}{\partial t} + \frac{\partial \varepsilon_{sk}}{\partial t} = 0. \qquad (1.6.55)$$

CASE B. We assume that

[A6.15]   the hydraulic conductivity, K, is not affected by variations in the density, ρ_f.

Then

$$-\nabla\cdot\{\rho_f(K\cdot\nabla\varphi^*)\} + \phi\beta_p\rho_f^2 g\frac{\partial \varphi^*}{\partial t} + \rho_f\frac{\partial \varepsilon_{sk}}{\partial t} = 0. \qquad (1.6.56)$$

CASE C. For a homogeneous porous medium, (1.6.53) can be rewritten in the form of the three equations

$$\overline{\mu'_s}\nabla^2 w_i + (\overline{\lambda''_s} + \overline{\mu'_s})\frac{\partial \varepsilon_{sk}}{\partial x_i} - \frac{\partial p^e}{\partial x_i} = 0, \qquad i = 1,2,3. \qquad (1.6.57)$$

Equations		Dependent Variables
Mass balance equation for the fluid, (1.6.41)–(1 eqn.)	$\nabla \cdot \rho_f \mathbf{q}_r^e + \phi \frac{\partial \rho_f}{\partial t}$ $+ \; \rho_f \frac{\partial \varepsilon_{sk}}{\partial t} = 0$	$\rho_f, \phi,$ $\mathbf{q}_r^e, \varepsilon_{sk}$ (6 vars.)
Equation of motion for the fluid (1.6.51)–(3 eqns.)	$\mathbf{q}_r^e = -\frac{k}{\mu}(\nabla p^e + \rho_f g \nabla z)$	p^e (1 var.)
Equilibrium relationships (1.6.48)–(3 eqns.)	$\nabla \cdot \sigma_s'^e - \nabla p^e = 0$	$\sigma_s'^e$ (6 vars.)
Stress-strain relationships for the solid matrix (1.6.49)–(6 eqns.)	$\sigma_s'^e = \overline{\mu_s'}\{\nabla \mathbf{w} + (\nabla \mathbf{w})^T\}$ $+ \overline{\lambda_s''}(\nabla \cdot \mathbf{w})\mathbf{I}$	$\mathbf{w}$ (3 vars.)
Dilatation - displacement relations (1.6.36)–(1 eqn.)	$\varepsilon_{sk} = \nabla \cdot \mathbf{w}$	(none)
Equation of state for the fluid–(1 eqn.)	$\rho_f = \rho_f(p)$	(none)
Dilatation– porosity relation –(1 eqn.)	$\dot{\varepsilon}_{sk} = \frac{1}{1-\phi}\dot{\phi}$	(none)
Total: 16 equations		16 (scalar) variables

Table 1.6.1: Summary of balance equations and constitutive relations for Darcian, saturated flow of a compressible Newtonian fluid in an isotropic linearly elastic porous medium

By differentiating each of these equations with respect to the corresponding x_i, and adding the resulting three equations, we obtain the single equation (Verruijt, 1969)

$$(\overline{\lambda_s''} + 2\overline{\mu_s'})\nabla^2\varepsilon_{sk} - \nabla^2 p^e = 0, \qquad (1.6.58)$$

which, together with (1.6.55), often simplified for a homogeneous isotropic porous medium to the form

$$-\frac{k}{\mu}\nabla^2 p^e + \phi\beta_p\frac{\partial p^e}{\partial t} + \frac{\partial\varepsilon_{sk}}{\partial t} = 0, \qquad (1.6.59)$$

constitute two equations in the variables p^e and ε_{sk}.

Following Verruijt (1969), we integrate (1.6.58), and obtain

$$(\overline{\lambda_s''} + 2\overline{\mu_s'})\varepsilon_{sk} = p^e + \Pi(\mathbf{x}, t), \qquad (1.6.60)$$

where Π is a function of position and time, that for every value of time, t, satisfies

$$\nabla^2\Pi = 0. \qquad (1.6.61)$$

When $\Pi \equiv 0$ (see below), we may insert

$$\varepsilon_{sk} = \frac{p^e}{\overline{\lambda_s''} + 2\overline{\mu_s'}} \qquad (1.6.62)$$

in (1.6.59), and obtain

$$\frac{k}{\mu}\nabla^2 p^e = \left(\phi\beta_p + \frac{1}{\overline{\lambda_s''} + 2\overline{\mu_s'}}\right)\frac{\partial p^e}{\partial t} \equiv (\phi\beta_p + \alpha)\frac{\partial p^e}{\partial t}, \qquad (1.6.63)$$

commonly employed in hydraulics of groundwater. In it

$$\alpha = \frac{1}{\overline{\lambda_s''} + 2\overline{\mu_s'}}$$

may be interpreted as a *coefficient of porous medium compressibility*.

Thus, we obtain

$$\frac{\partial\varepsilon_{sk}}{\partial t} \equiv \frac{1}{U_{sk}}\frac{\partial U_{sk}}{\partial t} = \alpha\frac{\partial p^e}{\partial t}, \qquad \alpha = \frac{1}{U_{sk}}\frac{\partial U_{sk}}{\partial p^e},$$

to be used in (1.6.41), (1.6.43), (1.6.52) and (1.6.56).

In soil mechanics, a *coefficient of compressibility*, a_v, is used, defined as

$$a_v = \frac{\partial e}{\partial \sigma'_s},$$

(in the case of vertical compaction only), making use of the definition of the *void ratio*

$$e = \frac{\phi}{(1 - \phi)},$$

as the volume of void space per unit volume of solid matrix. Then, for $d\sigma = 0$, and $d\sigma'_s = dp$, we have

$$a_v = \frac{\partial e}{\partial \sigma'_s} = \frac{\partial e}{\partial p} = \frac{\alpha}{1 - \phi} = \alpha(1 + e),$$

to be used in (1.6.63).

CASE D. Following Verruijt (1969), we assume that

[A6.16]   displacements occur in the vertical direction only, i.e.

$$w_z \neq 0, \quad w_x = w_y = 0,$$

and the total stress remains unchanged, i.e.

$$\sigma \equiv \sigma^o, \qquad \text{and} \quad \sigma^e = 0.$$

Then, (1.6.49) reduces to

$$\begin{aligned}
(\sigma'^{e}_s)_{xx} &= \overline{\lambda''_s}\frac{\partial w_z}{\partial z}, & (\sigma'^{e}_s)_{xy} &= (\sigma'^{e}_s)_{yx} = 0, \\
(\sigma'^{e}_s)_{xz} &= (\sigma'^{e}_s)_{zx} = \overline{\mu''_s}\frac{\partial w_z}{\partial x}, & & \\
(\sigma'^{e}_s)_{yy} &= \overline{\lambda''_s}\frac{\partial w_z}{\partial z}, & (\sigma'^{e}_s)_{zz} &= (\overline{\lambda''_s} + 2\overline{\mu'_s})\frac{\partial w_z}{\partial z}, & \varepsilon_{sk} &= \frac{\partial w_z}{\partial z}.
\end{aligned}$$

$$(1.6.64)$$

Since $\sigma^e \equiv 0$, it follows from (1.6.46) that

$$\sigma'^e_s = p^e \mathbf{I}. \tag{1.6.65}$$

Hence, from the last two equations in (1.6.64), we obtain

$$(\sigma'^e_s)_{zz} = p^e = (\overline{\lambda''_s} + 2\overline{\mu'_s})\frac{\partial w_z}{\partial z} = (\overline{\lambda''_s} + 2\overline{\mu'_s})\varepsilon_{sk}. \tag{1.6.66}$$

By comparing (1.6.60) with (1.6.66), Verruijt concludes that *for the simplifying assumptions leading to the latter equation, the function* Π *when assumption [A6.16], of* vertical consolidation only, *vanishes.* Hence, (1.6.66), which is based on (1.6.64), is valid only is valid. However, in general, horizontal displacements, w_x and w_y, do take place, in addition to vertical displacements, w_z.

By applying assumption [A6.16] to (1.6.56), we obtain

$$-\nabla\cdot(\rho_f \mathbf{K}\cdot\nabla\varphi^*) + \rho_f^2 g(\alpha + \phi\beta_p)\frac{\partial\varphi^*}{\partial t} = 0, \qquad (1.6.67)$$

where

$$\alpha = 1/(\overline{\lambda_s''} + 2\overline{\mu_s'})$$

is the *coefficient of compressibility of the porous medium.* For $\alpha = 0$, equation (1.6.67) reduces to (1.6.17).

In hydrology of groundwater, starting with Jacob (1940, 1950), assumption [A6.16] has usually been invoked (although, as mentioned above, significant horizontal displacements may occur under certain conditions). Then, (1.6.67) is written as

$$\nabla\cdot(\rho_f \mathbf{K}\cdot\nabla\varphi^*) = \rho_f S_o \frac{\partial\varphi^*}{\partial t}. \qquad (1.6.68)$$

Or, with the approximation introduced by [A6.5], and with a source $Q = Q(\mathbf{x},t)$

$$\nabla\cdot(\mathbf{K}\cdot\nabla\varphi^*) + Q = S_o \frac{\partial\varphi^*}{\partial t}, \qquad (1.6.69)$$

where

$$S_o = \rho_f g(\alpha + \phi\beta_p) \qquad (1.6.70)$$

is usually referred to as the *specific (volume) storativity* of a porous medium. It may be defined as the volume of water released from (or added to) storage per unit volume of porous medium per unit decline (or rise) of the piezometric head, φ^*. The use of (1.6.70) is associated with a test, called *undrained test*, common in soil mechanics. In such a test, the application of stress is so fast that, practically, no motion of the water relative to the solid takes place, i.e., $\mathbf{q}_r = 0$. Then, it is obvious from (1.6.69) that the added water goes only into elastic storage, producing a rise in the piezometric head. Thus, (1.6.70) corresponds to the verbal definition of S_o given above only under conditions equivalent to those prevailing during an undrained test.

Bear (1979) comments that, actually, the term specific (volume) storativity, with the above verbal definition, should be reserved to

$$S_o' = \rho_f g[\alpha(1 - \phi) + \beta_p \phi], \qquad (1.6.71)$$

as $-\nabla \cdot \mathbf{q}_r$, which constitutes the l.h.s. of (1.6.69), does not define excess of inflow of volume of water over outflow, per unit volume and unit time. This definition corresponds to $-\nabla \cdot \mathbf{q}$.

The verbal definition of specific storativity, S_o, given above is obtained from

$$
\begin{aligned}
\frac{\partial \phi \rho_f}{\partial t} &= \rho_f \frac{\partial \phi}{\partial t} + \phi \frac{\partial \rho_f}{\partial t} = \rho_f(1 - \phi)\frac{\partial \varepsilon_{sk}}{\partial t} + \phi \rho_f \beta_p \frac{\partial p^e}{\partial t} \\
&= \rho_f(1 - \phi)\alpha \frac{\partial p^e}{\partial t} + \phi \rho_f \beta_p \frac{\partial p^e}{\partial t} = \rho_f[(1 - \phi)\alpha + \phi \beta_p]\frac{\partial p^e}{\partial t} \\
&= \rho_f^2 g[(1 - \phi)\alpha + \phi \beta_p]\frac{\partial \varphi^*}{\partial t} = \rho_f S_o \frac{\partial \varphi^*}{\partial t}.
\end{aligned}
$$

$$(1.6.72)$$

The comment on point sources and sinks, made after (1.6.31)), is appropriate also here. For example, for sinks of strength, $Q^{(m)}$, located at points $\mathbf{x}^{(m)}$, equation (1.6.72) should read

$$\nabla \cdot \mathbf{K} \cdot \nabla \varphi^* - \sum_{(m)} Q^{(m)} \delta(\mathbf{x} - \mathbf{x}^{(m)}) = S_o \frac{\partial \varphi^*}{\partial t}. \qquad (1.6.73)$$

1.6.4 Boundary conditions

The need for including boundary conditions as part of any mathematical model, was explained in detail in LECTURE 5, and need not be repeated here. We have learned there that in order to obtain a solution to a specific problem of interest, it is necessary to provide additional information in the form of conditions that prevail, or are expected to prevail, on the boundary of the considered domain.

Let us consider a number of boundary conditions of practical interest, relevant to the problem of mass transport considered in the present lecture. The conditions are based on the discussion on boundary conditions presented in LECTURE 5. Sides 1 and 2 of a boundary will denote the considered porous medium domain and its environment, respectively.

We wish to emphasize again, that the selection of the conditions to be used for each problem depend on how, *in reality*, the considered domain interacts, or is expected to interact in the future, with its environment across the boundary. Sometimes, for lack of better information, we shall introduce a condition that only *approximately* represents the anticipated mode of interaction.

In all cases, the shape of a (possibly moving) boundary, or a segment, $\mathcal{B}$, of one, is described by the equation

$$F(\mathbf{x},t) = 0.$$

The unit vector, $\boldsymbol{\nu}$, normal to the boundary, is given by (1.5.3). We'll also make use of (1.5.2).

Whenever it is obvious that we refer to a single fluid that occupies the entire void space, the subscript f will be omitted.

Boundary of prescribed pressure, or head

When pressure is specified as a known function of position and time, say, $f_1(\mathbf{x},t)$, at all points on the external side of a boundary segment, $\mathcal{B}$, we employ (1.5.28) to write the boundary condition

$$p(\mathbf{x},t) \equiv p(\mathbf{x},t)|_1 = f_1(\mathbf{x},t) \qquad \text{on} \quad \mathcal{B}, \tag{1.6.74}$$

where $f_1(\mathbf{x},t) \equiv p(\mathbf{x},t)|_2$ is determined by phenomena that take place in the environment of the considered domain, independent of whatever happens within the latter.

When the piezometric head, $\varphi = z + p/\gamma$, used as the dependent variable of a considered problem, is a known function, $f_2(\mathbf{x},t)$, on the external side of a boundary, the boundary condition to be used is

$$\varphi(\mathbf{x},t) = f_2(\mathbf{x},t) \qquad \text{on} \quad \mathcal{B}. \tag{1.6.75}$$

In the theory of partial differential equations, the condition which specifies the values of a dependent variable at every point of a boundary is referred to as a *boundary condition of the first kind*, or a *Dirichlet condition*.

Boundary of Prescribed Mass Flux

Here, we employ (1.5.11), written for $\alpha = f$ and $\theta_\alpha = \phi$, which expresses the continuity of the total mass flux normal to a boundary. When this flux is known, on the external side of a boundary, say, $f_3(\mathbf{x},t)$, the boundary condition takes the form

$$\phi[\rho(\mathbf{V} - \mathbf{u}) + (\mathbf{J}^m + \mathbf{J}^{*m})]\cdot\boldsymbol{\nu} = f_3(\mathbf{x},t) \qquad \text{on} \quad \mathcal{B}. \qquad (1.6.76)$$

In view of (1.5.3), for a known surface $F(\mathbf{x},t) = 0$, this condition can be written in the form

$$\phi[\rho(\mathbf{V} - \mathbf{u}) + (\mathbf{J}^m + \mathbf{J}^{*m})]\cdot\nabla F = f_3'(\mathbf{x},t) \qquad \text{on} \quad \mathcal{B}, \qquad (1.6.77)$$

where $f_3'(\mathbf{x},t) = |\nabla F|f_3(\mathbf{x},t)$. For an *impervious boundary*, $f_3(\mathbf{x},t) = 0$. We may also use (1.5.3) to replace $\mathbf{u}\cdot\nabla F$ by $\partial F/\partial t$ in (1.6.76) and (1.6.77).

Equation (1.6.77), neglecting the diffusive and dispersive fluxes of the fluid mass, $(\mathbf{J}^m + \mathbf{J}^{*m})$, as we usually do in the mass balance equation itself, may also be written in the form

$$\rho[\mathbf{q}_r + \phi(\mathbf{V}_s - \mathbf{u})]\cdot\nabla F = f_3'(\mathbf{x},t) \qquad \text{on} \quad \mathcal{B}, \qquad (1.6.78)$$

where $\mathbf{q}_r$ is the specific discharge of the fluid, relative to the possibly moving solid matrix. If we now assume that [A5.3], and hence (1.5.14), are valid, the condition (1.6.78) becomes

$$\mathbf{q}_r\cdot\nabla F = f_3'(\mathbf{x},t)/\rho_f \equiv f_3''(\mathbf{x},t)/\rho_f \qquad \text{on} \quad \mathcal{B}, \qquad (1.6.79)$$

where, following assumption [A5.4] of LECTURE 5, $\rho|_1 = \rho|_2$ on $\mathcal{B}$.

The relative specific flux, $\mathbf{q}_r$, in (1.6.79), may now be expressed by any of the motion equations developed in LECTURE 4. For example, with (1.4.48), and $\mathbf{V}^m \simeq \mathbf{V}$ (i.e., neglecting molecular diffusion), we obtain

$$-\left\{\frac{\mathbf{k}}{\mu}\cdot(\nabla p + \rho g\nabla z)\right\}\cdot\nabla F = f_3''(\mathbf{x},t) \qquad \text{on} \quad \mathcal{B}. \qquad (1.6.80)$$

For $\rho = \rho(p)$, we employ (1.4.55) to obtain

$$-(\mathbf{K}\cdot\nabla\varphi^*)\cdot\nabla F = f_3''(\mathbf{x},t) \qquad \text{on} \quad \mathcal{B}. \qquad (1.6.81)$$

The boundary conditions (1.6.80) and (1.6.81) specify the values of ∇p and $\nabla\varphi^*$, respectively, on the boundary. Such boundary conditions are known as *boundary conditions of the second kind*, also referred to as *Neumann conditions*.

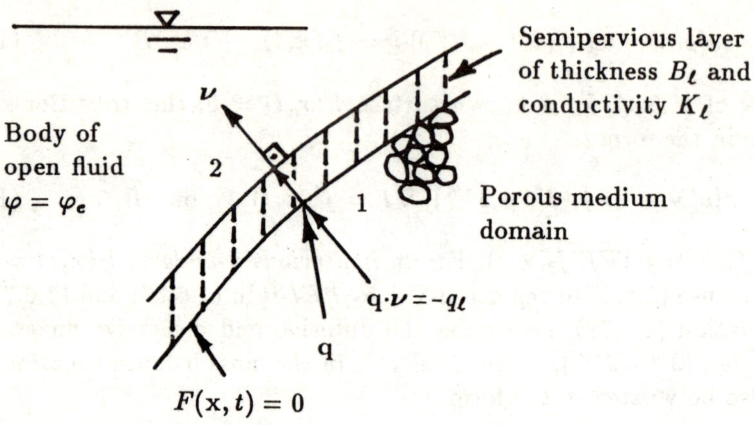

Figure 1.6.1: A semipervious boundary.

Semipervious Boundary

Figure 1.6.1 shows a semipervious layer that separates a porous medium domain from an external body of open fluid at known piezometric head, φ_e. When this layer is much thinner that the length dimensions of the porous medium domain, and its permeability much smaller than that of the latter, we may *approximate* it as a 'thin semipervious membrane' that permits the leakage of fluid through it. In this form, it serves as a boundary to the porous medium domain. Denoting the flux through this 'membrane' by q_ℓ, we may use Darcy's law, for example (1.4.52), and approximate $\nabla\varphi$ by $(\varphi_e - \varphi)/B_\ell$, so that the leakage is expressed by

$$q_\ell = K_\ell \frac{\varphi - \varphi_e}{B_\ell} = \frac{\varphi|_1 - \varphi_e}{c_r}, \qquad c_r = \frac{B_\ell}{K_\ell}, \qquad (1.6.82)$$

where K_ℓ, B_ℓ and c_r are the hydraulic conductivity, thickness and *resistance* (dims. T), respectively, of the semipervious layer. A similar expression may be written in terms of pressure.

Assuming, as an approximation, that as the piezometric head varies within the membrane, no changes in storage of fluid within it take place, the condition of no jump, in the volume flux across this membrane is expressed by

$$K_\ell \frac{\varphi - \varphi_c}{B_\ell} = (-\mathbf{K} \cdot \nabla \varphi)\Big|_1 \cdot \boldsymbol{\nu}, \tag{1.6.83}$$

where $\boldsymbol{\nu} = \nabla F / |\nabla F|$. A similar condition can be written in terms of pressure.

Condition (1.6.83) involves both the state variable, φ, and its gradient, $\nabla \varphi$. A boundary condition of this kind is called *a boundary condition of the third kind*, *a mixed boundary condition*, or *a Cauchy boundary condition*.

Discontinuity in solid matrix properties

This boundary is discussed in detail in LECTURE 5. Equation (1.5.11) serves as a basis for developing boundary conditions for mass transport of a single fluid that occupies the entire void space.

The condition of no-jump in the normal component of the specific discharge is (1.5.19), rewritten here for convenience in the form

$$[\phi(\mathbf{V} - \mathbf{u})]_{1,2} \cdot \boldsymbol{\nu} = 0 \qquad \text{on} \quad \mathcal{B}. \tag{1.6.84}$$

In view of (1.5.14), equation (1.6.84) reduces to

$$[\mathbf{q}_r]_{1,2} \cdot \boldsymbol{\nu} = 0 \qquad \text{on} \quad \mathcal{B}, \tag{1.6.85}$$

into which we may insert an appropriate equation for $\mathbf{q}_r$. For example, by employing (1.4.52), we obtain

$$[\mathbf{K} \cdot \nabla \varphi]_{1,2} \cdot \boldsymbol{\nu} = 0. \tag{1.6.86}$$

Another condition is that of no-jump in pressure, (1.5.28), rewritten here for convenience

$$[p]_{1,2} = 0 \qquad \text{on} \quad \mathcal{B}, \tag{1.6.87}$$

or, in terms of piezometric head

$$[\varphi]_{1,2} = 0 \qquad \text{on} \quad \mathcal{B}. \tag{1.6.88}$$

For a model that describes flow in a deformable porous medium, as the one summarized in Table 1.6.1, we need also conditions on stresses and displacements on the boundary.

Before leaving this subject, we should note that when surfaces of discontinuity in solid matrix properties occur within a considered porous medium domain, the latter is divided into a number of subdomains, such that no discontinuity occurs within any of them. Then, these surfaces of discontinuity serve as part of the boundary of each individual subdomain. Actually, each surface of discontinuity serves simultaneously as a boundary to the two subdomains separated by it. Therefore, *two* conditions are required on such a boundary–one for each subdomain. In the case considered here, the conditions to be employed are (1.6.86) and (1.6.88), or equivalent conditions in terms of pressure.

Sharp interface between two fluids

Here we consider the (assumed) sharp interface between adjacent domains, each occupied by a different fluid. As explained in LECTURE 5, such surface is an approximate model of reality, whether the two fluids are miscible, or immiscible. To simplify the presentation, we shall refer, as an example, to the (assumed sharp) interface between fresh water and salt water in a coastal aquifer.

An inherent difficulty in treating this kind of boundary, stems from the fact that the shape and position of the interface are a-priori unknown. In fact, in many cases, the determination of the latter constitutes the objective of the model. As explained in LECTURE 5, we find ourselves in a kind of a vicious circle: in order to solve the problem, we have to know the location of the boundary, but in order to determine the shape of the boundary, we have first to solve the problem.

In LECTURE 5, we have shown that the shape of the interface can be obtained from the boundary condition (1.5.28), viz.

$$F(\mathbf{x},t) = p_1(\mathbf{x},t) - p_2(\mathbf{x},t) = 0, \qquad (1.6.89)$$

where $p_1(\mathbf{x},t)$ and $p_2(\mathbf{x},t)$ denote the pressures in the lighter fresh (symbol 1) and in the heavier salt water (symbol 2) domains, respectively.

Assuming that in each domain, the fluid density is constant (or approximately so), we may introduce and employ the piezometric head, $\varphi_i(\mathbf{x},t)$, $i = 1,2$, as the state variable, instead of pressure, in each domain. Then, (1.6.89) becomes

$$F(\mathbf{x},t) = z + \delta\varphi_1(\mathbf{x},t) - (1+\delta)\varphi_2(\mathbf{x},t) = 0, \qquad (1.6.90)$$

where

$$\delta = \rho_1/(\rho_2 - \rho_1).$$

Approaching the interface from within the fresh water domain, we use (1.5.40), (1.6.90), (1.5.2) and (1.5.3) to obtain the condition

$$(\mathbf{u}\cdot\nabla F)|_1 \equiv \mathbf{V}_1\cdot\{\nabla z + \delta\nabla\varphi_1 - (1+\delta)\nabla\varphi_2\} = -\delta\frac{\partial\varphi_1}{\partial t} + (1+\delta)\frac{\partial\varphi_2}{\partial t}.$$

$$(1.6.91)$$

When the interface is approached from within the salt water domain, we obtain the condition

$$(\mathbf{u}\cdot\nabla F)|_2 \equiv \mathbf{V}_2\cdot[\nabla z + \delta\nabla\varphi_1 - (1+\delta)\nabla\varphi_2] = -\delta\frac{\partial\varphi_1}{\partial t} + (1+\delta)\frac{\partial\varphi_2}{\partial t}.$$

$$(1.6.92)$$

In the last two equations, the velocity in each domain is expressed by an appropriate motion equation, in terms of the piezometric head within that domain. For example, using (1.4.52), we obtain from (1.6.91)

$$\left(\frac{\mathbf{K}_1}{\phi}\cdot\nabla\varphi_1\right)\cdot[\nabla z + \delta\nabla\varphi_1 - (1+\delta)\nabla\varphi_2] = \delta\frac{\partial\varphi_1}{\partial t} - (1+\delta)\frac{\partial\varphi_2}{\partial t},$$

$$(1.6.93)$$

and from (1.6.92)

$$\left(\frac{\mathbf{K}_2}{\phi}\cdot\nabla\varphi_2\right)\cdot[\nabla z + \delta\nabla\varphi_1 - (1+\delta)\nabla\varphi_2] = \delta\frac{\partial\varphi_1}{\partial t} - (1+\delta)\frac{\partial\varphi_2}{\partial t},$$

$$(1.6.94)$$

where $\mathbf{K}_i = k\rho_i g/\mu_i$, $i = 1,2$, and we have assumed $\mathbf{V}_s = 0$ in both domains.

The last two equations constitute the (coupled) boundary conditions of the problem. One may think of each equation as serving as a boundary condition for one of the domains. Because of the coupling of the boundary conditions, the two problems: solving for φ_1 in domain 1 and for φ_2 in domain 2, have to be solved simultaneously.

We note the nonlinear nature of these boundary conditions. The shape and position of the interface, $F(\mathbf{x},t) = 0$, can be determined, using (1.6.90), once the problem has been solved and the piezometric head distributions $\varphi_1 = \varphi_1(\mathbf{x},t)$ and $\varphi_2 = \varphi_2(\mathbf{x},t)$ have been determined for the respective domains. This means that some kind of an iterative method of solution is required.

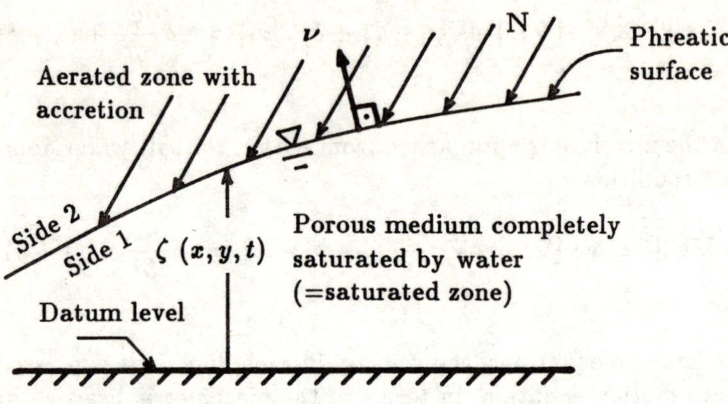

Figure 1.6.2: Nomenclature for a phreatic surface with accretion.

Phreatic surface

A particular case of the (assumed sharp) boundary between two fluids, discussed above, is that of an air-water interface, encountered in ground water hydrology. It may, therefore, be worthwhile to add a few comments on this kind of boundary.

Here we assume that a sharp interface–called *free surface*–separates an air occupied zone from a water occupied one. The water zone may occupy either the domain above, or below the air occupied one.

The free surface is an approximation, as in reality these two fluids are immiscible, and hence both air and water are present in the air occupied (= unsaturated) zone. When everywhere along a free surface, the air zone is *above* the water zone, (Fig. 1.6.2), the free surface is also called a *phreatic surface*, or a *water table*.

In principle, the entire discussion on a sharp interface between two fluids can be applied also to the phreatic surface that acts as a boundary to the saturated domain. However, in soil mechanics and hydrology of groundwater, the air is often assumed to be *stationary and at atmospheric pressure*. We shall consider this special case, taking into account the possibility of accretion (e.g., from precipitation) on the upper side of the phreatic surface.

We shall take into account accretion, $\mathbf{N}$, e.g., from precipitation, on the upper side of the phreatic surface.

Making use of (1.5.11), with $(\alpha = w)$, the condition of continuity of total mass flux at the phreatic surface, neglecting the diffusive and dispersive mass fluxes, may be expressed by

$$\phi\{\rho_w(\mathbf{V}_w - \mathbf{u})\}|_{\text{sat}}\cdot\boldsymbol{\nu} = (\rho'(\mathbf{N}')|_{\text{unsat}}, \qquad (1.6.95)$$

where the l.h.s. represents the water mass flux evaluated at the phreatic surface, as the latter is approached from within the saturated zone, while ρ' and $\mathbf{N}'$ are the density and the specific discharge of the fluid at the phreatic surface, respectively, as the latter is approached from above. The flux $\rho'\mathbf{N}'$ is independent of movement of the phreatic surface and the moisture content (if present) in the void space above it.

Assuming that the moisture in the unsaturated zone is everywhere at the *irreducible moisture content* θ_{wo} $(= \phi - \theta_a)$, the specific discharge, $\mathbf{N}'$, may be expressed in the form

$$\mathbf{N}' = \theta_{wo}(\mathbf{V}_w - \mathbf{u})|_{\text{unsat}} = (\mathbf{N} - \theta_{wo}\mathbf{u})|_{\text{unsat}}, \qquad (1.6.96)$$

where $\mathbf{N} = \theta_{wo}\mathbf{V}_w|_{\text{unsat}}$.

For a downward accretion at a rate N, we have $\mathbf{N} = -N\nabla z$.

Using (1.5.2) and (1.5.3), with $F(\mathbf{x},t) = 0$, to denote the shape of the phreatic surface, (1.6.95) becomes

$$(\rho_w\mathbf{q}_w - \rho'\mathbf{N})\cdot\nabla F + (\phi\rho_w - \theta_{wo}\rho')\frac{\partial F}{\partial t}\bigg|_{\text{sat}}\cdot\nabla F = 0, \qquad (1.6.97)$$

where $\mathbf{q}_w = \phi\mathbf{V}_w|_{\text{sat}}$.

When $\rho|_{\text{sat}} = \rho'$, equation (1.6.97) reduces to

$$(\mathbf{q}_w - \mathbf{N})\cdot\nabla F + \phi_{\text{eff}}\frac{\partial F}{\partial t} = 0, \qquad (1.6.98)$$

where $\phi_{\text{eff}}(= \phi - \theta_{wo})$ is often called the *effective porosity*.

In the model considered here, we assume that

[A6.17] the air in the unsaturated zone is everywhere at atmospheric pressure, taken as $p_a = 0$.

Hence, at all points of the phreatic surface, we also have $p_w(\mathbf{x}, t) = 0$. Making use of (1.5.55), the shape of the phreatic surface is given by

$$F(\mathbf{x}, t) \equiv p_w(\mathbf{x}, t) = 0, \tag{1.6.99}$$

or, in terms of piezometric head, $\varphi_w = z + p_w/\gamma_w$, for $\gamma_w = $ const.

$$F(\mathbf{x}, t) = \varphi_w(\mathbf{x}, t) - z = 0. \tag{1.6.100}$$

Using (1.6.100) to define ∇F and $\partial F/\partial t$ in (1.6.98), we obtain

$$(\phi \mathbf{V}_w|_{\text{sat}} - \mathbf{N})(\nabla \varphi_w - \nabla z) = -\phi_{\text{eff}}\frac{\partial \varphi_w}{\partial t}. \tag{1.6.101}$$

Neglecting the solid's velocity, i.e., $\mathbf{V}_s = 0$, hence, $\mathbf{q}_w \equiv \mathbf{q}_{wr} = \phi \mathbf{V}_w|_{\text{sat}}$, expressed by (1.4.52), equation (1.6.101) becomes

$$(\mathbf{K} \cdot \nabla \varphi_w + \mathbf{N})(\nabla \varphi_w - \nabla z) = \phi_{\text{eff}}\frac{\partial \varphi_w}{\partial t}. \tag{1.6.102}$$

This equation expresses the condition on a nonstationary phreatic surface with accretion that serves as a boundary to a flow domain. Actually, it is nothing but an expression of continuity of fluid flux across this boundary. A similar condition can be written in terms of pressure.

We note that (1.6.102) is *nonlinear* in φ_w. Also, since the specification of $F(\mathbf{x}, t)$ by (1.6.99) requires the knowledge of $\varphi_w(\mathbf{x}, t)$, which is actually the sought solution of the problem, we have here the vicious circle mentioned in the discussion on the interface between two fluids, presented above. Hence, an analytical solution of a free surface problem is impossible, except for a very limited number of special cases of steady flow. In numerical methods, iterative techniques are employed to obtain a solution.

Seepage Face

Whenever a phreatic surface approaches a body of open water which serves as part of the boundary of a flow domain (Fig. 1.6.3), it will *always* terminate on that (known) boundary at a point located at some elevation above the water table of the body of open water (Point A in Fig. 1.6.3). The segment AB is called the *seepage face* Through it, water seeps out of the porous medium domain.

The theoretical proof for the existence of a seepage face is given, among others, by Muskat (1937) and Bear (1972).

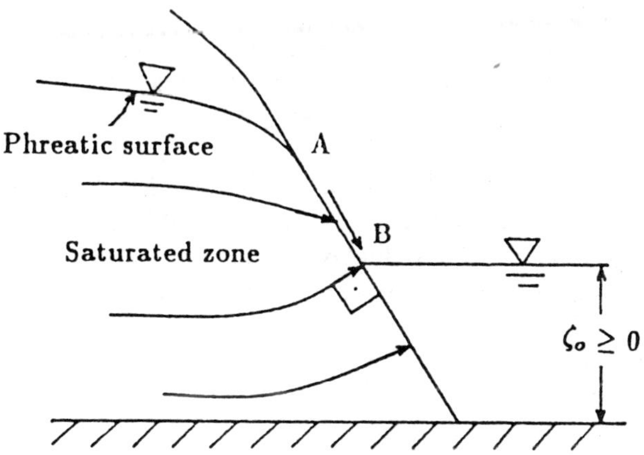

Figure 1.6.3: The seepage face, AB.

Since the seepage face is exposed to the atmosphere, the pressure on it
is $p_w = 0$. Hence, the condition on it is

$$\varphi_w(\mathbf{x}, t) = z, \qquad (1.6.103)$$

i.e., a condition of specified head (equal at each point to the known elevation
of the point).

The geometry of the seepage face is, thus, known, except for the location
of its end point A, which is also a point on the phreatic surface (whose
location, as we recall, is a-priori unknown).

1.6.5   Complete mathematical model

Following the discussion in Lecture 5, the complete mathematical model of
a problem of mass transport of a single component fluid phase that occupies
the entire void space of a porous medium, consists of five statements.

- Statement of the configuration of the surface bounding the domain,
  within which the considered mass transport takes place. The boundary is, usually, made up of a number of segments on each of which a
  different boundary condition may prevail.

- Statement of the dependent variable(s). In general, the fluid pressure,
  $p = p(\mathbf{x}, t)$, is the dependent variable. For a fluid of constant density,

we often use $\varphi = \varphi(\mathbf{x}, t)$.

If the porous medium is deformable, we may have to solve simultaneously for the fluid pressure, p, and for the strain, ε_{sk} (or displacement, $\mathbf{w}$, and effective stress, σ'_s) in the solid skeleton.

- Statement of the partial differential (balance) equation(s) to be satisfied by the dependent variable(s) at all points within the considered domain. Additional dependent variables (e.g., strain, ε_{sk}), require additional equations to be solved simultaneously.

- Statement of *initial conditions*, i.e., the values of all dependent variables at some initial time, taken as $t = 0$, at all points *within* the considered domain.

- Statement of boundary conditions to be satisfied by the dependent variable(s) at all points of the boundary of the considered domain at all times, $t > 0$.

To summarize, we have demonstrated how the general theory, presented in LECTURES 1 through 5, is applied to the case of mass transport of a single fluid that saturates a porous medium. It is important to emphasize that much attention should be given in each case to the assumptions that underlie the development of the partial differential equations and the boundary conditions. These assumptions, that together constitute the *conceptual model* of the problem, transform the (very complex) reality into the mathematical model, which we accept as describing the particular transport problem on hand. *Different assumptions lead to different models and, hence, to different solutions.*

1.7 Lecture Seven: Diffusive Flux

In this lecture, we consider the macroscopic diffusive fluxes of mass of a component of a phase, in single and multiphase flow, and of heat, where the latter flux is usually referred to as heat conduction. Following the usual methodology, we shall always start from the microscopic flux expressions and average them to obtain their macroscopic counterparts.

We recall that in LECTURE 2, the diffusive flux of any extensive quantity, E, of a phase, at the microscopic level, was expressed by

$$\mathbf{j}^{EU} = e(\mathbf{V}^E - \mathbf{V}), \qquad (1.7.1)$$

or

$$\mathbf{j}^{Em} = e(\mathbf{V}^E - \mathbf{V}^m). \qquad (1.7.2)$$

The corresponding macroscopic diffusive fluxes, $\overline{\mathbf{j}^{EU}}^f$, and $\overline{\mathbf{j}^{Em}}^f$, appear in the macroscopic balance equations (1.3.1) and (1.3.2), respectively.

The average diffusive flux (per unit area of fluid) is given by

$$\overline{\mathbf{j}^E}^f \equiv \overline{\mathbf{j}^{EU}}^f = \overline{e(\mathbf{V}^E - \mathbf{V})}^f. \qquad (1.7.3)$$

We note that

$$\overline{\mathbf{j}^E}^f \neq \overline{e}^f(\overline{\mathbf{V}^E}^f - \overline{\mathbf{V}}^f). \qquad (1.7.4)$$

We shall limit the discussion to *linear diffusive flux equations*, i.e., equations in which the flux of E at the microscopic level is proportional to a *driving force* that takes the form of a gradient of a single intensive quantity ($\equiv$ function of state), without *coupling* to other extensive quantities. Also, we shall assume no coupling among components, in the case of a multicomponent phase. Diffusive flux equations that include the effect of coupling among components and/or among extensive quantities, can be developed by following the procedure outlined here. When considering only a single phase, no symbol will be used to indicate this phase.

Rather than discuss a general case, we shall consider two particular diffusive fluxes: of mass of a component of a phase and of heat of a phase, from which other ones may be derived.

As already stated on several occasions in this series of lectures, the diffusive flux of momentum is identical to the stress. This flux is related to the gradient of the velocity (= momentum per unit mass) through an appropriate constitutive relation. In LECTURE 4, we have already developed the macroscopic expression for this flux.

1.7.1　Diffusive mass flux

Let us start by considering the diffusive flux of the mass of a γ-component in a single fluid phase that occupies the entire void space. This flux, $\mathbf{j}^\gamma$, is often referred to as *molecular diffusion*. It was defined in LECTURE 2 by (1.2.6), and (1.2.9).

At the microscopic level, this flux is expressed by *Fick's law*, that takes the form

$$\mathbf{j}^\gamma \equiv \mathbf{j}^{m\gamma U} \;=\; c^\gamma(\mathbf{V}^{m\gamma} - \mathbf{V})$$
$$\;=\; -\mathcal{D}^\gamma \cdot \nabla c^\gamma, \tag{1.7.5}$$

where $\mathcal{D}^\gamma$ is the (scalar) coefficient of molecular diffusion of the γ-component in a fluid phase, and c^γ denotes the concentration of the γ-component in the fluid. We note that

$$\sum_{(\gamma)} \mathbf{j}^\gamma \equiv \rho(\mathbf{V}^m - \mathbf{V}) = 0$$

only when $\mathbf{V}^m \equiv \mathbf{V}$, i.e., in a single component fluid. Otherwise, $\sum_{(\gamma)} \mathbf{j}^\gamma \neq 0$. However, $\sum_{(\gamma)} \mathbf{j}^{m\gamma m} = 0$, in a multicomponent fluid.

Equation (1.7.5) is the simplest form of Fick's law; it expresses the flux of a single component in a single fluid that is a *binary system*, assuming that $\mathcal{D}^\gamma$ is independent of c^γ.

Other forms of Fick's law for the diffusive mass flux of a γ-component are

$$\mathbf{j}^{m\gamma m} \;=\; c^\gamma(\mathbf{V}^{m\gamma} - \mathbf{V}^m) = -\rho\mathcal{D}^\gamma\nabla\left(\frac{c^\gamma}{\rho}\right), \tag{1.7.6}$$

$$\mathbf{j}^{\gamma mol} \;=\; c^{\gamma mol}(\mathbf{V}^{m\gamma} - \mathbf{V}^m) = -\rho^{mol}\mathcal{D}^\gamma\nabla\left(\frac{c^{\gamma mol}}{\rho^{mol}}\right), \tag{1.7.7}$$

where $c^{\gamma mol}(= c^\gamma/M^\gamma)$ is the *molar concentration* of γ (= number of moles of γ per unit volume of solution), $\rho^{mol}(= \sum_{(\gamma)} c^{\gamma mol})$ is the total *molar density* of the solution, and $\mathbf{j}^{\gamma mol}$ is the *molar (diffusive) flux* of γ (say in gr - moles, per cm^2 per sec.), relative to the fluid moving at the mass weighted velocity.

Actually, the diffusive mass flux of a γ-component should be expressed in the form

$$\mathbf{j}^{m\gamma m} = -\mathcal{D}'^\gamma\nabla\mu^\gamma, \tag{1.7.8}$$

where $\mu^\gamma = \mu^\gamma(p,\omega^\gamma,T)$ is the *chemical potential* of the γ-component. For *dilute solutions*, $\mu^\gamma = Cp\omega^\gamma T$, where C is a coefficient, so that (1.7.8) reduces to (1.7.7), with $\mathcal{D}^\gamma \sim C\mathcal{D}'^\gamma pT/\rho$ for constant C, ρ, T, and p.

In the passage from (1.7.5) to its macroscopic counterpart, the configuration of the solid - fluid interface surface, and conditions on it, affect the transformation of the (local) gradient of concentration, appearing in (1.7.5), into a gradient of the average concentration, which serves as the state variable at the macroscopic level.

Following our averaging routine, we attempt to derive the averaged diffusive flux, say, for a fluid that occupies the entire void space, by applying (1.2.57) to (1.7.5), obtaining, for a constant $\mathcal{D}^\gamma$

$$\overline{\mathbf{j}^\gamma} \equiv \phi\overline{\mathbf{j}^{\gamma f}} \;=\; -\phi\mathcal{D}^\gamma\overline{\nabla c^{\gamma f}}$$

$$=\; -\phi\mathcal{D}^\gamma(\nabla\overline{c^{\gamma f}} + \frac{1}{U_o}\int_{\mathcal{S}_{fs}} \overset{\circ}{c}^\gamma \nu\, dS), \tag{1.7.9}$$

where

$$\overset{\circ}{c}^\gamma = c^\gamma - \overline{c^{\gamma f}},$$

denotes the deviation of ρ at a point within a phase from the average over the REV.

Obviously, (1.7.9) is of no practical value to us, as we have no information on the values of c^γ on the solid-fluid surface. We note that this information constitutes a first type boundary condition for c^γ on the surface bounding the fluid phase. The configuration of $\mathcal{S}_{fs}$ is, obviously, also not known.

Since, as explained above, our objective is to study the influence of the configuration of the solid-fluid boundary at every instant of time, it is sufficient to investigate the concentration distribution *at* that instant of time, assuming no γ-sources or sinks within the fluid. Under such assumption, a monotonous distribution of c^γ takes place within $\mathcal{U}_{ov}$, satisfying

$$\nabla^2 c^\gamma = 0 \quad \text{in} \quad \mathcal{U}_{ov}. \tag{1.7.10}$$

A more rigorous justification of this assumption of quasi-steady state within the void space, can be obtained by employing the methodology of deletion of nondominant terms already demonstrated several times in earlier lectures. The starting point is the *diffusion equation* (= mass balance equation for a component), written for the fluid in the void space. In a one-dimensional domain (the same procedure can easily be applied to a three dimensional domain), this equation takes the form

$$\frac{\partial c^\gamma}{\partial t} = \mathcal{D}^\gamma\frac{\partial^2 c^\gamma}{\partial x^2}, \tag{1.7.11}$$

where we assume that $\mathcal{D}^\gamma$ is constant within $\mathcal{U}_{ov}$. Then, following the methodology of nondimensionalization, with the objective of comparing order of magnitude of terms, we introduce

$$c^\gamma = c^{\gamma*}c_c^\gamma \equiv c^{\gamma*}c^\gamma|_{\max},$$

$$x = x^*L_c,$$

$$\frac{\partial c^\gamma}{\partial t} = \left(\frac{\partial c^\gamma}{\partial t}\right)^*\left(\frac{\partial c^\gamma}{\partial t}\right)_c = \left(\frac{\partial c^\gamma}{\partial t}\right)^*\left(\frac{\partial c^\gamma}{\partial t}\right)\Big|_{\max} = \left(\frac{\partial c^\gamma}{\partial t}\right)^*\frac{c^\gamma|_{\max}}{t_c},$$

$$\frac{\partial c^\gamma}{\partial x} = \left(\frac{\partial c^\gamma}{\partial x}\right)^*\left(\frac{\partial c^\gamma}{\partial x}\right)_c = \left(\frac{\partial c^\gamma}{\partial x}\right)^*\left|\frac{\partial c^\gamma}{\partial x}\right|_{\max},$$

$$L_c = \frac{c^\gamma|_{\max}}{|\partial c^\gamma/\partial x|_{\max}}.$$

where subscript c denotes a characteristic value, and L_c is a distance (within the void space) over which significant changes in c^γ take place.

With these relations, the diffusion equation, (1.7.11), can be written in the form

$$\frac{L_c^2}{t_c\mathcal{D}^\gamma}\left(\frac{\partial c^\gamma}{\partial t}\right)^* = \frac{\partial^2 c^{\gamma*}}{\partial(x^*)^2}. \tag{1.7.12}$$

For component diffusion within the pore space, we may use the hydraulic radius Δ_f $(= U_{of}/S_{fs})$ as the characteristic length, L_c. Thus, when the condition

$$\mathrm{Fo}^{\mathcal{D}} \equiv \frac{t_c}{\Delta_f^2/\mathcal{D}^\gamma} \gg 1$$

prevails, where the dimensionless number $\mathrm{Fo}^{\mathcal{D}}$ is the *Fourier number* associated with the fluid's diffusivity within the void space, the diffusion process can be described by the Laplace equation (1.7.10), as a quasi-steady state one. In most cases, this condition is indeed valid within the void space. Here, the Fourier number, $\mathrm{Fo}^{\mathcal{D}}$, gives the ratio between the time interval during which a significant change in concentration occurs and the time required for smoothing out spatial concentration differences by diffusion.

When the solid-fluid interface, $\mathcal{S}_{fs}$, acts as a material surface to both the fluid as a whole and to the γ-component in it, i.e., when there is no transfer of mass of both of them across it, then

$$j_i^\gamma\nu_i \equiv -\mathcal{D}^\gamma\frac{\partial c^\gamma}{\partial x_i}\nu_i = 0 \qquad \text{on} \qquad \mathcal{S}_{fs}, \tag{1.7.13}$$

where ν is the outward normal unit vector on the $\mathcal{S}_{fs}$-surface.

With $\mathcal{D}^\gamma = \text{const.}$ within $\mathcal{U}_{ov}$, equations (1.7.10) and (1.7.13) are identical to (1.2.59) and (1.2.70), respectively, that describe **CASE A** in **LECTURE 2**, with $G_\alpha \equiv c^\gamma$. Hence, by averaging (1.7.5), making use of (1.2.71), we obtain

$$\overline{j_j^\gamma} = -\mathcal{D}^\gamma \overline{\frac{\partial c^\gamma}{\partial x_j}} = -\phi \mathcal{D}^\gamma T_{ji}^* \frac{\partial \overline{c^\gamma}^f}{\partial x_i} = -\phi (\mathcal{D}^{*\gamma})_{ij} \frac{\partial \overline{c^\gamma}^f}{\partial x_i}, \qquad (1.7.14)$$

where ϕ is the porosity and $\mathcal{D}^{*\gamma} = \mathcal{D}^\gamma \mathbf{T}^*$, a second rank symmetric tensor, is the *coefficient of molecular diffusion in a (saturated) porous medium*. The definition of $\mathbf{T}^*$ is presented in (1.2.63) and discussed in detail following that equation.

Consider a liquid α-phase that occupies only part of the void space, and a component that does not interact with the solid, nor does it cross the (microscopic) interfaces between the α-phase and any of the other phases present in $\mathcal{U}_o$. Using β to denote all other phases within $\mathcal{U}_o$, and $\mathcal{S}_{\alpha\beta}$ to denote the α–*beta*-surface, and when condition (1.7.13) prevails on $\mathcal{S}_{\alpha\beta}$, equation (1.7.14) becomes

$$\overline{j_j^\gamma} = -\mathcal{D}^\gamma \overline{\frac{\partial c^\gamma}{\partial x_j}} = -\theta \mathcal{D}^\gamma T_{ji}^* \frac{\partial \overline{c^\gamma}^\alpha}{\partial x_i}$$

$$= -\theta (\mathcal{D}^{*\gamma})_{ji} \frac{\partial \overline{c^\gamma}^\alpha}{\partial x_i}, \qquad (1.7.15)$$

where $\mathcal{D}^{*\gamma}$ depends on θ and on all geometrical factors which determine the value of $\mathbf{T}^*$.

Thus, subject to the conditions (1.7.10) in $\mathcal{U}_{o\alpha}$ and (1.7.13) on $\mathcal{S}_{\alpha\beta}$, equation (1.7.15) expresses the diffusive mass flux of a γ-component in an α-phase that occupies only part of the void space.

Let us consider the case when the γ-**component can be adsorbed on the solid surface.** For a liquid phase that completely occupies the void space, and with the assumptions introduced above with respect to the γ-distribution within $\mathcal{U}_{ov}$, equation (1.7.10) remains valid.

Because the solid-fluid surface is a material surface with respect to fluid mass, *the adsorbed component can reach the solid wall only by diffusion.* Let us assume that this diffusive flux, normal to the solid, can be approximated by

$$j_i^\gamma \nu_i = -\mathcal{D}^\gamma \frac{\partial c^\gamma}{\partial x_i} \nu_i = \mathcal{D}_\gamma \frac{\overline{c^\gamma}^f - c^\gamma|_{S_{fs}}}{\Delta}, \qquad (1.7.16)$$

where Δ is a length characterizing the distance between the $\mathcal{S}_{fs}$-surface and the interior of the fluid phase within $\mathcal{U}_o$. From (1.7.16), it follows that

$$\frac{\partial c^\gamma}{\partial x_i}\nu_i = \frac{c^\gamma|_{\mathcal{S}_{fs}} - \overline{c^\gamma}^f}{\Delta} \qquad \text{on} \qquad \mathcal{S}_{fs}. \qquad (1.7.17)$$

By comparing (1.7.17) with (1.2.80), we conclude that the case under consideration is identical to **CASE C** of **LECTURE 2**, with $G_\alpha \equiv c^\gamma$. Hence, using (1.2.82), we obtain

$$\overline{\mathbf{j}^\gamma} = -\phi(\mathcal{D}^{\gamma*})_{ji}\frac{\partial \overline{c^\gamma}^f}{\partial x_i} - \phi M_j \mathcal{D}^\gamma \frac{\widetilde{c^\gamma}^{fs} - \overline{c^\gamma}^f}{\Delta_f^2}, \qquad (1.7.18)$$

where $\mathbf{M}$ is a (vector) coefficient defined by (1.2.81), $\widetilde{c^\gamma}^{fs}$ is the average value of c^γ on the surface $\mathcal{S}_{fs}$, and Δ_f is the hydraulic radius of the void space. In LECTURE 8, we shall see that $\widetilde{c^\gamma}^{fs}$ is another state variable in a diffusion, or a diffusion–dispersion problem with adsorption.

For a fluid that occupies only part of the void space, we replace f by α and ϕ by θ_α in (1.7.18). Also, both $\mathcal{D}_\alpha^*$ and Δ_α depend on θ_α.

1.7.2   Diffusive heat flux

In what follows, we shall take into account the fact that heat can diffuse in both the fluid and the solid phases, with the possibility of heat exchange between them.

At the microscopic level, the diffusive heat flux (commonly referred to as *heat conduction*) is expressed by *Fourier's law*

$$\mathbf{j}_\alpha^H = -\lambda_\alpha \nabla T_\alpha, \qquad \alpha = f, s, \qquad (1.7.19)$$

where we have used subscript f to indicate that we consider only the case in which a single fluid phase occupies the entire void space.

As in the case of molecular diffusion discussed above, we assume that λ_f is constant and the distribution of T_f is quasi–steady within $\mathcal{U}_{of}$. In the present case, following the discussion presented above, for molecular diffusion, the assumption of a quasi–steady distribution of temperature is valid as long as

$$\text{Fo}^\lambda \equiv \frac{(\Delta t)_c}{\Delta_f^2/(\lambda_f/\rho_f C_{vf})} \gg 1, \qquad (1.7.20)$$

where $(\Delta t)_c$ is the characteristic time for temperature changes, and Fo^λ is the Fourier number associated with conductive heat transfer.

It is easy to verify that this problem of heat conduction in the fluid-solid system comprising the porous medium is described by **CASE B1** of LECTURE 2, with $G_\alpha \equiv T_f$ and $\lambda_\alpha \equiv \lambda_f$, representing the temperature and the thermal conductivity of the fluid phase, and $G_\beta = T_s$ and $\lambda_\beta = \lambda_s$, representing the temperature and the thermal conductivity of the solid phase. We may, therefore, average (1.7.19), employing (1.2.79), to obtain

$$
\begin{aligned}
\overline{\mathbf{j}_f^H} &= -\lambda_f \overline{\nabla T_f} = -\phi \lambda_f \overline{\nabla T_f}^f \\
&= -\frac{\phi \lambda_f}{\lambda_f - \lambda_s}\left\{ (\lambda_f \overline{\nabla T_f}^f - \lambda_s \overline{\nabla T_s}^s) \cdot \mathbf{T}_f^* - \frac{\lambda_s}{\phi} \nabla[\phi\left(\overline{T_f}^f - \overline{T_s}^s\right)] \right\}.
\end{aligned}
$$
$$(1.7.21)$$

An analogous expression, in terms of $\theta_s = 1 - \phi$ and $\theta_s \mathbf{T}_s^* = \mathbf{I} - \theta_f \mathbf{T}_f^*$ (see (1.2.77)) can be written for the macroscopic conductive heat flux, $\overline{\mathbf{j}_s^H}$ in the solid phase.

It is interesting to note the difference between (1.7.14) and (1.7.21). Because the fluid - solid interface, $\mathcal{S}_{fs}$, is 'impervious' to mass transfer, the relationship (1.7.14) shows that the macroscopic diffusive mass flux of the component depends on the concentration of the component within the fluid phase only. As the component stays always within the phase, the tortuosity coefficient, $\mathbf{T}_f^*$, that appears in the flux expression, depends only on the configuration of the fluid phase within the REV. However, with respect to heat, the solid-fluid interface is a 'permeable' surface. As a result, the macroscopic conductive heat flux depends on the temperatures in *both* the solid and the fluid phases. In this case, the $\mathcal{S}_{fs}$-surface permits heat to be transferred across it, with conditions (1.2.72) and (1.2.73) prevailing on it.

Equation (1.7.21) shows that coupling exists between the heat transported in the two domains, $\mathcal{U}_{of}$ and $\mathcal{U}_{os}$, with heat being continuously exchanged between the two phases.

When the two phases are in a state of thermal equilibrium, i.e., $\overline{T_f}^f \simeq \overline{T_s}^s$, equation (1.7.21) reduces to

$$
\overline{\mathbf{j}_f^H} = -\phi \lambda_f \mathbf{T}_f^* \cdot \overline{\nabla T_f}^f = -\phi \lambda_f^* \cdot \overline{\nabla T_f}^f, \tag{1.7.22}
$$

in which $\lambda_f^* = \lambda_f^* \mathbf{T}_f^*$ is the coefficient of heat conduction in a fluid that fully occupies the void space of a porous medium, with no heat exchange between the fluid phase and the solid one. We note the similarity between (1.7.22) and to (1.7.14). Actually, the form of (1.7.22) should have been expected, since, *on the average*, no heat crosses the interface between the two phases. Hence, (1.7.22) can also be obtained from (1.2.79), in which we insert $\lambda_s = 0$

The equation for heat flux through the porous medium as a whole, $\overline{\mathbf{j}}_{pm}^{H}$, is obtained by summing the averaged flux equations for the two phases, taking into account the condition of equality of normal flux on the boundary between them. We obtain

$$\overline{\mathbf{j}_{pm}^{H}} = \phi\overline{\mathbf{j}_{f}^{H}}^{f} + (1-\phi)\overline{\mathbf{j}_{s}^{H}}^{s} = -[\phi\lambda_{f}\overline{\nabla T_{f}}^{f} + (1-\phi)\lambda_{s}\overline{\nabla T_{s}}^{s}]$$

$$= -\phi\lambda_{f}\mathbf{T}_{f}^{*}\cdot\nabla\overline{T_{f}}^{f} + (1-\phi)\lambda_{s}\mathbf{T}_{s}^{*}\cdot\nabla\overline{T_{s}}^{s}.$$

$$(1.7.23)$$

When $\overline{T_{f}}^{f} = \overline{T_{s}}^{s}$, the total heat flux in the porous medium as a whole is expressed by

$$\overline{\mathbf{j}_{pm}^{H}} \equiv \overline{\mathbf{j}_{f}^{H}} + \overline{\mathbf{j}_{s}^{H}}$$

$$= -[\phi\lambda_{f}^{*} + (1-\phi)\lambda_{s}^{*}]\cdot\nabla\overline{T_{f}}^{f} = -\Lambda^{H}\cdot\nabla\overline{T_{f}}^{f}, \qquad (1.7.24)$$

where

$$\Lambda^{H} = \phi\lambda_{f}^{*} + (1-\phi)\lambda_{s}^{*}$$

is the thermal conductivity of a saturated porous medium.

To summarize, we have developed expressions for the macroscopic diffusive fluxes of mass and heat (recalling that we have developed such expression also for the diffusive flux of momentum in LECTURE 4) that appear in the respective macroscopic balance equation. We note the possibility of coupling between the two phases when the considered extensive quantity can be exchanged between adjacent phases. This possible exchange may affect our definition of tortuosity as a coefficient that reflects only the geometry of the domain in which a considered extensive quantity is being transported.

The methodology employed here can, in principle, be extended to cases in which more than one fluid phase are present in the void space.

1.8 Lecture Eight: Modelling Contaminant Transport

In this lecture, we focus our attention on the transport of a component, e.g., a solute, contained in a fluid phase that occupies the entire void space, or only part of. The extensive quantity under consideration is the mass of this component. We shall use the term *concentration*, and the symbol c^γ, to denote the mass density of a γ-component (= mass of component per unit volume of fluid).

The use of mass density as a measure of the amount of a component in a fluid phase, is not always convenient. When the chemical interactions among components are being considered, preferable measures are, for example, the component's *chemical potential*, or its *activity*. To simplify the presentation, we shall refer to concentration, as defined above, with the understanding that, when necessary, the symbol c^γ may also stand for other measures of the quantity of the component in the fluid.

As a fluid moves within a considered domain, it carries the component with it. As we have already seen in the discussion on the macroscopic mass balance equation for a component of a phase (Subs. 3.2(c)), the total flux of a component is made up of the sum of three fluxes:

- an advective flux (discussed in LECTURE 4),

- a dispersive flux (to be considered in this lecture), and

- a diffusive flux (discussed in LECTURE 7).

In addition, the continuous variation of the component's concentration is affected by various sources and sinks of the considered component within the domain of interest.

Our objective in this lecture is to construct the mathematical model that describes the transport of a component of a fluid phase in a porous medium domain. The solution of the model should provide information on the future spatial distribution of the component's concentration within a given domain. Problems that require such information arise in groundwater investigations, in connection with the movement of pollutants, in chemical engineering, in reservoir engineering (e.g., in connection with the use of solvents to enhance recovery), and in investigations associated with radioactive waste repositories in geological formations.

To obtain the component's advective flux (and as we shall soon see, also the dispersive one), we must know the fluid's (i.e., the 'carrier's') velocity. In other words, the solution of a problem of component transport, requires information concerning the velocity of the fluid phase that carries that component. This information is obtained by solving the fluid flow models discussed, for a single phase flow, in LECTURE 4. As we shall see below, sometimes these two problems, the fluid flow problem and the component transport one, are coupled and have to be solved simultaneously.

As long as we shall consider the transport of a component in a single fluid phase, no special symbols will be used to indicate the considered component and phase. Similarly, when only one component is being considered, we shall omit the superscript γ in c^γ.

1.8.1   The Phenomenon of dispersion

The term $\overline{\mathring{c}\mathring{\mathbf{V}}}^\alpha (\equiv \overline{c\mathring{\mathbf{V}}}^\alpha)$ appearing in the general macroscopic balance equation for a component of an α-phase, (1.3.12), represents the *dispersive flux* of the component (per unit area of fluid). It is a macroscopic flux of the latter, relative to the transport of the phase, at the average (volume weighted) velocity $\overline{\mathbf{V}}^\alpha$ of the phase. This flux results from the variation of both the microscopic velocity and the concentration, c, within the REV. We recall that (1.3.12) is an average of the balance equation (1.2.22), in which we preferred to express the total flux of the mass of a γ-component, m^γ, by $c^\gamma \mathbf{V}$ $+\mathbf{j}^{m^\gamma U}$. The dispersive flux of a component that is present in an α-phase that occupies only part of the void space, *per unit area of porous medium,* is given by $\theta_\alpha \overline{\mathring{c}_\alpha \mathring{\mathbf{V}}_\alpha}^\alpha$.

In order to make use of the dispersive flux, $\overline{\mathring{c}\mathring{\mathbf{V}}}^\alpha$, in a macroscopic transport model, we have to express it in terms of average variables, such as $\overline{c}^\alpha$ and $\overline{\mathbf{V}}^\alpha$. This is our objective in the first part of this lecture.

Before presenting an expression for the dispersive flux, in terms of the average velocity of the phase, and the average density of the component, let us attempt to gain some insight of the phenomena that produce the dispersion of a component of a fluid phase.

For the sake of simplicity, we shall refer to a single fluid phase that occupies the entire void space.

Let us conduct two field experiments. Figure 1.8.1a shows an (assumed) sharp front in two-dimensional flow in a porous medium domain at some initial time, $t = 0$. This front separates the subdomain occupied by the component labelled fluid, at a uniform concentration, $c = 1$, from the one occupied by a nonlabelled fluid, $c = 0$. If uniform flow (normal to the initial front) at an average velocity, $\overline{V}$, takes place in the entire domain, Darcy's law can be used to determine $\overline{V}$, which, in turn can be used to determine the new position of the (assumed) abrupt front at time t.The new position is at $x = \overline{V}t$. Thus, on the basis of Darcy's law alone, the two parts of the fluid would continue to occupy subdomains separated by an abrupt front. However, by measuring concentrations in a number of observation points scattered in the domain, we find that no such front exists. Instead, we observe a gradual transition from the subdomain containing fluid at $c = 1$, to that containing fluid at $c = 0$. Experience shows that as flow continues, the width of the transition zone increases. This spreading of the labelled fluid, beyond the zone it is supposed to occupy according to the description of fluid movement by Darcy's law, cannot be explained by considering only the averaged movement of the fluid.

As a second experiment, consider the injection of a certain quantity of labelled fluid at a point, $x = 0$, $y = 0$, at some initial time, $t = 0$ into a tracer free fluid that is in uniform flow in a two-dimensional porous medium domain. Making use of the (average) velocity as calculated by Darcy's law, we should expect the labelled fluid to move as a volume of fixed shape, reaching the distance $x = \overline{V}t$ at time t. Again, field observations (shown in Fig. 1.8.1b) reveal a completely different picture. We observe a spreading of the labelled fluid, not only in the direction of the uniform (average) flow, but also normal to it. The region occupied by the labelled fluid will continue to grow, both *longitudinally*, i.e., in the direction of the uniform flow, and *transversally*, i.e., normal to it. Curves of equal concentration have the shape of confocal ellipses. Again, this spreading cannot be explained by the average flow alone (especially, noting that spreading occurs also perpendicular to the direction of the uniform flow).

The spreading phenomenon described above is called *hydrodynamic dispersion* (or *miscible displacement*). It is a nonsteady, *irreversible*, process, (in the sense that the initial component distribution cannot be obtained by reversing the direction of the uniform flow), in which the mass of the component continuously spreads out, mixing with the nonlabelled fluid.

The dispersion phenomenon may be demonstrated also by a simple laboratory experiment. Consider steady flow of water in a column of homogeneous sand, at a constant discharge, Q. At a certain instant, $t = 0$, tracer-marked water (e.g., water with NaCl at a low concentration (so that the effect of density variations on the flow pattern is negligible) starts to dis-

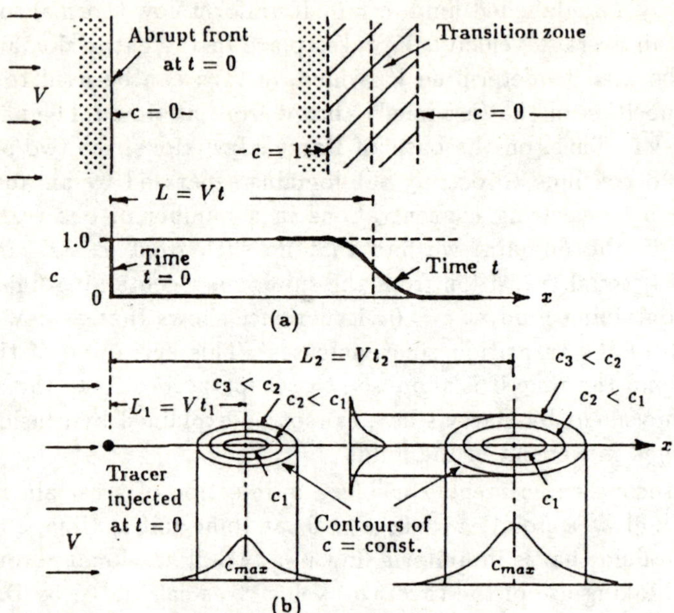

Figure 1.8.1: Longitudinal and transversal spreading of a tracer. (a) Longitudinal spreading of an initially sharp front. (b) Spreading of a tracer injected at a point.

place the original unmarked water in the column. The tracer concentration $c = c(t)$ is measured at the outflow end of the column and presented in the form of a graph, called a *breakthrough curve*, which relates the relative tracer concentration to time. In the absence of dispersion, the breakthrough curve would have taken the form of the broken line shown in Fig. 1.8.2, where U_{col} is the pore volume in the column, and Q is the constant discharge through the column. In reality, due to hydrodynamic dispersion, the curve will take the form of the S-shaped curve shown in full line in Fig. 1.8.2.

As stated above, we cannot explain all the above observations on the basis of the average (Darcy) flow. Following the methodology presented throughout these lectures, we must 'take our magnifying glass', and study what happens at the *microscopic level*, viz., inside the void space. There, we observe velocity variations in both magnitude and direction across any

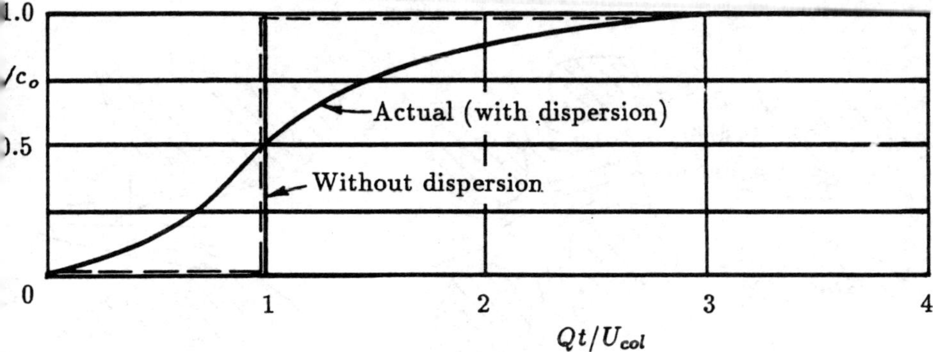

Figure 1.8.2: Breakthrough curve in a sand column.

pore cross section. We usually assume zero fluid velocity at the solid surface, and observe a maximum velocity at some internal point of every pore cross-section. The maximum velocity itself varies according to the size of the pore. We may also note stagnation points. Because of the shape of the interconnected pore space, the (microscopic) streamlines fluctuate in space, with respect to the direction of the average flow (Fig. 1.8.3a and b). This phenomenon, referred to as *mechanical dispersion*, causes the spreading of any initially close cloud of component particles. As flow continues, these particles will occupy an ever growing volume of the flow domain. The two basic factors that produce mechanical dispersion are, therefore, *flow* and the *presence of a pore system* through which flow takes place.

Although this spreading is in both the longitudinal direction, namely, that of the average flow, and in the direction transversal to the latter, it is primarily in the former direction. Very little spreading in a direction perpendicular to the average flow is produced by velocity variations alone. Also, such velocity variations alone cannot explain the ever-growing width of the zone occupied by component particles normal to the direction of flow. In order to explain the latter observed spreading, we must refer to an additional phenomenon that takes place in the void space, viz., *molecular diffusion*.

Molecular diffusion, caused by the random motion of molecules in a fluid, produces a flux of tracer particles (at the microscopic level) from regions of higher tracer concentrations to those of lower ones. Molecular diffusion will occur, independent of whether the fluid is at rest, or is moving. This means that inside every pore, as marked particles spread *along* microscopic stream-tubes, due to velocity variations, a concentration gradient of these particles

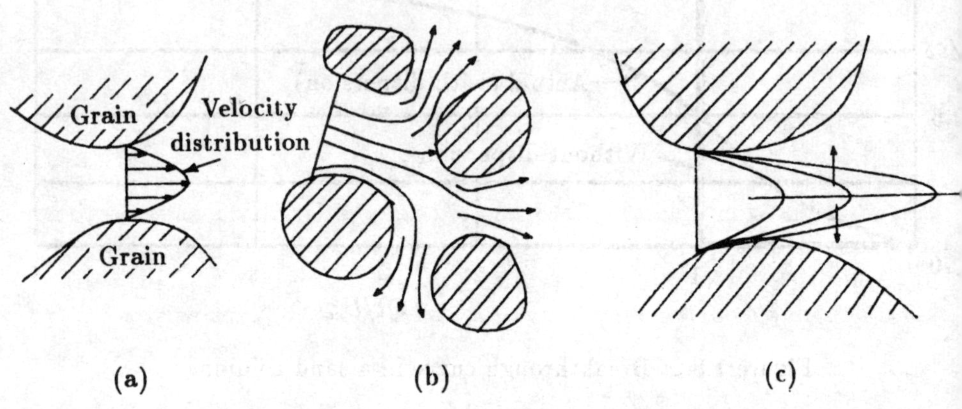

Figure 1.8.3: Spreading due to mechanical spreading (a,b) and molecular diffusion (c).

is produced, which, in turn, produces a flux of the component by the mechanism of molecular diffusion. The latter phenomenon tends to equalize the concentration along the stream tube. Relatively, this is a minor effect. However, at the same time, a concentration gradient is also produced between adjacent streamlines, causing *lateral molecular diffusion across streamtubes* (Fig. 1.8.3c), tending to equalize the concentration across pores. It is this phenomenon that explains the nature of the observed transversal dispersion.

In addition to the role played at the microscopic level by molecular diffusion, in enhancing the transversal component of mechanical dispersion, it also produces a macroscopic flux of its own. This is easily demonstrated by letting the velocity vanish. Then the component is transported by (macroscopic) molecular diffusion only.

We refer to the flux that causes mechanical dispersion (of a component) as *dispersive flux*. It is a *macroscopic* flux that expresses the effect of the microscopic variations of velocity in the vicinity of a considered point. However, as explained above, it is intrinsically affected by molecular diffusion within every pore.

The decomposition of the average of the total (local) advective fluxes within an REV, into an advective flux and a dispersive one, is a result of the averaging process. No dispersive flux exists at the microscopic level. In employing the dispersive flux, we lose information about the behavior at the microscopic level (which we do not have anyway). As we shall see below, this loss is compensated for by the introduction of a coefficient.

We use the term *hydrodynamic dispersion* to denote the spreading (at the macroscopic level) resulting from both mechanical dispersion and molecular diffusion. As explained above, the separation between the two processes is rather artificial, as they are inseparable. However, molecular diffusion alone does take place also in the absence of motion (both in a porous medium and in a fluid continuum). Because molecular diffusion depends on time, its effect on the overall spreading of a component is more significant at low velocities. It is molecular diffusion which makes the phenomenon of hydrodynamic dispersion irreversible, even in purely laminar flow.

In addition to inhomogeneity on a microscopic scale (i.e., the presence of pores and grains), we may also have inhomogeneity on a macroscopic scale, due to variations in permeability from one portion of a domain to the next. This inhomogeneity also produces dispersion of marked particles, but on a much larger scale. We shall briefly refer to this dispersion later in this lecture.

Dispersion may take place both in a microscopically laminar flow regime, where a fluid moves along definite paths and in a turbulent regime, where the turbulence may cause yet an additional mixing. In what follows, we shall focus our attention only on flow of the first type.

Variations in concentrations cause changes in the fluid's density and viscosity. These, in turn, affect the flow regime (i.e., velocity distribution) that depends on these properties. At low concentrations, the effect is relatively small, and may be overlooked for most practical purposes. However, in certain cases, for example in the problem of sea water intrusion into a region of fresh water, the density (and the viscosity) may vary appreciably. Concentration affected changes in density and in viscosity cannot be ignored. The flow and contaminant transport models are then coupled to each other through the relationship between density and concentration.

With the above comments on the spreading of a component carried by the fluid, it should be clear why we refer to the term average, $\overline{\overset{\circ}{c}\overset{\circ}{V}}^{\alpha}$, as the *dispersive flux* of the mass of a component in a fluid phase. Although in the discussion we referred to the mass of a component, the conclusions are equally valid for any extensive quantity (for example heat, with heat conduction playing the role assigned above to molecular diffusion). Not much is known about the dispersive flux of momentum.

In many cases of practical interest, the fluid contains more than one component of interest, possibly with chemical interactions taking place among these components and with the solid. We may also have a number of immiscible fluids that together occupy the void space, with components that can diffuse in more than one fluid. Some of these cases will also be discussed below.

In the macroscopic mass balance for a component of a phase, (1.3.12), we see the three fluxes that together comprise the total flux of a component. In the same balance equation, we also see a term that expresses the rate at which the mass of the component leaves the phase through the microscopic interface area between the phase and all other phases, and a term that represents the rate of production of the component *within* the phase. Both source terms are per unit volume of porous medium. In what follows, we shall first discuss the fluxes and then the various sources, so that eventually we may rewrite (1.3.12) in more convenient and practical forms.

1.8.2   Fluxes

The first flux term that appears in the macroscopic mass balance for a component of a phase, (1.3.12), is the advective flux of a component in the phase, $\overline{c}^{\alpha}\overline{\mathbf{V}_{\alpha}}^{\alpha}$. No further discussion is required with respect to this flux.

The second flux is a dispersive flux, to be discussed shortly, while the third is a diffusive flux, $\overline{\mathbf{j}_{\alpha}^{\gamma}}^{\alpha}$, discussed in detail in LECTURE 7.

The dispersive flux of a component in a fluid phase, $\overline{\overset{\circ}{c}^{\gamma}\overset{\circ}{\mathbf{V}}}^{\alpha}$, was introduced in the discussion on the macroscopic balance equation, (1.3.12), for a component of a fluid phase. It expresses the mass of a component passing through a unit area *of fluid* in a cross-section of a porous medium which is perpendicular to the direction of the flux.

Over the past three decades, especially in the 60's and early 70's, much research efforts have been devoted to the expression of this flux in terms of macroscopic parameters of the flow , the fluid, the transported component and the configuration of the void space (or the fluid occupied part of, it in multiphase flow). The works of Nikolaevski (1959), Bear (1961, 1972), Scheidegger (1961), Bear and Bachmat (1967, 1990) , de Josselin de Jong (1958), Saffman (1959), Bachmat (1972) and Bachmat and Bear (1983), among many others, may be mentioned in this respect. Although a variety of different approaches have been employed by these authors, most of the resulting flux equations were practically the same. In what follows, we shall start by presenting the commonly accepted equation that describes the dispersive flux in a porous medium.

In recent years, primarily in connection with ground water contamination in aquifers of large areal extent, many efforts have been devoted to the effect of large scale heterogeneities, e.g., in permeability, or due to the presence of a fracture network, on the dispersive flux equation. These efforts have led to a better understanding of large scale dispersion (often referred to as *macrodispersion*), to flux equations, and to coefficients of dispersion which modify the basic flux equation presented here. In the literature, the issue is often raised in the form of the question as to whether the dispersive flux takes the form of a Fickian-type law, or whether the dispersivity coefficient, to be discussed shortly, is a constant, or it depends on time of travel, or distance travelled by a plume of polluted water. We shall not discuss this topic here.

The conclusion of the various investigations that were mentioned above was that the dispersive flux, $\mathbf{J}_\alpha^{*m^\gamma} \equiv \overline{\overset{\circ}{c}\overset{\circ}{\mathbf{V}}}^\alpha$ is given by

$$\mathbf{J}^{*m^\gamma} \equiv \overline{\overset{\circ}{c}\overset{\circ}{\mathbf{V}}}^\alpha = -\mathbf{D}\cdot\nabla\overline{c}^\alpha, \qquad \mathbf{J}_i^{*m^\gamma} \equiv \overline{\overset{\circ}{c}\overset{\circ}{V}_i}^\alpha = -\mathbf{D}_{ij}\frac{\partial\overline{c}^\alpha}{\partial x_j}, \qquad (1.8.1)$$

where $\mathbf{D}$ is a coefficient called *coefficient of (mechanical, or advective) dispersion*. This coefficient is further expressed in the form

$$D_{ij} = a_{ijkm}\frac{\overline{V_k}^\alpha\overline{V_m}^\alpha}{\overline{V}^\alpha}f(\text{Pe},\delta), \qquad (1.8.2)$$

in which we note three factors that together contribute to its structure.

First we note the coefficient a_{ijkm} (dim. L), called *dispersivity of the porous medium*. This coefficient, that has the form of a fourth rank tensor, with components a_{ijkm}, represents, in single phase flow, the effect of the pore space configuration. It actually represents a measure of the heterogeneity that is due to the microscopic configuration of the pore space. In a phase in multiphase flow, the dispersivity components depend on the configuration of the phase occupied portion of the void space, which, in turn, depends on the saturation of the phase.

The second effect that we note is that of the (average) fluid velocity, $\overline{\mathbf{V}}^\alpha$. We note the linear relationship between the dispersion coefficient and the velocity.

The third factor is a function of the Peclet number, Pe, and δ, a param- eter that represents the ratio of the length and width characterizing a pore. As we recall, the Peclet number here is defined as

$$\text{Pe} = \frac{L_c V_c}{\mathcal{D}},$$

in which L_c is a characteristic length of the pores, e.g., the hydraulic radius of the pore space, Δ_c, V_c represents a characteristic average velocity and $\mathcal{D}$ is the coefficient of molecular diffusion. The function f introduces the effect of component transfer by molecular diffusion between adjacent (microscopic) streamlines inside the pores. Through this factor we have the effect of molecular diffusion on dispersion, as discussed earlier in this lecture. We recall that this effect is distinct from molecular diffusion at the macroscopic level. Once we recognize the need for such a factor, we may follow most authors who approximate it by one.

We note that (1.8.1) has the outward shape of Fick's law of molecular diffusion.

Henceforth we shall delete the average symbol and use c and $\mathbf{V}$ to denote $\overline{c}^{\alpha}$ and $\overline{\mathbf{V}_{\alpha}}^{\alpha}$, respectively.

When the fluid phase occupies only part of the void space, the dispersivity components, a_{ijkm}, depend on the volumetric fraction, θ, of the considered phase.

The coefficient of dispersion, $\mathbf{D}$, is a symmetric second rank tensor, with components $D_{ij}, = D_{ji}, \quad i, j = 1, 2, 3$. As such, it has *three principal directions*. Using these principal directions as Cartesian coordinate axes, x_1, x_2, x_3, we may write this coefficient in the matrix form

$$\mathbf{D} = \begin{bmatrix} D_{x_1 x_1} & O & O \\ O & D_{x_2 x_2} & O \\ O & O & D_{x_3 x_3} \end{bmatrix}. \tag{1.8.3}$$

However, unlike the second rank tensor of permeability, which is a property *only* of the microscopic configuration of the considered phase inside the void space, the tensor $\mathbf{D}$ (and this means its components and principal directions) depends also on the macroscopic velocity field. This is a consequence of (1.8.2).

Another basic difference between these two tensors, k_{ij} and D_{ij}, is the following. In an isotropic porous medium, any three mutually orthogonal directions in space may serve as principal directions for the permeability. However, due to the effect of velocity, the principal axes of the dispersion coefficient, D_{ij}, at a point in an isotropic porous medium, are always in the direction of the tangent to the streamline passing through that point and in the directions of the two main normals to that direction.

Thus, even in an isotropic porous medium, a distinct set of principal directions exists at every point in a flow domain. As the velocity varies from point to point, so do the principal directions. They may also vary at every point with time, in response to velocity variations.

If *locally*, we select the principal axes as a coordinate system, then **D** can be written in the matrix form (1.8.3).

In this case, the coefficient $D_{x_1 x_1}$ is called *coefficient of longitudinal dispersion*, while the coefficients $D_{x_2 x_2}$ and $D_{x_3 x_3}$ are called *coefficients of transversal dispersion*.

In a three-dimensional space, the dispersivity, a_{ijkm}, is a fourth rank tensor that has 81 components. However, because of various symmetry considerations (e.g., Spain, 1956, Bear, 1961, Scheidegger, 1961), all but 36 components are non-zero.

In an *isotropic porous medium*, only 21 non-zero components remain. They, in turn, depend only on *two* parameters: a_L and a_T, called the *longitudinal* and the *transversal dispersivities of the porous medium*, respectively. Both have a length dimension. As mentioned above, the longitudinal dispersivity, a_L, has been shown to be related to the length that characterizes the microscopic configuration of the phase within the REV. Thus, for a phase that completely fills the void space in a granular porous medium, a_L should be of the order of magnitude of a pore size. Experiments have shown that a_T is 8 to 24 times smaller than a_L.

We can express the components of the dispersivity of an isotropic porous medium in terms of a_L and a_T, in the form

$$a_{ijkm} = a_T \delta_{ij}\delta_{km} + \frac{a_L - a_T}{2}(\delta_{ik}\delta_{jm} + \delta_{im}\delta_{jk}). \qquad (1.8.4)$$

The coefficient of dispersion is then expressed by

$$D_{ij} = a_T V \delta_{ij} + (a_L - a_T)\frac{V_i V_j}{V}. \qquad (1.8.5)$$

where $\delta_{ij}(= 0$, for $i \neq j$, and $\delta_{ij} = 1$, for $i = j)$ is the *Kronecker delta*.

In Cartesian coordinates, and with velocity components V_x, V_y and V_z, we obtain from (1.8.5)

$$D_{xx} = a_T V + (a_L - a_T)\frac{V_x^2}{V} = \frac{a_T(V_y^2 + V_z^2) + a_L V_x^2}{V},$$

$$D_{xy} = \frac{(a_L - a_T)V_x V_y}{V} = D_{yx},$$

$$D_{xz} = \frac{(a_L - a_T)V_x V_z}{V} = D_{zx},$$

$$D_{yy} = a_T V + \frac{(a_L - a_T)V_y^2}{V} = \frac{a_T(V_x^2 + V_z^2) + a_L V_y^2}{V},$$

$$D_{yz} = \frac{(a_L - a_T)V_y V_z}{V} = D_{zy},$$

$$D_{zz} = a_T V + \frac{(a_L - a_T)V_z^2}{V} = \frac{a_T(V_x^2 + V_y^2) + a_L V_z^2}{V}.$$

$$(1.8.6)$$

In the special case of uniform flow, say $V_x = V \neq 0$, $V_y = V_z = 0$, (1.8.6) reduces to

$$D_{xx} = a_L V, \qquad D_{yy} = a_T V, \qquad D_{zz} = a_T V,$$

$$D_{xy} = D_{yx} = D_{xz} = D_{zx} = D_{yz} = D_{zy} = 0, \qquad (1.8.7)$$

or, in matrix form

$$\mathbf{D} = \begin{bmatrix} a_L V & O & O \\ O & a_T V & O \\ O & O & a_T V \end{bmatrix}. \qquad (1.8.8)$$

For an anisotropic porous medium, with axial symmetry around a unit vector $\boldsymbol{\lambda}$, with components λ_i, the dispersivity components are expressed by

$$a_{ijkm} = a_1 \delta_{ij}\delta_{km} + a_2(\delta_{ik}\delta_{jm} + \delta_{im}\delta_{jk}) + a_3(\delta_{ij}\lambda_k\lambda_m + \delta_{km}\lambda_i\lambda_j)$$
$$+ a_4(\delta_{ik}\lambda_j\lambda_m + \delta_{jk}\lambda_i\lambda_m + \delta_{im}\lambda_j\lambda_k + \delta_{jm}\lambda_i\lambda_k) + a_5\lambda_i\lambda_j\lambda_k\lambda_m.$$

$$(1.8.9)$$

This means that *five* independent coefficients of dispersivity are required in order to fully define the dispersivity tensor. We may understand this by noting that in flow normal to the layers, we need *one longitudinal* coefficient and *one transversal* one to describe the dispersion, while in flow parallel to the layers, we need *one (different) longitudinal* coefficient and *two additional transversal* ones. For an isotropic porous medium, (1.8.9) reduces to (1.8.4), with $a_T = a_1$ and $a_L = a_1 + 2a_2$.

For an anisotropic porous medium, with three mutually orthogonal planes of symmetry, we need nine independent dispersivity coefficients.

To summarize, the dispersive flux (per unit area of the fluid), $\mathbf{J}^{*m^\gamma}$, is expressed by (1.8.1), in which the dispersion coefficient, $\mathbf{D}$, is described by (1.8.2), with the dispersivity, $\mathbf{a}$, described, *for an isotropic porous medium*, by (1.8.4).

When we consider the transport of a component in a phase that occupies only part of the void space, the above discussion remains valid, except that the dispersivity components, that reflect the geometry of the phase occupied domain within an REV, are functions of the volumetric fraction, θ, of the phase.

The diffusive flux has already been discussed in LECTURE 7. Here, we limit the discussion to a single component that does not diffuse into adjacent components, when such components are present in the void space.

Following the discussion, in LECTURE 7, we use (1.7.14) to express the macroscopic flux due to molecular diffusion in a *binary system*. For convenience, we rewrite here this equation, without the γ superscript, and the average symbol over c, in the form

$$\overline{j_j} = -\phi \mathcal{D} T^*_{ji} \frac{\partial c}{\partial x_i} = -\phi (\mathcal{D}^*)_{ji} \frac{\partial c}{\partial x_i}, \tag{1.8.10}$$

where the second rank symmetric tensor $\mathcal{D}^* = \mathcal{D} T^*$ denotes the *coefficient of molecular diffusion in a porous medium* and T^* is defined by (1.2.63)). In an isotropic medium, the components $\mathcal{D}_{ij}$ reduce to $\mathcal{D}^* \delta_{ij}$.

We recall that in a multiphase flow in a porous medium, T^* is a function of the volumetric fraction, θ, of the phase.

Because both the diffusive and the dispersive fluxes of a component are proportional to ∇c, it is common to sum up the two coefficients, $\mathbf{D}$ and $\mathcal{D}^*$, and refer to the sum

$$\mathbf{D}_h = \mathbf{D} + \mathcal{D}^*, \tag{1.8.11}$$

as the *coefficient of hydrodynamic dispersion*.

Because $\mathbf{D}$ is linearly proportional to the velocity, in regions of high velocity, the effect of molecular diffusion on the spreading of a component may be negligible, relative to the effect of mechanical dispersion, while at a very low velocity, and in the limit for a stagnant fluid phase, molecular diffusion predominates. Say, in one-dimensional flow, dispersion dominates as long as

$$|a_l V| \gg \mathcal{D},$$

or

$$\mathrm{Pe} \equiv \frac{L_c V_c}{\mathcal{D}} \gg 1.$$

1.8.3   Sources and Sinks

Sources and sinks at the fluid - solid interface

We now refer to the surface integral term in the balance equation (1.3.12), rewritten here in the form

$$f_{f\to s} = \overline{\{c(\mathbf{V}-\mathbf{u})+\mathbf{j}_f\}\cdot\boldsymbol{\nu}_f}^{\,fs}\;\Sigma_{fs}$$

$$\equiv \frac{1}{U_o}\int_{S_{fs}}\{c(\mathbf{V}_f-\mathbf{u})+\mathbf{j}_f\}\cdot\boldsymbol{\nu}_f\,dS. \qquad (1.8.12)$$

It expresses the rate, per unit volume of porous medium, at which a considered component of a fluid phase, (f), reaches the (possibly moving) solid - fluid interface, thus leaving the phase, by two mechanisms: advection and molecular diffusion. We shall use the symbol $f_{f\to s}$ to denote this transfer from the fluid onto the solid. When the solid - fluid interface is a material surface for the fluid, $(\mathbf{V}_f-\mathbf{u})\cdot\boldsymbol{\nu}_f \equiv 0$, so that the component can reach this interface only by molecular diffusion.

Upon leaving the fluid phase, the considered component adheres to the (microscopic) walls of the solid, thus building up a concentration of the component on the solid's surface. The component may also leave the solid, where it has been previously accumulated, and reenter the fluid phase by the mechanism of molecular diffusion.

Several mechanisms account for these phenomena, primarily *adsorption*, and *ion exchange*. Let us briefly describe them.

Adsorption is the phenomenon of accumulation of a substance (*component of a fluid; adsorbate*) on the solid (*adsorbent*) at a fluid-solid interface. In *desorption*, the quantity of the substance on the solid decreases. The component's affinity to the solid surface is due to electrical attraction, *van der Waals attraction*, i.e., intermolecular forces of attraction between molecules of the solid and those of the adsorbed component, and *chemisorption*, i.e., chemical interaction between the solid and the adsorbed substance. Hence, the main factors affecting the adsorption and desorption of chemicals to or from a solid, are the physical and chemical characteristics of the considered component and of the solid's surface. Additional factors are temperature, and the presence of other components in the fluid phase (e.g., through the pH that results from their presence in the fluid phase).

Ion exchange is a process of adsorption of electrolytes present in a solution, where a certain reaction occurs between the ions in the solution and localized sites on the solid's surface. The reaction involved is essentially an exchange between ions in solution and ions fixed on the solid. For example, reversible ion exchange for a univalent component, may take the form

$$X^+ + AS \rightleftharpoons A^+ + XS,$$

in which X^+ is a cationic component in solution, XS is the component in the adsorbed state on the solid, denoted by S, and A^+ is the cation, initially on the solid, remaining in the fluid phase. Upon reaching a state of equilibrium, according to the *law of mass action*, we obtain

$$K_A^X = \frac{[XS][A^+]}{[AS][X^+]},$$

in which K_A^X is the *equilibrium coefficient* for the exchange reaction, and the bracketed terms represent thermodynamic concentrations, or activities (denoted for a γ-component by a^γ). These activities can be related to their *molar concentrations*, expressed as the number of moles per liter of solution, by $[\beta] \equiv a^\beta = \gamma^\beta c^\beta$, where γ^β is the *activity coefficient* of β, and c^β is the concentration of the β-component. For dilute solutions, $\gamma^\beta \cong 1$ and $a^\beta \cong c^\beta$. Usually $\gamma^\beta < 1$.

At a low concentration of a considered component, X^+, relative to that of the exchangable cation, A^+, and when the exchange between the two does not cause a significant change in the activity ratio, $[A^+]/[AS]$, the partitioning coefficient, K_d, can be obtained from

$$K_d = \frac{K_A^X}{[A^+]/[AS]} = \frac{(XS)}{(X^+)},$$

where the simple brackets denote component concentrations.

An *adsorption isotherm* is an expression that relates the quantity of an adsorbed component to its quantity (expressed as concentration) in the fluid phase, at constant temperature. Let us denote the former by F (= mass of adsorbed component per unit mass of solid (adsorbate)). Although it would have seemed more natural to define the quantity of the component on the solid 'per unit surface area of the solid', the definition, 'per unit mass of solid' stems from the way this quantity is measured in the laboratory. Thus, an isotherm relates F to the concentration in the fluid, c.

Different adsorbate–adsorbent pairs have different isotherms. However, in general we may distinguish two classes of isotherms:

- **Equilibrium isotherms** that are based on the assumption (substantiated, of course, by experiments), that the quantities of the component on the solid and in the adjacent solution, are continuously at equilibrium. Any change in the concentration of one of them produces an *instantaneous* change in the other.

- **Nonequilibrium isotherms**, which assume that equilibrium is not achieved instantaneously, but rather that it is approached at a certain rate, which, in general, depends on both F and c.

Following are a number of examples of the more commonly encountered isotherms.

(i) Freundlich (1926), suggested the *nonlinear equilibrium isotherm*

$$F = bc^m, \tag{1.8.13}$$

where b and m are constant coefficients.

(ii) For $m = 1$, and replacing the symbol b by the more commonly used symbol K_d, (1.8.13) reduces to

$$F = K_d c, \tag{1.8.14}$$

known as the *linear equilibrium isotherm*. It assumes that adsorption is instantaneous, reversible and linear. The coefficient K_d is called the *distribution coefficient*, or *partitioning coefficient*. From (1.8.14), it follows that K_d gives at every instant the mass of a component on the solid (per unit mass of the latter), per unit concentration of the component in the fluid phase. It describes the partitioning of the total amount of the component, say in a unit volume of porous medium, between the part adsorbed on the solid and the part remaining in the fluid phase. Sometimes, K_d for the adsorption process differs from that of the desorption one. This means that the process is not completely reversible. Another observation is that there is often a limit to the adsorptive capacity of the solid surface. This requires a modification of the isotherm (1.8.14).

Consider a fluid phase which is a liquid that occupies only part of the void space, at a volumetric fraction θ_f (or saturation $S_f = \theta_f/\phi$). Then, overlooking the presence of a thin stagnant liquid film that covers the solid in the gas occupied zone, only part of the total area of the solid is exposed to adsorption, or ion - exchange phenomena which occur at the solid-liquid interface. By overlooking the presence of the thin film, we also accept the assumption that this film does not provide for the passage of the adsorbate, by molecular diffusion, from the liquid phase to the solid surface adjacent to it. The portion of the total surface of the solid that is in contact with the liquid phase, depends on θ_f. Let us *assume* (as one of many possibilities, to be verified by experiments for a particular porous medium) that the ratio of the area of the solid - liquid interface to the total surface area of the solid, is equal to the ratio of active solid mass (i.e., solid mass participating in the surface phenomena) to the total mass of the solid, and that each of these ratios, in turn, is equal to the ratio of the liquid occupied portion of the void space to the total void space volume, i.e., equal to S_f. Then, leaving the definition of F unchanged, (= mass of adsorbate per total mass of adsorbent), the linear equilibrium isotherm (1.8.14) reduces to

$$F = S_f K_d c, \qquad (1.8.15)$$

where K_d has the same definition as in (1.8.14). In other models, S_f in (1.8.15) may be replaced by some other function of S_f. Obviously, it is also possible to assume that, because molecular diffusion does take place through the film the presence of the latter enables the entire area of the solid to be available for adsorption. In this case, the isotherm (1.8.14) remains valid also for $S_f < 1$.

Another possible model is that the component's concentration in the immobile film is different from that in the mobile phase. For example, we may regard the film as a separate 'apparent phase'.

Considerations similar to those discussed above, are applicable also to the isotherms considered below.

(iii)   A more general form of the linear equilibrium isotherm is given by

$$F = k_1 c + k_2, \qquad (1.8.16)$$

where k_1 and k_2 are constant coefficients.

(iv)   Langmuir (1915, 1918) suggested the *nonlinear equilibrium isotherm*

$$F = \frac{k_3 c}{1 + k_4 c}, \qquad (1.8.17)$$

where k_3 and k_4 are constant coefficients.

(**v**)   Lindstrom *et al.* (1971) and Van Genuchten (1974), mention the *non-linear isotherm*

$$F = k_5 c \exp(-2k_6 F), \tag{1.8.18}$$

where k_5 and k_6 are constant coefficients.

(**vi**)   The simplest *non-equilibrium isotherm* for an irreversible system (Langmuir, in Adamson, 1967) is

$$\frac{\partial F}{\partial t} = k_r c, \tag{1.8.19}$$

where k_r is a *kinetic rate coefficient*.

(**vii**)   Lapidus and Amundson (1952) proposed the *non-equilibrium isotherm*

$$\frac{\partial F}{\partial t} = k_r(k_7 c + k_8 - F), \tag{1.8.20}$$

where k_7 and k_8 are constant coefficients.

(**viii**)   A *non-equilibrium Langmuir isotherm* is (e.g., Hendricks, 1972)

$$\frac{\partial F}{\partial t} = k_r \left(\frac{k_9 c}{1 + k_{10} c} - F \right). \tag{1.8.21}$$

Only one of the above isotherms would apply for any particular case. The selection of the appropriate isotherm, and the determination of the value of the various coefficients appearing in it, should be based on the study of the thermodynamics of the interacting components and on experimental work with the particular soil and particular component of interest.

We have still to find a convenient expression for the sink term $f_{f \to s}$ defined by (1.8.12). In order to derive such an expression, we turn now to the equation of balance of the considered component *on the solid phase*. We obtain this equation from (1.3.12) in which the α subscript is replaced by s, $\theta_\alpha c$ is replaced by $\theta_s c_s F$ $(= (1 - \phi)\rho_s F)$, $\theta_s \equiv 1 - \phi$, and we assume no flux, whether advective, dispersive, or diffusive, of the component within the solid phase and/or on the solid surface. Since at every point on $\mathcal{S}_{fs}, \nu_s = -\nu_f$, the mass balance for the component on the solid surface, takes the form

$$\frac{\partial(\theta_s \rho_s F)}{\partial t} = -f_{s \to f} + (1 - \phi)\rho_s \Gamma_s, \tag{1.8.22}$$

where $(1 - \phi)\rho_s \Gamma_s$ is the strength of the component's source on the solid surface, per unit volume of porous medium (e.g., due to growth, or decay processes), and $f_{s\to f}$ replaces the surface integral that expresses the quantity of the component leaving the solid surface, per unit volume of porous medium.

Conservation requires that $f_{f\to s} = -f_{s\to f}$. Hence, the sought rate of transfer from the fluid phase to the solid one can be expressed by

$$f_{f\to s} = \frac{\partial(1-\phi)\rho_s F}{\partial t} - (1-\phi)\rho_s\Gamma_s, \qquad (1.8.23)$$

in which F, or $\partial F/\partial t$ may be expressed by the appropriate isotherm.

Sources and sinks within the fluid phase

Sources and sinks (= negative sources) of a component in a liquid phase, occupying the entire void space (ϕ), or part of it (θ_f), are expressed by the term $\theta_f\overline{\rho_f \Gamma_f^{m\gamma}}^f$, approximated as $\theta_f\overline{\rho_f}^f\overline{\Gamma_f^{m\gamma}}^f$, appearing on the r.h.s. of the balance equation (1.3.12). These may be due to the actual withdrawal or injection of liquid (containing the component) from the porous medium, or into it, either at selected points, or as distributed sources or sinks over portions of a two-dimensional domain. Sources and sinks of a component may also result from various processes, e.g., radioactive decay, biodegradation and growth due to bacterial activities, that cause the quantity of the considered component within the liquid phase to decrease, or increase. Chemical reactions among components within the liquid phase, may also cause the quantity of a considered component to increase or decrease.

Let the liquid phase containing a considered component at a known concentration c_R, be added as a distributed source at a rate $R_{ex} = R_{ex}(\mathbf{x},t)$ (= volume of liquid added per unit volume of porous medium per unit time). Then, the component source term $\theta_f\overline{\rho_f\Gamma_f^{m\gamma}}^f$, appearing in the component balance equation, is expressed by

$$\theta_f\overline{\rho_f\Gamma_f^{m\gamma}}^f - R_{ex}(\mathbf{x},t)c_R. \qquad (1.8.24)$$

When sources exist at N isolated points, $\mathbf{x}^{(i)}$, $i = 1,2,...N$, the source term takes the form

$$\theta_f\overline{\rho_f\Gamma_f^{m\gamma}}^f = \sum_{(i)} R_{ex}^{(i)}(\mathbf{x}^{(i)},t)\delta(\mathbf{x}-\mathbf{x}^{(i)})c_R^{(i)}, \qquad (1.8.25)$$

where $R_{ex}^{(i)}$ represents the rate of injection (in terms of volume of liquid per unit volume of porous medium per unit time) at point $\mathbf{x}^{(i)}$ at time t, of liquid at a known concentration, $c_R^{(i)}$.

For sinks, $Q_{ex}^{(r)}$, (e.g., by pumping), the corresponding expression is

$$\theta_f\overline{\rho_f\Gamma_f^{m\gamma}}^f = -\sum_{(r)} Q_{ex}^{(r)}(\mathbf{x}^{(r)}, t)\delta(\mathbf{x} - \mathbf{x}^{(r)})c(\mathbf{x}, t), \qquad (1.8.26)$$

where we note that the withdrawn liquid is at the unknown concentration, $c(\mathbf{x}, t)$.

When the component in the fluid phase undergoes radioactive decay, the sink term, expressing the rate of disappearance of the component, is given by

$$\theta_f\overline{\rho_f\Gamma_f^{m\gamma}}^f = -\theta_f\lambda c, \qquad (1.8.27)$$

where $\lambda = 1/T$, and T is the *half life* of the radioactive component.

When Γ_s represents a rate of radioactive decay of a radioactive adsorbed component, it can be expressed as

$$\Gamma_s = -\lambda F \qquad (1.8.28)$$

For other decay, or degradation phenomena of a component in the fluid phase, we may write the sink term in the form

$$\theta_f\overline{\rho_f\Gamma_f^{m\gamma}}^f = -\theta_f k_f c, \qquad (1.8.29)$$

where $k_f\ (\equiv k_f^\gamma)$ is a *degradation rate constant* for the component in the fluid phase.

When the considered γ-component participates in chemical reactions which cause its concentration to increase (or decrease), we may express the source by

$$\theta_f\overline{\rho_f\Gamma_f^{m\gamma}}^f = \theta_f\sum_{(j)} R^{\gamma j}, \qquad (1.8.30)$$

where $R^{\gamma j}$ is the rate of production of the mass of the γ-component by the jth reaction, per unit volume of porous medium. In general $R^{\gamma j} = R^{\gamma j}(c^{\gamma 1}, c^{\gamma 2}, c^{\gamma 3}\ldots)$, i.e., a function of the concentrations of the various components that are present within the fluid phase.

The rate of reaction, $R^{\gamma j}$, is also defined by the *rate equation*

$$R^{\gamma j} = \left.\frac{dc^\gamma}{dt}\right|_{\text{chem}, j}, \qquad (1.8.31)$$

which gives the amount of a component, measured as mass per unit volume of porous medium, that is added within the fluid phase by the jth chemical reaction, per unit time. This addition should not to be mixed with additions from other sources, hence the symbol 'chem'.

The general *chemical balance equation* of a reversible reaction, can be expressed by the *stoichiometric equation*

$$e\mathrm{E} + f\mathrm{F} \underset{k_r}{\overset{k_f}{\rightleftharpoons}} g\mathrm{G} + h\mathrm{H}, \tag{1.8.32}$$

in which e, f, g, h are *stoichiometric coefficients.* In this case

$$\frac{1}{e}\frac{d[\mathrm{E}]}{dt} = \frac{1}{f}\frac{d[\mathrm{F}]}{dt} = \frac{1}{g}\frac{d[\mathrm{G}]}{dt} = \frac{1}{h}\frac{d[\mathrm{H}]}{dt}, \tag{1.8.33}$$

where a stoichiometric coefficient is negative for reactants and positive for products of the reaction. Often, the rate law, as determined by experiments, takes the form

$$R = k[\mathrm{E}]^{\nu_e}[\mathrm{F}]^{\nu_f}[\mathrm{G}]^{\nu_g}[\mathrm{H}]^{\nu_h}, \tag{1.8.34}$$

where the exponents, negative, or positive, are determined experimentally. The reaction is said to be of order ν_e in E, ν_f in F, etc. The reaction is then said to be of the order of the sum of the exponents.

In general, the rate of a reaction is a function of the concentrations of the various components that interact within the fluid phase.

Under conditions of *non-equilibrium*, a situation often encountered in groundwater contamination, we have to introduce the appropriate expressions for the rate of production (or disappearance) of every considered component.

For example, for a simple reaction of the form

$$\mathrm{E} \overset{k}{\rightarrow} \mathrm{G}, \tag{1.8.35}$$

we have

$$\frac{R}{M^E} = \frac{d[\mathrm{E}]}{dt} = -k[\mathrm{E}], \tag{1.8.36}$$

where [E] is the molar concentration of E, and k is called the *first-order rate constant* (dims. T^{-1}).

When the reaction is

$$\mathrm{E} + \mathrm{F} \overset{k}{\rightarrow} \mathrm{G} + \mathrm{H} + ..., \tag{1.8.37}$$

the rate of change in E may be given by

$$\frac{R}{M^E} = \frac{d[\mathrm{E}]}{dt} = -k[\mathrm{E}][\mathrm{F}], \tag{1.8.38}$$

in which k (dims. $M^{-1}T^{-1}$) is referred to as a *second-order rate constant*.

When the reaction is

$$E \underset{k_r}{\overset{k_f}{\rightleftharpoons}} F, \qquad (1.8.39)$$

the rate expression for E may take the form

$$\frac{R}{M^E} = \frac{d[E]}{dt} = -k_f[E] + k_r[F]. \qquad (1.8.40)$$

When the reaction takes the form

$$E + F \underset{k_r}{\overset{k_f}{\rightleftharpoons}} G + H, \qquad (1.8.41)$$

the rate expression for E may be

$$\frac{R}{M^E} = \frac{d[E]}{dt} = -k_f[E][F] + k_r[G][H]. \qquad (1.8.42)$$

At equilibrium, and for a closed system, the rates of the various components are related to each other, for the case (1.8.41), by

$$\frac{d[E]}{dt} = \frac{d[F]}{dt} = \frac{d[G]}{dt} = \frac{d[H]}{dt} = 0. \qquad (1.8.43)$$

At equilibrium, with the forward rate of reaction (described by k_f) equalling the reverse one (described by k_r), the *law of mass action* must be satisfied. Thus, referring to the reaction described by (1.8.32), the law of mass action takes the form

$$K = \frac{[G]^g[H]^h}{[E]^e[F]^f}, \qquad (1.8.44)$$

in which the symbol K represents a (known) *thermodynamic equilibrium constant* that depends on the temperature, and the quantities in the square brackets denote thermodynamic concentrations, or *activities*, a^β, $\beta = $ E, F, G, H.

Suppose we consider the simultaneous transport of the four components E, F, G, H, which, as they are transported, undergo a chemical reaction according to (1.8.32). As we shall see below, in each of the four component mass balance equations, a term appears that describes the rate of production of the considered component. We shall, therefore, have four unknowns (in addition to the four unknown component concentrations, for the solution of which we have the four component balance equations). Equation (1.8.33) contains three relations between the rates of production of the four individual phases. Under equilibrium conditions, (1.8.44) provides the fourth additional relationship required to obtain a solution for the four concentrations and the four rates of production. Under non-equilibrium conditions, the four relations are obtained from (1.8.33) and (1.8.34).

1.8.4 Mass balance equation for a single component

Let the considered component be one that is adsorbed to the solid surfaces and, in addition, undergoes degradation, but at rates that are different for the adsorbate (on the solid) and for the component within the fluid phase. In addition, point sources and sinks of the component exist within the considered domain, due to recharge and extraction of the fluid. Expressing:

- the advective flux by $\theta_f c \mathbf{V}_f$,

- the dispersive flux by (1.8.1),

- the diffusive flux by (1.8.10),

- the sink due to adsorption by (1.8.23),

- the sink due to degradation within the fluid phase e.g., by (1.8.27), or by (1.8.29),

- the sink due to degradation of the adsorbate,

- the sources due to fluid point injection by (1.8.25),

- the sinks due to point extraction by (1.8.26),

and inserting these expressions in the balance equation (1.3.12), we obtain

$$
\begin{aligned}
\frac{\partial \theta_f c}{\partial t} = \; & - \; \nabla \cdot \theta_f (c\mathbf{V}_f - \mathbf{D} \cdot \nabla c - \mathcal{D}^* \cdot \nabla c) - \frac{\partial \theta_s \rho_s F}{\partial t} - \theta_s \rho_s k_s F \\
& - \; \theta_f k_f c + \sum_{(i)} R_{ex}^{(i)}(\mathbf{x}^{(i)}, t)\delta(\mathbf{x} - \mathbf{x}^{(i)})c_R^{(i)}(\mathbf{x}^{(i)}, t) \\
& - \; \sum_{(r)} Q_{ex}^{(r)}(\mathbf{x}^{(r)}, t)\delta(\mathbf{x} - \mathbf{x}^{(r)})c(\mathbf{x}, t),
\end{aligned}
\tag{1.8.45}
$$

where we recall that for a single fluid that occupies the entire void space, $\theta_f \equiv \phi$, $c \equiv \overline{c}^f$, $\mathbf{V}_f \equiv \overline{\mathbf{V}_f}^f$, etc. We also recall that for a fluid that fills only part of the void space, $\mathbf{D} = \mathbf{D}(\theta_f)$ and $\mathcal{D}^* = \mathcal{D}^*(\theta_f)$. The symbols $\mathbf{x}^{(i)}$ and $\mathbf{x}^{(r)}$ denote the locations of point sources and sinks, respectively.

Equation (1.8.45) contains the two variables, $c(\mathbf{x},t)$ and $F(\mathbf{x},t)$. We have to supplement this equation by an appropriate isotherm that gives the relationship between them. For example, with (1.8.15), in which the fluid's saturation, S_f, is replaced by a more general term, $f_a(\theta_f)$, to express the effect of saturation on the adsorption isotherm, we obtain from (1.8.45)

$$
\frac{\partial}{\partial t}\{\theta_f + \theta_s \rho_s f_a(\theta_f) K_d\}c = -\nabla \cdot \theta_f(c\mathbf{V}_f - \mathbf{D}\cdot\nabla c - \mathcal{D}^*\cdot\nabla c)
$$

$$
-\{\theta_f k_f + \theta_s \rho_s f_a(\theta_f) k_s K_d\}c + \sum_{(i)} R_{ex}^{(i)}(\mathbf{x}^{(i)},t)\delta(\mathbf{x} - \mathbf{x}^{(i)})c_R^{(i)}(\mathbf{x}^{(i)},t)
$$

$$
-\sum_{(r)} Q_{ex}^{(r)}(\mathbf{x}^{(r)},t)\delta(\mathbf{x} - \mathbf{x}^{(r)})c(\mathbf{x},t). \qquad (1.8.46)
$$

This equation now involves only the single variable $c(\mathbf{x},t)$, expressing the spatial concentration distribution of the considered component. Actually, (1.8.45) and (1.8.46) contain two additional variables, viz., $\theta_f(\mathbf{x},t)$ and $\mathbf{V}_f(\mathbf{x},t)$. For a deformable porous medium, $\theta_s (= 1 - \phi))$ is another variable. Hence, the complete model describing the distribution $c(\mathbf{x},t)$, will contain additional equations (see below, in the summary on the complete model).

In the absence of adsorption, and other sink/source phenomena at the solid-fluid interface, (1.8.45) reduces to

$$
\frac{\partial \theta_f c}{\partial t} = -\nabla \cdot \theta_f(c\mathbf{V}_f - \mathbf{D}\cdot\nabla c - \mathcal{D}^*\cdot\nabla c) - \theta_f k_f c
$$

$$
+ \sum_{(i)} R_{ex}^{(i)}(\mathbf{x}^{(i)},t)\delta(\mathbf{x} - \mathbf{x}^{(i)})c_R^{(i)}(\mathbf{x}^{(i)},t)
$$

$$
- \sum_{(r)} Q_{ex}^{(r)}(\mathbf{x}^{(r)},t)\delta(\mathbf{x} - \mathbf{x}^{(r)})c(\mathbf{x},t). \qquad (1.8.47)
$$

Let us consider the case of a single fluid (= liquid) that completely occupies the void space, i.e., $\theta_f = \phi$, $\theta_s = 1 - \phi$, and $f_a(\theta_f) = 1$. In addition, we assume that no external sources or sinks exist, that $\partial\phi/\partial t = 0$, i.e., nondeformable solid matrix, $\partial\rho_s/\partial t = 0$, and $\partial K_d/\partial t = 0$. Then, (1.8.46) reduces to the form

$$
R_d \phi \frac{\partial c}{\partial t} = -\nabla \cdot \phi(c\mathbf{V}_f - \mathbf{D}\cdot\nabla c - \mathcal{D}^*\cdot\nabla c)
$$

$$
- k_f \left[\phi + \frac{(1-\phi)\rho_s K_d k_s}{k_f}\right]c, \qquad (1.8.48)
$$

where

$$R_d = 1 + \frac{(1-\phi)\rho_s K_d}{\phi} \quad (>1) \tag{1.8.49}$$

is called the *coefficient of retardation*, or *retardation factor*.

To understand the significance of R_d, and the reason for calling it a 'retardation factor', let us further simplify (1.8.48) by assuming $k_s = k_f$, and that the porous medium is homogeneous, i.e., $\nabla\phi = 0$. Then (1.8.48) may be rewritten as

$$\phi\frac{\partial c}{\partial t} = -\nabla\cdot\phi\left(c\frac{\mathbf{V}_f}{R_d} - \frac{\mathbf{D}}{R_d}\cdot\nabla c - \frac{\mathcal{D}^*}{R_d}\cdot\nabla c\right) - \phi k_f R_d c. \tag{1.8.50}$$

For comparison, let us consider the case without adsorption, i.e., $K_d = 0$, and rewrite (1.8.46) in which θ_f is replaced by ϕ (with $\partial\phi/\partial t = 0$), and the external source and sink terms have been deleted. We obtain

$$\phi\frac{\partial c}{\partial t} = -\nabla\cdot\phi(c\mathbf{V}_f - \mathbf{D}_h\cdot\nabla c) - \phi k_f c. \tag{1.8.51}$$

We note that (1.8.50) and (1.8.51) are similar, except that in (1.8.50), the average fluid velocity *seems to be* $\mathbf{V}_f/R_d$ and the coefficient of hydrodynamic dispersion in (1.8.50) is reduced to $\mathbf{D}_h/R_d$. Thus, the effect of adsorption (and similar activities) is to *retard* the advance of the component (as part of it is adsorbed to the solid surface rather than advance with the fluid moving at an average velocity, $\mathbf{V}_f$). At the same time, the coefficient of mechanical dispersion, $\mathbf{D}$, which is shown in (1.8.2) to be proportional to the average velocity, is also reduced by the factor R_d. The coefficient of molecular diffusion in a porous medium, $\mathcal{D}^*$, is also reduced by the factor R_d.

Figure 1.8.4 shows the effect of retardation in an example of a semi infinite column. We note that the point of $c = 0.5$ advances at a speed V/R_d, and that the curve is steeper, as produced by a smaller coefficient of hydrodynamic dispersion.

Although we have reduced (1.8.46) to the simpler form (1.8.50), in order to explain the phenomenon of retardation, this phenomenon obviously exists also in the more general case expressed by (1.8.46), which can be rewritten in the form

$$\frac{\partial}{\partial t}\{\theta_f R_d(\theta_f)c\}$$

$$= -\nabla\cdot\theta_f(c\mathbf{V}_f - \mathbf{D}_h\cdot\nabla c) - \theta_f k_f \left(1 + \frac{\theta_s\rho_s f_a(\theta_f)k_s K_d}{\theta_f k_f}\right)c$$

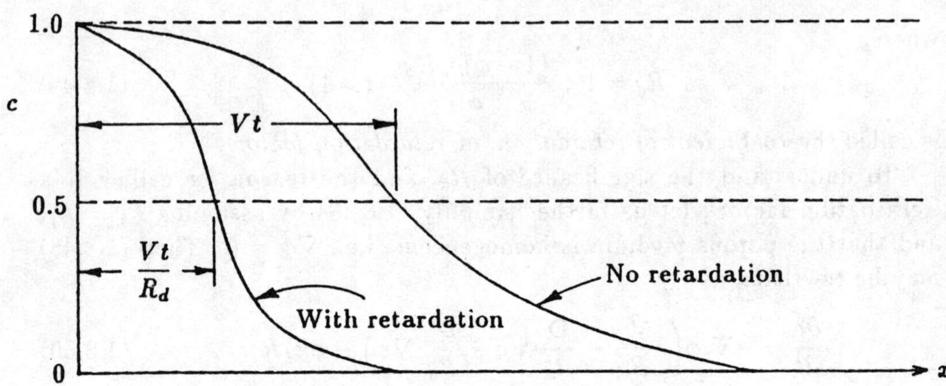

Figure 1.8.4: Effect of retardation

$$+ \sum_{(i)} R_{ex}^{(i)}(\mathbf{x}^{(i)}, t)\delta(\mathbf{x} - \mathbf{x}^{(i)})c_R^{(i)}(\mathbf{x}^{(i)}, t)$$

$$- \sum_{(r)} Q_{ex}^{(r)}(\mathbf{x}^{(r)}, t)\delta(\mathbf{x} - \mathbf{x}^{(r)})c(\mathbf{x}, t), \qquad (1.8.52)$$

where

$$R_d(\theta_f) = 1 + \frac{\theta_s \rho_s f_a(\theta_f) K_d}{\theta_f} \qquad (1.8.53)$$

depends now on $\theta_f(\mathbf{x}, t)$.

The phenomenon of retardation also exists when F in (1.8.45) is expressed by any other, not necessarily linear equilibrium isotherms. The structure of R_d will then depend on the selected isotherm.

By combining (1.8.45) with the volume balance equation (1.3.10), to which we add the point sources and sinks introduced above, we obtain

$$\theta_f \frac{\partial c}{\partial t} = - \theta_f \mathbf{V}_f \cdot \nabla c + \nabla \cdot \theta_f \mathbf{D}_h \cdot \nabla c - \frac{\partial \theta_s \rho_s F}{\partial t} - \theta_s \rho_s k_s F - \theta_f k_f c$$

$$+ \sum_{(i)} R_{ex}^{(i)}(\mathbf{x}^{(i)}, t)\delta(\mathbf{x} - \mathbf{x}^{(i)})(c_R^{(i)} - c). \qquad (1.8.54)$$

It may be of interest to note another way of expressing the balance equation. So far, all the balance equations in this lecture have been written as an *Eulerian* formulation, i.e., from the point of view of what happens at a fixed point in space. As an example of a second formulation, we note that (1.8.54) can be rewritten as

$$\theta_f \frac{\mathrm{D}_f c}{\mathrm{D} t} \equiv \theta_f \left(\frac{\partial c}{\partial t} + \mathbf{V}_f \cdot \nabla c \right) = \nabla \cdot \theta_f \mathbf{D}_h \cdot \nabla c - \frac{\partial \theta_s \rho_s F}{\partial t} - \theta_s \rho_s k_s F$$

$$-\theta_f k_f c + \sum_{(i)} R_{ex}^{(i)}(\mathbf{x}^{(i)}, t)\delta(\mathbf{x} - \mathbf{x}^{(i)})(c_R^{(i)} - c). \qquad (1.8.55)$$

The l.h.s. of (1.8.55) is a *Lagrangian expression* (following a particle of fixed identity as it travels with the fluid velocity, $\mathbf{V}_f$). The r.h.s. is an Eulerian format. In the absence of sources, sinks, adsorption and decay, (1.8.55) reduces to

$$\theta_f \frac{\mathrm{D}_f c}{\mathrm{D}t} = \nabla\cdot\theta_f \mathbf{D}_h\cdot\nabla c. \qquad (1.8.56)$$

If we further assume $|c\mathbf{V}_f| \gg |\mathbf{D}_h\cdot\nabla c|$, i.e., advection dominates the component's transport, (1.8.56) further reduces to

$$\frac{\mathrm{D}_f c}{\mathrm{D}t} = 0. \qquad (1.8.57)$$

Let us rewrite (1.8.46) for a fluid phase that occupies part of the void space, in the form

$$\frac{\partial}{\partial t}\left[\phi S_w + (1-\phi)\rho_s K_d\right] c =$$
$$-\nabla\cdot\left[\mathbf{q}_{rw}c + \phi S_w(\mathbf{V}_s c - \mathbf{D}_h\cdot\nabla c)\right] - \left[\phi S_w k_f + (1-\phi)\rho_s k_s K_d\right]c$$
$$+R_{ex}(\mathbf{x},t)c_R - Q_{ex}(\mathbf{x},t)c, \qquad (1.8.58)$$

where subscript w is used to indicate the fluid, S_w denotes the fluid's saturation, we have assumed that $f_a(\theta_f) \cong 1$ and have used the symbols $R_{ex}(\mathbf{x},t)$ and $Q_{ex}(\mathbf{x},t)$ to represent fluid external sources and sinks, respectively. In this equation, the relative specific discharge, $\mathbf{q}_{rw}$, can be expressed by any of the motion equations for a fluid in multiphase flow. Obviously, in most practical cases, $\mathbf{V}_s \simeq 0$

Again, in the complete model, S_w, and $\mathbf{q}_{rw}$ constitute additional variables of the problem.

1.8.5   Balance equations with immobile liquid

In the conceptual model employed for describing multiphase flow in porous media, it is usually assumed that the wetting phase is continuous and mobile as long as its saturation is above the irreducible wetting phase saturation, S_{wo}. Similarly, it is implicitly assumed that the nonwetting fluid is continuous and mobile as long as its saturation is above the residual nonwetting saturation, S_{no}. However, depending on the pore sizes and shapes, e.g., the presence of *dead-end* pores and pendular rings, part of a liquid phase may be discontinuous and *immobile*, or stagnant, or practically so, even at saturations above these lower bounds. For example, water in pores with narrow throats may be practically stagnant in the unsaturated zone in the soil. Similarly, part of the wetting phase may be present in the form of isolated pendular rings, even at relatively high saturations. The fraction of the void space which contains an *immobile wetting liquid* is not a constant, but rather some function of the saturation, reaching its maximum when $S_w = S_{wo}$. However, it does not necessarily vanish at full saturation, as even then, a wetting liquid phase in part of the void space may be practically immobile.

It is of interest to note that once we accept the notion that only part of an α-phase is actually in motion, and that this part is not identical to the irreducible saturation, $S_{\alpha o}$, of that phase, but is a function of the saturation, then all the coefficients that represent properties of the phase occupied portion of the void space, and hence were, hitherto, regarded as functions of the saturation, should actually be regarded as functions of the mobile part of the saturation only. Mass balance equations may have to be written separately for the mobile and the immobile portions of each participating phase, each portion regarded as a continuum within the entire domain.

The immobile portion of a phase is, usually, in direct contact with the mobile portion of it, thus enabling the transfer of components from one to the other by the mechanism of molecular diffusion. Even when part of the immobile portion of a wetting fluid phase is in the form of isolated pendular rings, transport of components between them may take place by molecular diffusion through the fluid film that covers the solid surface in the larger pores, occupied by the nonwetting phase.

In what follows, we consider an example with mobile and immobile water in the unsaturated zone in the soil. The conceptual model is comprised of the following assumptions:

- Following the continuum approach to multiphase flow in porous media, we regard the immobile portion of a wetting fluid phase, or *apparent phase*, as a continuum that fills the entire considered domain. The mobile water constitutes a separate continuum. We denote the water contents of these two continua by θ_{im} and θ_m, respectively, with

$$\theta_{im} + \theta_m = \theta_w.$$

- A considered component, say, a solute, is transported in the mobile water by advection, dispersion and diffusion, while in the immobile phase, it can be transported only by diffusion. We denote the concentrations of a considered component in the two phases by c_m and c_{im}, respectively.

- Adsorption is assumed to take place on the entire surface of the solid matrix, as this surface is everywhere in contact with water. However, we have to refer separately to the contact with each of the two phases. We use $p(< 1)$ and $1 - p$ to denote the areal fractions of the total area of the solid matrix, or the fractions of the total number of adsorption sites, that are in contact with the mobile and immobile water phases, respectively. We shall assume that the same fractions represent also the corresponding fractions of the solid mass that interacts with the two (apparent) water phases. Obviously, this last assumption may be a very poor one for certain solid matrix configurations.

- Let F_m and F_{im} denote the concentrations of the component on the two portions of the solid's surface. Then, assuming that a linear equilibrium isotherm is applicable, we have

$$F_m = pK_d c_m, \qquad \text{and} \quad F_{im} = (1 - p)K_d c_{im}. \qquad (1.8.59)$$

Other isotherms may also be used.

- The considered component undergoes decay described by (1.8.14).

- The considered component can be exchanged between the two fluid continua. We shall assume that the rate at which this exchange, say, from the mobile phase to the immobile one (in terms of mass of component per unit volume of porous medium per unit time) is proportional to the difference between their respective concentrations. Using the symbol $\alpha^\star$ to denote the coefficient of proportionality, this rate can be expressed as $\alpha^\star(c_m - c_{im})$. The *transfer coefficient*, $\alpha^\star$ ($\equiv \alpha^{\star\gamma}$), is proportional to the molecular diffusivity of the component in the water, to the total area of the surface of contact between the two fluid phases, and inversely proportional to some length characterizing the distance between the centroids of the subdomains occupied by the two phases. In principle, this coefficient need not be a constant, although it is often approximated as such, rather than make it depend on the saturations of the two phases.

- The fluid's density is constant, independent of changes in component concentration.

- The solid matrix is rigid and stationary.

For this conceptual model, the balance equation for a component in the mobile phase, takes the form

$$\frac{\partial \theta_m c_m}{\partial t} = -\nabla \cdot [c_m \mathbf{q} - \theta_m \mathbf{D}_{hm}(\theta_m) \cdot \nabla c_m] - f_{m \to s}$$

$$+ \alpha^*(c_{im} - c_m) - \theta_m \lambda c_m, \tag{1.8.60}$$

where $\mathbf{q}$ denotes the specific discharge of the mobile phase, and $f_{m \to s}$ denotes the rate at which the component leaves the mobile phase, to be adsorbed on the solid. When combined with the mass balance equation for the mobile phase

$$\frac{\partial \theta_m}{\partial t} = -\nabla \cdot \mathbf{q}, \tag{1.8.61}$$

the balance equation for a component in the **mobile phase** takes the form

$$\theta_m \frac{\partial c_m}{\partial t} = -\mathbf{q} \cdot \nabla c_m + \nabla \cdot [\theta_m \mathbf{D}_{hm}(\theta_m) \cdot \nabla c_m] - f_{m \to s}$$

$$+ \alpha^*(c_{im} - c_m) - \theta_m \lambda c_m. \tag{1.8.62}$$

The balance equation for the component in the **immobile phase** is

$$\frac{\partial \theta_{im} c_{im}}{\partial t} = \nabla \cdot [\theta_{im} \mathcal{D}^*_{im}(\theta_{im}) \cdot \nabla c_{im}] - f_{im \to s}$$

$$+ \alpha^*(c_m - c_{im}) - \theta_{im} \lambda c_{im}. \tag{1.8.63}$$

The balance equation for the component on the portion of the **solid surface that is in contact with the mobile water phase**, is

$$\frac{\partial \theta_s \rho_s F_m}{\partial t} = + f_{m \to s} - \theta_s \rho_s \lambda F_m. \tag{1.8.64}$$

For the portion of the **solid surface that is in contact with the immobile water phase**, we write

$$\frac{\partial \theta_s \rho_s F_{im}}{\partial t} = + f_{im \to s} - \theta_s \rho_s \lambda F_{im}, \tag{1.8.65}$$

where $\theta_s = 1 - \phi$.

Following the methodology presented in Subs. 6.1.5, in order to eliminate the terms that represent interphase transfer, we sum up the balance equations for the component in the mobile phase and its adsorbed part on the solid. We obtain

$$\frac{\partial \theta_m R_{dm} c_m}{\partial t} = -\nabla \cdot [c_m \mathbf{q} - \theta_m \mathbf{D}_{hm}(\theta_m) \cdot \nabla c_m]$$
$$+\alpha^\star (c_{im} - c_m) - \theta_m R_{dm} \lambda c_m, \qquad (1.8.66)$$

where

$$R_{dm} = R_{dm}(\theta_m) = 1 + \frac{\theta_s \rho_s p K_d}{\theta_m}$$

is the *retardation factor* for the mobile water.

For the component in the immobile water, we obtain

$$\frac{\partial \theta_{im} R_{dim} c_{im}}{\partial t} = \nabla \cdot [\theta_{im} \mathcal{D}^*_{im}(\theta_{im}) \cdot \nabla c_{im}]$$
$$+\alpha^\star (c_m - c_{im}) - \theta_{im} R_{dim} \lambda c_{im}, \qquad (1.8.67)$$

where

$$R_{dim} = R_{dim}(\theta_{im}) = 1 + \frac{\theta_s \rho_s (1 - p) K_d}{\theta_{im}}.$$

As an approximation, some authors use an average retardation factor in the form

$$R_d = \frac{\rho_s (1 - \phi) K_d}{\theta_w}.$$

Thus, we have two equations to be solved simultaneously for c_m and c_{im}. In saturated flow, $\theta_w = \phi$, and $\theta_{im} =$const., and (1.8.67) reduces to

$$\theta_{im} R_{dim} \frac{\partial c_{im}}{\partial t} = +\alpha^\star (c_m - c_{im}) + \nabla \cdot (\theta_{im} \mathcal{D}^*_{im} \cdot \nabla c_{im}) - \theta_{im} R_{dim} \lambda c_{im}$$
$$(1.8.68)$$

If we neglect diffusion in the immobile phase, (1.8.68) reduces to

$$\theta_{im} R_{dim} \frac{\partial c_{im}}{\partial t} = +\alpha^\star (c_m - c_{im}) - \theta_{im} R_{dim} \lambda c_{im}. \qquad (1.8.69)$$

Finally, consider saturated flow of a component carrying fluid in a granular porous medium domain in which the solid matrix, say, in the form of grains, or part of it, is porous. However, the size of the pores in the porous solid matrix is such that the latter's permeability to water flow is very low.

Clay aggregates may also serve as an example of such porous solid matrix. The transport of a component in such a porous medium can be described by a conceptual model that envisions mobile water in the larger pores, say between clay aggregates, and immobile water in the porous solid matrix, or clay aggregates. The considered component may be exchanged by molecular diffusion between the two apparent water phases: the mobile water in the void space, and the immobile water in the porous solid matrix. In this way, the porous solid matrix acts as distributed sinks for the considered component. Adsorption may take place on the relatively large surface area of the solid inside the porous solid matrix. The component reaches these sites by diffusion through the void space inside the solid matrix. However, molecular diffusion does not necessarily constitute a (macroscopic) transport mechanism in the balance equation written for the immobile water.

So far, we have considered only a single component. Let us consider two examples of multicomponent phases: a single multicomponent phase, and two multicomponent phases.

1.8.6　Balance equations for radionuclide decay chain

Consider the case of a single liquid (e.g., water) phase that occupies the entire void space, or part of it, at a liquid volumetric fraction, θ. The remaining part is occupied by a gas (e.g., air), or by another liquid, immiscible with the first. Let a decay chain of elements, e.g., due to radioactive disintegration,

$$A_1 \rightarrow A_2 \rightarrow A_3 \rightarrow \ldots \rightarrow A_N,$$

be dissolved only in the considered liquid, such that none of them can cross the (microscopic) interphase boundaries. The rate of decay (or growth) of an element (= component) is expressed as some function of its concentration. As a simple example, consider the case

$$\left. \frac{\partial c^\gamma}{\partial t} \right|_{\text{decay}} = -\lambda^\gamma c^\gamma, \tag{1.8.70}$$

i.e., the rate of decay of a γ-component is proportional to the concentration of that component. For the entire chain, this means

$$c^2 = c^1 \exp(-\lambda^1 t), \quad c^3 = c^2 \exp(-\lambda^2 t), \ldots, c^N = c^{N-1} \exp(-\lambda^{N-1} t), \tag{1.8.71}$$

where $c^1, c^2, c^3, \ldots, c_N$, each a function of space and time, denote the concentrations of the respective N elements of the chain. Thus, (1.8.70), for

$\gamma = 1, 2, \ldots, N$, represent a source (= negative sink) for the γth element, due to the decay phenomenon. Usually, $\lambda^N = 0$, i.e., a non-decaying element.

Let us assume that each of these elements can also be adsorbed to the solid matrix, according to a linear equilibrium isotherm, say (1.8.14), with $K_d^1, \quad K_d^2, \ldots, K_d^N$ denoting the respective distribution coefficients. Obviously, for an element that cannot be adsorbed, the corresponding coefficient is equal to zero.

In the absence of external sources or sinks, and with no chemical reactions among the elements, the only source results from the radioactive decay and growth of the elements. Accordingly the source term for γth component in the liquid phase takes the form

$$\rho\Gamma_f^\gamma = -\lambda^\gamma c^\gamma + \lambda^{\gamma-1} c^{\gamma-1}, \tag{1.8.72}$$

and for γth component on the solid, by

$$\Gamma_s^\gamma = -K_d \lambda^\gamma c^\gamma + K_d^{\gamma-1} \lambda^{\gamma-1} c^{\gamma-1}, \tag{1.8.73}$$

with $\gamma = 1, 2, \ldots N$, and $c^o = 0$.

Inserting these expressions into the mass balance equation for the γth component, leads to

$$\frac{\partial}{\partial t}(\theta R_d^\gamma c^\gamma) = \quad - \quad \nabla\cdot(c^\gamma \mathbf{q} - \theta \mathbf{D}_h \cdot \nabla c^\gamma)$$
$$- \quad \theta(\lambda^\gamma R_d^\gamma c^\gamma - \lambda^{\gamma-1} R_d^{\gamma-1} c^{\gamma-1}), \tag{1.8.74}$$

where

$$R_d^\gamma = 1 + \frac{1-\phi}{\theta}\rho_s K_d^\gamma$$

is the retardation factor for the γth component (= element). For saturated flow, we replace the moisture content, θ, by the porosity, ϕ.

Equations (1.8.74), with $\gamma = 1, \ldots, N$ represent N equations to be solved simultaneously for the N concentrations, c^γ.

1.8.7 Two multicomponent phases

Consider the case of two fluid phases, a gas, g (= air) and a liquid, ℓ (= water), that together occupy the entire void space. Let the gaseous phase be radioactive Radon, R, produced by the solid matrix at a rate Γ^R, expressed as mass of Radon per unit mass of solid matrix per unit time. The Radon can dissolve in the water. We wish to write the model that describes the spatial and temporal distributions of the Radon in the two phases, i.e., $c_\ell^R(\mathbf{x}, t)$ and $c_g^R(\mathbf{x}, t)$.

For the sake of simplicity, we shall assume that the solid matrix is rigid and stationary. We shall also neglect the difference between mass and volume weighted velocities of the two fluids. We shall assume that the considered component can move from one fluid to the next by diffusion, across the microscopic interfaces.

The core of the model consists of two mass balance equations. For the Radon in the gaseous phase, we write

$$\frac{\partial}{\partial t}(\phi S_g c_g^R) = \quad - \quad \nabla \cdot \phi S_g (c_g^R \mathbf{V}_g - \mathbf{D}_{gh} \cdot \nabla c_g^R)$$
$$- \quad \lambda \phi S_g c_g^R + S_g (1 - \phi) \rho_s \Gamma^R - f_{g \to \ell}^R, \qquad (1.8.75)$$

where, assuming that the fraction of the total surface area of the solid matrix that is in contact with the gas is expressed by the gas saturation, S_g, the term $S_g \Gamma^R$ denotes the strength of the Radon source entering from the rock into the gaseous phase. In a similar way, the strength of the Radon source entering the liquid phase will be given by $S_w \Gamma^R$. Obviously, different weights may be assigned to each fluid phase. The symbol $f_{g \to \ell}^R$ denotes the rate (in mass per unit volume of porous medium per unit time) at which Radon is transferred from the gaseous phase to the liquid one (to be dissolved in the latter), across their (microscopic) common interface.

The mass balance equation for the Radon in the liquid phase. takes the form

$$\frac{\partial}{\partial t}(\phi S_\ell c_\ell^R) = -\nabla \cdot \phi S_\ell (c_\ell^R \mathbf{V}_\ell - \mathbf{D}_{\ell h} \cdot \nabla c_\ell^R)$$
$$-\lambda \phi S_\ell c_\ell^R + S_\ell (1 - \phi) \rho_s \Gamma^R + f_{g \to \ell}^R, \qquad (1.8.76)$$

with $f_{g \to \ell}^R = -f_{\ell \to g}^R$.

We have here two balance equations in three variables: c_ℓ^R, c_g^R and $f_{g \to \ell}^R$. The input information concerning the phase velocities, $\mathbf{V}_\ell$, $\mathbf{V}_g$, and the phase saturations, S_ℓ, S_g, should be obtained by writing and solving the two phase flow model.

The rate of transfer of a component from one phase to the other, can be expressed by employing the concept of a *transfer coefficient*, $\alpha^\star$, and the assumption that this rate is proportional to the difference in the component's concentration between the two phases, viz.

$$f_{g \to \ell}^R = \alpha^{\star R}(c_g - c_\ell). \qquad (1.8.77)$$

Since the transfer mechanism is molecular diffusion, the transfer coefficient is proportional to some weighted average of the diffusion of the component in the two phases, and to the contact area between them, and inversely proportional to the distance between the centroids of the partial volumes occupied by them within the void space. This distance is equal to some weighted average of the characteristic distances between each phase and the common boundary between them. In both cases, the weights are functions of the (time and space dependent) phase saturations.

Altogether, it means that, even if we do accept the (questionable) notion of a constant transfer coefficient in the case of single phase flow, in two and three-phase flow, the transfer coefficient is saturation dependent.

If we employ (1.8.77), we have three equations to be solved for the three variables of the problem.

Instead, we may make use a methodology that eliminates the rates of transfer between the two phases by summing up the two balance equations. In the more general case, and whenever an extensive quantity, e.g., heat, is exchanged between phases, By summing up all the balance equations for the same component in all the phase, we obtain a single equation for the considered component in all the phases present in the system. The rates of interphase transfers do not appear in this equation. However, the 'cost' of eliminating the (unknown) rates of interphase transfers is that we end up with a smaller number of equations. We have now to make use of various thermodynamic relations between the densities of the considered extensive quantity in adjacent phases. Actually, we have employed exactly this methodology when dealing with adsorption.

Here, by summing up the two balance equations, we obtain a single balance equation for the component, R, in the entire system

$$\frac{\partial}{\partial t}\left[\phi(S_\ell c_\ell^{\mathrm{R}} + S_g c_g^{\mathrm{R}})\right]$$
$$= -\nabla\cdot\phi\{S_\ell(c_\ell^{\mathrm{R}}\mathbf{V}_\ell - \mathbf{D}_{\ell h}\cdot\nabla c_\ell^{\mathrm{R}}) + S_g(c_g^{\mathrm{R}}\mathbf{V}_g - \mathbf{D}_{gh}\cdot\nabla c_g^{\mathrm{R}})\}$$
$$-\lambda\phi(S_\ell c_\ell^{\mathrm{R}} + S_g c_g^{\mathrm{R}}) + (1-\phi)\rho_s\Gamma^{\mathrm{R}}, \tag{1.8.78}$$

where we have made use of $S_\ell + S_g = 1$.

At this stage, we have a single equation in two variables. The additional information required in order to obtain a complete set of equations is that of the partitioning of the Radon between the two phases. *Assuming* that a local phase equilibrium exists between the two phases (based on the assumption of equilibrium at the microscopic level between adjacent phases), this information takes the form of *Henry's law*, which can be written as

$$H_{g,\ell}^{\mathrm{R}} = H_{g,\ell}^{\mathrm{R}}(T) = \frac{c_g^{\mathrm{R}}}{c_\ell^{\mathrm{R}}}, \qquad (1.8.79)$$

in which the value of *Henry's coefficient*, $H_{g,\ell}^{\mathrm{R}}$, is known. With this relationship, the combined balance equation can be written in terms of a single state variable, say, c_ℓ^R, in the form

$$\frac{\partial}{\partial t}\left[\phi(S_\ell + S_g H_{g,\ell}^{\mathrm{R}})c_\ell^{\mathrm{R}}\right]$$

$$= -\nabla\cdot\phi\left[H_{g,\ell}^{\mathrm{R}}S_g(c_\ell^{\mathrm{R}}\mathbf{V}_g - \mathbf{D}_{gh}\cdot\nabla c_\ell^{\mathrm{R}}) + S_\ell(c_\ell^{\mathrm{R}}\mathbf{V}_\ell - \mathbf{D}_{\ell h}\cdot\nabla c_\ell^{\mathrm{R}})\right]$$

$$- \lambda\phi(S_\ell + S_g H_{g,\ell}^{\mathrm{R}})c_\ell^{\mathrm{R}} + (1-\phi)\rho_s\Gamma^{\mathrm{R}}. \qquad (1.8.80)$$

As mentioned above, the complete model should include also the part that involves the mass transport of the two phases. Its solution will provide information on $\mathbf{V}_\ell$, $\mathbf{V}_g$, S_ℓ and S_g.

1.8.8   Boundary Conditions

There is no need to elaborate on the need to supplement the balance equation(s) by boundary conditions, in order to write a well posed model of component transport. The conditions presented here are based on the discussion presented in LECTURE 5, with sides 1 and 2 denoting the considered domain and its environment, respectively. The dependent variable in the model is the concentration, $c(\mathbf{x},t)$, of a component.

The discussion here is similar to that presented in LECTURE 6. We continue to assume that [A5.3] and hence also (1.5.14) are valid. The symbols f and γ, that denote the fluid and the component will be omitted.

Boundary of prescribed concentration

When the concentration, $c(\mathbf{x},t)$, can be specified as a known function, $g_1(\mathbf{x},t)$, at all points of a boundary segment, $\mathcal{B}$, say due to phenomena that take place in the environment, independent of what happens within the considered domain, we employ (1.5.21) to write the boundary condition

$$c(\mathbf{x},t) = g_1(\mathbf{x},t), \qquad \text{on} \qquad \mathcal{B}, \qquad (1.8.81)$$

where $c(\mathbf{x},t) \equiv c(\mathbf{x},t)|_1$ and $g_1(\mathbf{x},t) \equiv c(\mathbf{x},t)|_2$. This is a *Dirichlet boundary condition*.

Boundary of prescribed flux

When phenomena occurring in the external domain impose a known flux of the considered component, say $g_2(\mathbf{x},t)$, normal to a boundary segment, $\mathcal{B}$, at all points of that segment, independent of what happens within the considered domain itself, we use (1.5.22) to write the condition

$$(c\mathbf{q}_r - \theta\mathbf{D}_h\cdot\nabla c)\cdot\boldsymbol{\nu} = g_2(\mathbf{x},t), \qquad \text{on} \quad \mathcal{B}, \qquad (1.8.82)$$

or, in view of (1.5.3)

$$(c\mathbf{q}_r - \theta\mathbf{D}_h\cdot\nabla c)\cdot\nabla F = g_2(\mathbf{x},t)|\nabla F| \qquad \text{on} \quad \mathcal{B}, \qquad (1.8.83)$$

where $F(\mathbf{x},t) = 0$ specifies the configuration of the known surface, $\mathcal{B}$. For saturated flow, θ is replaced by the porosity, ϕ.

Since both c and ∇c are involved in (1.8.83), this is a *Cauchy boundary condition*.

A particular boundary of this kind, is the impervious boundary. For such boundary, $g_2 = 0$, and $\mathbf{q}_r\cdot\boldsymbol{\nu} = 0$. Then, (1.8.83) reduces to

$$(\mathbf{D}_h\cdot\nabla c)\cdot\nabla F = 0, \qquad \text{on} \quad \mathcal{B}, \qquad (1.8.84)$$

i.e., a *Neumann boundary condition*.

Boundary between two porous media

In this case, we assume the existence of discontinuities in all porous matrix characteristics, e.g., ϕ, k_{ij} and $a_{ijk\ell}$. Neither the concentration, nor the flux are known a-priori on the boundary. However, *both* (1.5.21) and (1.5.22) must be satisfied, i.e.,

$$c|_1(\mathbf{x},t) - c|_2(\mathbf{x},t), \qquad \text{on} \quad \mathcal{B}, \qquad (1.8.85)$$

and

$$(c\mathbf{q}_r - \theta\mathbf{D}_h\cdot\nabla c)|_1\cdot\nabla F = (c\mathbf{q}_r - \theta\mathbf{D}_h\cdot\nabla c)|_2\cdot\nabla F, \qquad (1.8.86)$$

where $F(\mathbf{x},t) = 0$ defines the (known) shape of the boundary, and $\mathbf{q}_r \equiv \theta(\mathbf{V} - \mathbf{V}_s)$ can be expressed by one of the motion equations. Under the

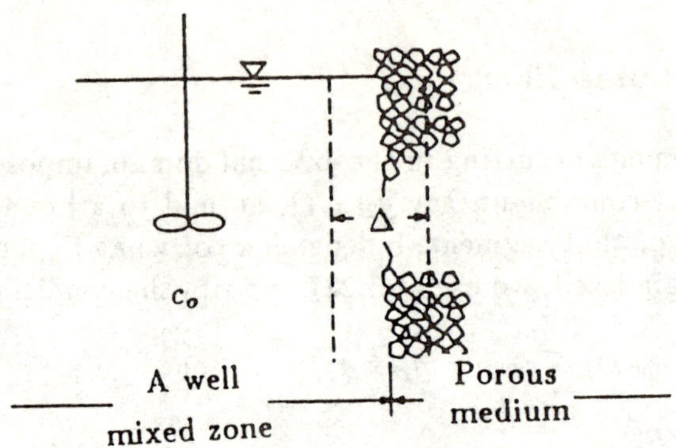

Figure 1.8.5: Nomenclature for a well mixed zone.

conditions that lead to (1.5.19), which means no-jump in advective flux, (1.8.86) reduces to

$$(\theta \mathbf{D}_h \cdot \nabla c)|_1 \cdot \nabla F = \{\theta \mathbf{D}_h \cdot \nabla c\}|_2 \cdot \nabla F. \qquad (1.8.87)$$

Boundary with a body of 'well mixed' fluid

Although, here the considered extensive quantity is the mass of a component, with the concentration, $c\ (\equiv c^\gamma)$, as state variable, the development and results presented below can also be applied to cases where temperature and pressure serve as state variables, as well as to cases where the external domain is a solid body.

Consider a saturated porous medium domain in hydraulic contact with a body of water (e.g., a river, or lake), and let the water contain a component which will serve as the considered extensive quantity (Fig. 1.8.5). No subscript will be used to denote the water. We assume that

[A8.1] The external body of water is 'well mixed' i.e., it is everywhere at a constant value of the concentration.

The condition of no-jump in the total flux of the considered component, takes the form of (1.5.42), in which $\theta_f|_{pm}$ is replaced by the porosity, ϕ, while $\phi|_{fb} = 1$, so that $\mathbf{V}|_{fb} \equiv \mathbf{q}|_{fb}$. This condition may also be written in the form

$$[c''(\mathbf{V} - \mathbf{u})]|_{fb}\cdot\boldsymbol{\nu} - [c(\mathbf{q} - \phi\mathbf{u}) + \phi(\mathbf{J} + \mathbf{J}^*)]|_{pm}\cdot\boldsymbol{\nu} = 0$$

$$(1.8.88)$$

where $\mathbf{q} = \phi\mathbf{V}$ is the specific discharge of water in the porous medium, subscripts fb and pm denote the fluid (= water) body and the porous medium, respectively, $\theta|_{fb} = 1$, and c'' denotes the (*assumed constant*) concentration of the component in the 'well mixed' water body. Equation (1.8.88) expresses the continuity of the component's mass flux across the boundary, where the total flux in the water body, because it is well mixed, consists of advection only, while in the porous medium it consists of advection, diffusion and dispersion.

Assumption [A5.1] underlies (1.8.88). In view of (1.5.14), we could replace $(\mathbf{q} - \phi\mathbf{u})$ in (1.8.88) by $\mathbf{q}_r$.

To simplify the presentation, let us assume that the boundary is stationary, i.e., $\mathbf{u} = 0$. Then, (1.8.88) reduces to

$$c''\mathbf{V}|_{fb}\cdot\boldsymbol{\nu} - (c\mathbf{q} - \phi\mathbf{D}_h\cdot\nabla c)|_{pm}\cdot\boldsymbol{\nu} = 0,\qquad(1.8.89)$$

where the sum $\mathbf{J} + \mathbf{J}^*$ has been expressed in terms of ∇c.

Consequently, when no advection takes place across the boundary, $\mathbf{V}|_{fb}\cdot\boldsymbol{\nu} \equiv \mathbf{q}\cdot\boldsymbol{\nu} = 0$, $\mathbf{J}^* = 0$, and (1.8.89) reduces to

$$(\boldsymbol{\mathcal{D}}^*\cdot\nabla c)|_{pm}\cdot\boldsymbol{\nu} = 0.\qquad(1.8.90)$$

This implies that no mass is transferred by molecular diffusion across such a boundary, even when $c|_{pm} \neq c''$. This conclusion is unacceptable, as under the physical conditions of this case, we would expect transport of the component by molecular diffusion to take place between the porous medium domain and the adjacent water body, as molecular diffusion remains the only possible mode of transport.

The error in the conclusion which follows from (1.8.90), stems from the assumption that a 'well mixed' zone exists on the external side of the boundary. This assumption, in the absence of advection, combined with the sharp boundary approximation, yields no mass flux by diffusion across it. In order to reinstate the diffusive–dispersive flux, which, as we know should take place in reality, we introduce the concept of a *transition*, or *buffer zone*, at the boundary (Fig. 1.8.5). We may associate the width of this transition

zone, Δ, with the magnitude of an REV, assuming that the abrupt boundary passes through its midpoint. Instead of the boundary between the body
of water and the porous medium, we now consider the boundary between
the latter and the transition zone. Assuming that the sum of dispersive and
diffusive fluxes through the transition zone is proportional to the average
concentration gradient, and that the latter is proportional to the concentration difference $c - c''$, we express the condition of continuity of flux at the
boundary by

$$c''\mathbf{V}|_{fb}\cdot\boldsymbol{\nu} - \alpha(c - c'') = (c\mathbf{q} - \phi\mathbf{D}_h\cdot\nabla c)|_{pm}\cdot\boldsymbol{\nu}, \qquad (1.8.91)$$

where α is a coefficient such that $\alpha(c'' - c)$ represents the sum of diffusive
and dispersive fluxes through the transition zone.

Since, with $[\rho_f]_{1,2} = 0$, we have $\mathbf{V}|_{fb}\cdot\boldsymbol{\nu} = \mathbf{q}\cdot\boldsymbol{\nu}$, equation (1.8.91) reduces
to

$$(c'' - c|_{pm})(\mathbf{q}\cdot\boldsymbol{\nu} + \alpha) = -\phi\mathbf{D}_h\cdot\nabla c|_{pm}\cdot\boldsymbol{\nu}, \qquad (1.8.92)$$

which now serves as the boundary condition.

In the absence of advection, or when $|\mathbf{q}| \ll \alpha$, equation (1.8.92) reduces
to

$$\alpha(c'' - c|_{pm}) = -\phi\mathbf{D}_h\cdot\nabla c|_{pm}\cdot\boldsymbol{\nu}. \qquad (1.8.93)$$

We note that if we accept (1.8.92), then $c''|_{fb} \neq c|_{pm}$ on the boundary, i.e.,
a jump in concentration takes place on the boundary. This is a consequence
of introducing the transition zone and the 'well mixed zone' approximation.

When $|\mathbf{q}\cdot\boldsymbol{\nu}| \gg \alpha$, equation (1.8.92) reduces to

$$(c'' - c|_{pm})\mathbf{q}\cdot\boldsymbol{\nu} = -\phi\mathbf{D}_h\cdot\nabla c|_{pm}\cdot\boldsymbol{\nu}, \qquad (1.8.94)$$

which is identical to (1.8.89), yet is based on different reasoning.

Phreatic surface

The phreatic surface boundary condition for fluid mass transport was
presented in LECTURE 6. Here we consider the phreatic surface boundary
in a problem of mass transport of a component of a fluid phase. The concentration of a considered component in the infiltrating water is denoted by
c', and we assume that, in spite of concentration changes, the mass density
of the fluid remains constant. The boundary condition is derived from the
requirement of no jump in the component's flux normal to the phreatic surface. With this in mind, the boundary condition is similar to that expressed

by (1.6.97), except that c' replaces ρ', $\mathbf{q}$ replaces $\mathbf{q}_w$, and c replaces ρ_w, viz.

$$(c\mathbf{q} - c'\mathbf{N})\cdot\nabla F + (\phi c - \theta_{wo}c')\frac{\partial F}{\partial t} - \phi(\mathbf{D}_h\cdot\nabla c)|_{\text{sat}}\cdot\nabla F = 0. \qquad (1.8.95)$$

If we assume $\rho_w = \rho'$, equation (1.6.98) is valid and (1.8.95) reduces to

$$(c - c')\left(\mathbf{N}\cdot\nabla F + \theta_{wo}\frac{\partial F}{\partial t}\right) - \phi(\mathbf{D}_h\cdot\nabla c)|_{\text{sat}}\cdot\nabla F = 0, \qquad (1.8.96)$$

where we note that $[c]_{\text{sat, unsat}} \neq 0$. Thus the zone above the phreatic surface represents a *well mixed zone* in the sense discussed earlier.

Seepage face

The seepage face was discussed in LECTURE 5. In this case, fluid containing a component leaves a porous medium domain through the boundary into an environment which is free of porous medium.

Because no porous medium is present in the environment, and $\mathbf{u} \equiv 0$ for a stationary seepage face, the condition of continuity of flux of a component, takes the form

$$(c\phi\mathbf{V} - \phi\mathbf{D}_h\cdot\nabla c)|_{pm}\cdot\boldsymbol{\nu} = (c\mathbf{V})|_{en} \qquad (1.8.97)$$

where symbols pm and en denote the porous medium domain and its environment, respectively. With (1.5.19) and (1.5.20), equation (1.8.97) reduces to

$$-(\mathbf{D}_h\cdot\nabla c)\cdot\boldsymbol{\nu} = 0. \qquad (1.8.98)$$

1.8.9   Complete Mathematical Model

Because the velocity within a considered domain is a necessary input information to a component transport problem, it is always necessary to solve the (total) mass transport problem (LECTURE 5), before solving a component transport one. However, when solving a case in which the component's concentration affect's the fluid's density (and possibly also its viscosity), the two problems are coupled and must be solved simultaneously. As an example, consider the case of *sea water intrusion* into a coastal aquifer. The total dissolved solids may be used as the considered component. With ρ denoting the water's density, the core of the model (= balance equations and constitutive relations), is comprised of the following equations:

Mass balance of the component. We assume no adsorption, no decay or growth. From (1.8.54), we obtain

$$\phi \frac{\partial c}{\partial t} \;=\; -\mathbf{q}\cdot\nabla c + \nabla\cdot\phi\mathbf{D}_h\cdot\nabla c$$
$$+ \sum_{(i)} R_{ex}^{(i)}(\mathbf{x}^{(i)},t)\delta(\mathbf{x}-\mathbf{x}^{(i)})(c_r^{(i)}-c). \qquad (1.8.99)$$

Mass balance of the fluid. Assuming $\partial\phi/\partial t = 0$, and $\rho = \rho(p,c)$, we may use (1.6.16) to write

$$\phi \frac{\partial \rho}{\partial t} = -\nabla\cdot\rho\mathbf{q} - \sum_{(r)} Q_{ex}^{(r)}(\mathbf{x}^{(r)},t)\delta(\mathbf{x}-\mathbf{x}^{(r)}), \qquad (1.8.100)$$

noting, of course, all other assumptions that underlie (1.6.16).

Motion equation. We use Darcy's law (= simplified momentum balance equation)

$$\mathbf{q} = -\frac{\mathbf{k}}{\mu}(\nabla p + \rho g\nabla z), \qquad (1.8.101)$$

again, noting all the underlying assumptions.

Constitutive relations. We need information on $\mu = \mu(p,c)$ and $\rho = \rho(p,c)$. For example

$$\rho = \rho_o[1 + \beta_p'(p-p_o) + \beta_c'(c-c_o)]. \qquad (1.8.102)$$

Altogether, we have here four variables: c, ρ, p, and $\mathbf{q}$ (or five, had we added μ), and the same number of equations.

Actually, inserting (1.8.101) and (1.8.102) into (1.8.99) and (1.8.100), recalling that $\mathbf{D}_h$ also depends on $\mathbf{q}$, we have to solve the two partial differential equations, (1.8.99) and (1.8.100), *simultaneously.*

With this example of a coupled flow–component transport problem in mind, we can now return to the general content of a component transport problem. As in the case of total mass transport, following the discussion in LECTURE 5, the statement, or *model* of a problem of component transport, consists of the following statements:

- Statement of the configuration of the surface that bounds the porous medium domain within which the transport of the considered component takes place.

- Statement of the dependent variable–the concentration of the considered component. When two or more components, which may interact with each other chemically, or which affect each other in any other way, are present in the fluid phase, the models that describe the transport of each component, separately, are coupled to each other. The coupling is produced, for example, by the term $\sum_{(j)} R^{\gamma j}$ defined by (1.8.30). In such a case, we have a number of state variables which have to be solved for simultaneously.

- Statement of the partial differential equation that describes the balance of the considered component (or equations, in the case of a number of interacting components).

- Statement of initial conditions for each of the relevant state variables and

- Statement of boundary conditions for each of the relevant state variables.

We need information on all relevant porous medium properties, such as porosity, permeability, dispersivity, decay and growth coefficients, partitioning coefficient, etc. As discussed earlier, we also need information on the spatial and temporal distribution of the velocity.

As in the case of the flow model, we wish to emphasize also here that a *conceptual model* ($\equiv$ set of assumptions that simplify the real world) must be established as a basis for the *mathematical model*. This completes the series of eight lectures on transport phenomena in porous media. We have presented a general methodology for constructing a well posed model at the macroscopic level for any extensive quantity. We have applied the methodology, in some more detail, to two particular extensive quantities: mass of a phase and mass of a component of a phase.

References

Adamson, A. W. *Physical Chemistry of Surfaces, 2nd edn.* Interscience, New York , 1967.

Aitchison, G. D. and Donald, I. B. Effective Stresses in Unsaturated Soils. In *Proc. 2nd Australian-New Zealand Conference on Soil Mechanics Fdn. Engnrng.*, Inst. Engrs., 1956.

Bachmat, Y. Spatial macroscopization of processes in heterogeneous systems. *Israel J. of Tech.*, 10:391–403, 1972.

Bachmat, Y. and Bear, J. The dispersive flux in transport phenomena. *Adv. Water Resour.*, **6**:169–174, 1983.

Bachmat, Y. and Bear, J. Macroscopic modelling of transport phenomena in porous media. Part 1: the continuum approach. *Transport in Porous Media*, **1**:213–240, 1986.

Bear, J. *Dynamics of Fluids in Porous Media*. American Elsevier (also Dover publ., 1988), 1972.

Bear, J. *Hydraulics of Groundwater*. McGraw-Hill Book Co., London, 1979.

Bear, J. On the tensor form of dispersion. *J. Geophys. Res.*, **66**:1185–1197, 1961.

Bear, J. and Bachmat, Y. A generalized theory on hydrodynamic dispersion. In *Proc. I.A.S.H. Symposium on Artifical Recharge and Management of Aquifers*, p. 7, 1967.

Bear, J. and Bachmat, Y. *Introduction to Modelling of Transport Phenomena in Porous Media*. Kluwer Academic Publishers, 1990.

Bear, J., Corapcioglu, Y. M. and Bulkarishna, J. Modeling of centrifugal filtration in unsaturated deformable porous medium. *Adv. Water Resour.*, **7**:150–167, 1984.

Bensabat, J. *Heat and Mass Transfer in Unsaturated Porous Media with Application to an Energy Storage Problem*. D.Sc. dissertation, Technion–Israel Inst. of Technology, Haifa, Israel, 1986.

Biot, M. A. General theory of three-dimensional consolidation. *J. Appl. Phys.*, **12**:155–164, 1941.

Brinkman, H. C. Calculations of the flow of heterogeneous mixture through porous media. *Appl. Sci. Soc.*, **2**:81–86, 1948.

De Josselin de Jong, G. Longitudinal and transverse diffusion in granular deposits. *Trans. Amer. Geophys. Union*, **39**:67–74, 1958.

Debye, H. R. Jr., Anderson, P. and Brumberger, H. Scattering by an inhomogeneous solid ii: the correlation function and its application. *Appl. Phys.*, **28**:679–83, 1957.

Denbigh, K. *The Principles of Chemiccls Equilibrium*. Cambridge University Press, 1955.

Freundlich, C. G. L. *Colloid and Capillary Chemistry*. Methuen, London, 1926.

Hendricks, D. W. Sorption in flow through porous media. In Bear, J., (ed.) *Fundamentals of Transport Phenomena in Porous Media*, pp. 384–392, Elsevier, Amsterdam, 1972.

Hubbert, M. K. The theory of ground water motion. *J. Geol.*, **48**:785–944, 1940.

Jacob, C. E. On the flow of water in an elastic artesian aquifer. *Trans. Amer. Geophys.*, **21**:547–586, 1940.

Jacob, C. E. Flow of ground water. In Rouse, H. (ed.) *Engineering Hydraulics*, pp. 321-386. John Wiley, New York, 1950.

Landau, L. and Lifshitz. E. M. *Fluid Mechanics.* Addison-Wesley, Reading, 1960.

Langmuir, I. Chemical reactions at low temperatures. *J. Amer. Chem. Soc.*, **37**:1139, 1915.

Langmuir, I. The adsorption of gases on plane surfaces of glass, mica and platinum. *J. Amer. Chem. Soc.*, **40**:1361–1403, 1918.

Lapidus, L. and Amundson, N. R. Mathematics of absorption in beds vi: the effect of longitudinal diffusion in ion exchange and chromatographic columns. *J. Phys. Chem.*, **56**:984–988, 1952.

Lindstrom, L., Boersma, F. T. and Stockard, D. A theory on the mass transport of previously distributed chemicals in a water saturated sorbing porous medium. *Soil Sci.*, **112**:291–300, 1971.

Muskat, M. *The Flow of Homogeneous Fluids Through Porous Media.* McGraw-Hill, New York, 1937.

Nikolaevski, V. N. Convective diffusion in porous media. *J. Appl. Math. Mech. (P.M.M.)*, **23**:1042–1050, 1959.

Saffman, P. G. A theory of dispersion in a porous medium. *J. Fluid Mech.*, **6**:321–349, 1959.

Santalo, L. A. *Integral Geometry and Geometric Probability.* Addison Wesley, 1976.

Scheidegger, A. E. General theory of dispersion in porous media. *J. Geophys. Res.*, **66**:3273–3278, 1961.

Spain, B. *Tensor Calculus.* Oliver and Boyd, London, 1960.

Terzaghi, K. *Erdbaumechanik auf Bodenphysikalische Grundlage.* Franz Deuticke, Leipzig, 1925.

Van Genuchten, M. Th. *Mass Transfer Studies of Sorbing Porous Media.* New Mexico State Univ., La Cruz, NM, 1974.

Verruijt, A. *Elastic Storage of Aquifers*, pp. 331-376. In De Wiest, R. J. M., (ed.) Flow through Porous Media. Academic Press, New York, 1969.

List of Main Symbols

a As subscript, symbol denoting air.

$\mathbf{a}$ Dispersivity of a porous medium.

a_{ijkl} Components of $\mathbf{a}$.

a_L Longitudinal dispersivity of an isotropic porous medium.

a_T Transversal dispersivity of an isotropic porous medium.

c As subscript, symbol denoting a characteristic value.

c_r Hydraulic resistance of a semipervious layer ($=$ ratio of thickness to hydraulic conductivity).

c_α^γ Concentration of a γ-component in an α-phase ($\equiv \rho_\alpha^\gamma$ $=$ mass of component per unit volume of phase).

C_s Specific heat of solid at constant strain.

C_V Specific heat of fluid at constant volume.

d^* Length characterizing macroscopic heterogeneity.

$\mathbf{D}_\alpha^E$ Coefficient of dispersion of E in an α-phase.

$\mathbf{D}_\alpha^\gamma$ Coefficient of dispersion of mass of a γ-component in an α-phase.

$\mathcal{D}_\alpha^\gamma$ coefficient of molecular diffusion of a γ-component in an α-phase.

$\mathcal{D}_\alpha^{*\gamma}$ Coefficient of molecular diffusion of a γ-component in a porous medium.

$\mathbf{D}_{\alpha h}^\gamma$ Coefficient of hydrodynamic dispersion of a γ-component in an α-phase ($= \mathcal{D}_\alpha^{*\gamma} + \mathbf{D}_\alpha^\gamma$).

e Void ratio ($= U_{ov}/U_{os}$). Density of an extensive quantity, E.

e_α^γ Density of E_α^γ.

E_α An extensive quantity, E, of an α-phase, e.g., $E = m, m^\gamma, M, H$.

f As subscript, symbol denoting fluid.

F Equation of a surface, $F(\mathbf{x}, t) = 0$.

F Concentration of a component adsorbed on a solid ($=$ mass of a component per unit mass of solid).

g Gravity acceleration. As subscript, symbol denoting gas.

h Enthalpy. As subscript, symbol denoting hydrodynamic dispersion.

H Heat. As superscript, a synbol denoting heat.

$\mathbf{I}$ Unit tensor.

I_α Specific internal energy of an $\alpha - phase$.

$\mathbf{j}^{E_1 E_2}$ Microscopic diffusive flux of E_1 with respect to E_2 ($= e_1(\mathbf{V}^{E_1} -\mathbf{V}^{E_2})$).

$\mathbf{j}^H$ — Microscopic conductive heat flux $(= \rho I(\mathbf{V}^H - \mathbf{V}^m))$.

$\mathbf{j}^m$ — Microscopic diffusive flux of total mass $(\equiv \mathbf{j}^{mU} = \rho(\mathbf{V}^m - \mathbf{V}))$.

$\mathbf{j}^\gamma$ — Microscopic diffusive mass flux of a γ-component $(\equiv \mathbf{j}^{mU} = \rho(\mathbf{V}^\gamma - \mathbf{V}))$.

$\mathbf{j}^{tE_\alpha^\gamma}$ — Total microscopic flux of E_α^γ $(= e_\alpha^\gamma \mathbf{V}^{E_\alpha^\gamma})$.

$\mathbf{J}^{*E}$ — Dispersive flux of E.

$\mathbf{J}_h^E$ — Sum of diffusive and dispersive fluxes of E.

$\mathbf{J}^H$ — Macroscopic conductive heat flux $(\equiv \overline{\mathbf{J}^H}^\alpha)$.

$\mathbf{J}^\gamma$ — Macroscopic diffusive flux of a γ-component $(\overline{\mathbf{j}^\gamma}^\alpha)$.

$\mathbf{J}^{tE_\alpha^\gamma}$ — Total macroscopic flux of E_α^γ.

$\mathbf{k}$ — Permeability.

$\mathbf{k}_\alpha$ — Effective permeability of an α-phase.

$\mathbf{K}$ — Hydraulic conductivity.

K_d — Partitioning coefficient.

L^* — Characteristic size of a domain.

ℓ — Characteristic size of an REV.

m — Mass; also as superscript.

m_α^γ — Mass of a γ-component of an α-phase.

$\mathbf{M}$ — Momentum. Also as superscript.

M — Molecular weight.

n — As subscript, symbol denoting a nonwetting fluid.

p — Pressure.

p_α — Pressure in an α-phase.

pm — As subscript, symbol denoting a porous medium.

$\mathbf{q}_\alpha$ — Specific discharge of an α-phase.

$\mathbf{q}_r$ — Specific discharge of an α-phase, relative to the solid.

$\mathbf{q}^m$ — Mass weighted specific discharge.

R — Universal gas constant. Rate of recharge of an aquifer.

R — Retardation factor.

S_α — Saturation of an α-phase (e.g., $\alpha = a, w$).

S_o — Specific storativity of a porous medium.

$S_{\alpha\beta}$ — Area of $S_{\alpha\beta}$-surface (similarly, $S_{\alpha\alpha}$, etc.).

$\mathcal{S}_{\alpha\beta}$ — Surface of contact of an α-phase with all other phases (denoted by β) within $\mathcal{U}_o$.

t — Time.

T Temperature.

$\mathbf{T}^*$ Tortuosity.

$\mathbf{u}_\alpha$ Velocity of a surface (e.g., of $\mathcal{S}_{\alpha\beta}$.

U Volume. Volume of a domain $\mathcal{U}$.

U^E Volume occupied by a quantity E.

U_o Volume of the domain of an REV ($\mathcal{U}_o$).

$\mathcal{U}$ Domain of volume U. Similarly, domains $\mathcal{U}_{o\alpha}$, $\mathcal{U}_{os}$, etc.

v As subscript, symbol indicating void space. Specific volume (of mass $=1/\rho$).

$\mathbf{V}$ Velocity.

$\mathbf{V}_s$ Velocity of the solid.

$\mathbf{V}_\alpha$ Volume weighted velocity of an α-phase ($\equiv \mathbf{V}^{U_\alpha}$).

$\mathbf{V}^{E_\alpha^\gamma}$ Velocity of E_α^γ-continuum.

w As subscript, symbol denoting water, or wetting fluid.

$\mathbf{w}$ Displacement.

x Horizontal coordinate.

$\mathbf{x}$ Position vector.

$\overset{\circ}{\mathbf{x}}$ $= \mathbf{x} - \mathbf{x}_o$.

$\mathbf{x}'$ Position vector of a point at the microscopic level.

$\mathbf{x_o}$ Position vector of the centroid of an REV.

$\mathbf{x}_{o\alpha}$ Position vector of the centroid of $\mathcal{U}_{o\alpha}$.

y Horizontal coordinate.

z Vertical coordinate (positive upward).

Z Compressibility factor.

Greek Letters

α As subscript, symbol for an α-phase.

β As subscript, symbol for a β-phase. A symbol for all other phases, except α).

β_p Coefficient of fluid compressibility at constant pressure.

β_T Coefficient of fluid compressibility at constant temperature.

γ As superscript, symbol denoting a γ-component.

$\gamma(\mathbf{x})$ Characteristic function of the void space.

$\gamma_\alpha(\mathbf{x})$ Characteristic function of an α-phase.

$\Gamma^{E_\alpha^\gamma}$ Rate of production of E_α^γ, per unit mass of phase.

Γ^{SE} Rate of production of E on a boundary.

δ_{ij} Components of Kronecker delta.

Δ	Characteristic distance from the solid surface to the fluid within the void space.
Δ_α	Hydraulic radius of an α-phase.
ε	Volumetric strain. First invariant of ε
ε	Strain tensor.
$\dot{\varepsilon}$	Rate of strain.
ε_{ij}	Components of ε.
ε_{sk}	Dilatation of the solid matrix.
ε_s	Dilatation of the solid phase.
η	Coefficient of thermoelasticity.
θ_α	Volumetric fraction of an α-phase.
θ_α^S	Fraction of the α–α-surface in S_o.
λ_α	Thermal conductivity of an α-phase. A coefficient of an α-phase.
λ_α^*	Coefficient of thermal conductivity of an α-phase in a porous medium.
Λ^H	Combined thermal conductivity in the fluid and the solid in a saturated porous medium.
Λ_h	Combined thermal dispersion in the fluid and thermal conductivity in the fluid and the solid, in a saturated porous medium.
λ''_s	Lamè constant of elastic solid.
μ_α	Dynamic viscosity of an α-phase.
μ'_s	Lamè constant of an elastic solid.
ν	Poisson's ratio.
ν_α	Kinematic viscosity of an α-phase $(=\mu_\alpha/\rho_\alpha)$.
$\boldsymbol{\nu}_\alpha$	Outward unit vector to $\mathcal{U}_{\alpha\beta}$ on $S_{\alpha\beta}$.
ρ_α	Mass density of an α-phase.
ρ_α^γ	Mass density of a γ-component in an α-phase.
σ	Stress tensor.
σ'_s	Effective stress $(\equiv \overline{\sigma'_s})$.
$\Sigma_{\alpha\beta}$	Specific area of $S_{\alpha\beta}$. Similarly, $\Sigma_{\alpha alpha}$ for $S_{\alpha\alpha}$, etc.
τ	Shear stress.
τ_γ	Correlation coefficient of $\gamma(\mathbf{x})$.
ϕ	Porosity.
ϕ_{eff}	Effective porosity.
φ	Piezometric head.
$\chi(\theta)$	Function of moisture content.
ω_α^γ	Mass fraction of a γ-component in an α-phase.

Special Symbols

$\overline{(..)}$	Average, volume average, or phase average of $(..)$ $\left(=\frac{1}{U_o}\int_{\mathcal{U}_o}(..)\,dU\right)$.
$\overline{(..)}^\alpha$	Intrinsic phase average of $(..)$ $\left(=\frac{1}{U_{o\alpha}}\int_{\mathcal{U}_o}(..)dU\right)$.

$\overset{\circ}{G}$ Deviation of G from its intrinsic phase average, $\overline{G}^{\alpha}$, over an REV.

$\text{Cov}(..)$ Covariance of $(..)$.

$\text{Var}(..)$ Variance of $(..)$.

$E(..)$ Expected value of $(..)$.

$\widetilde{(..)}^{\alpha\beta}$ Average of $(..)$ over the $\mathcal{S}_{\alpha\beta}$-surface.

$\dot{G}$ Material derivative of G.

$\nabla \cdot \mathbf{A}$ Divergence of a vector $\mathbf{A}$ ($=$ div $\mathbf{A}$).

∇A Gradient of a scalar, A ($\equiv$ grad A).

$\dfrac{D_E(..)}{Dt}$ Material derivative of $(..)$, as observed by the E-continuum.

Chapter 2

MULTIPHASE FLOW IN POROUS MEDIA

Th. DRACOS
Swiss Federal Inst. of Technology (E.T.H.)
Zurich, Switzerland.

In this chapter, we consider the simultaneous flow of liquids and gases through porous media. We shall assume that cohesive forces in these fluids exceed the adhesion ones, and that the fluids are *immiscible* and separated from each other by stable abrupt interfaces. Air, water and oil are immiscible under normal conditions (atmospheric pressure, room temperature, etc.). This does not mean that solubility (or volatility) is zero, but that they are so small that a sharp interface forms and that the physical properties of the fluids are not affected.

2.1 Capillary Pressure

2.1.1 Interfacial tension, contact angle and wettability

The origin of the interfacial tension is the difference between the cohesive and adhesive forces in the two fluids. This difference, in turn, depends on the thermal energy of the molecules. Within a phase, cohesive forces acting on a molecule are statisticaly in equilibrium. At the interface between different phases (liquid and its vapor, liquid and gas, immiscible liquids, liquid and solid body, etc.), the equilibrium is disturbed because of the unbalance between cohesive and adhesive forces of molecules of different type, or because of a difference in the thermal excitation level of the molecules at the

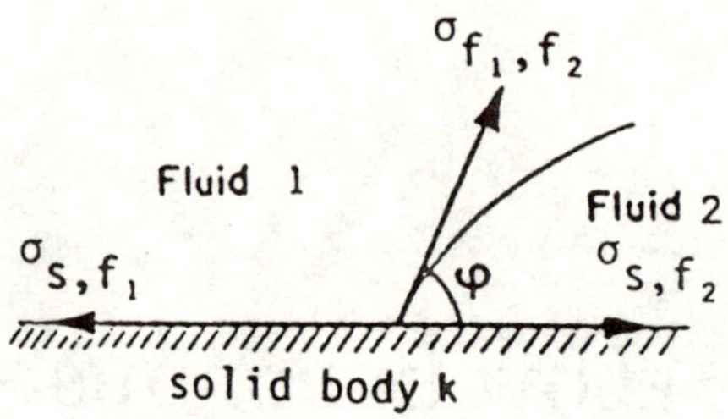

Figure 2.1.1: Equilibrium of the interfacial tensions at the contact point of two fluids at the surface of a solid body.

interface. The unequilibrated forces manifest themselves as *surface-* or *interfacial tension*, $\sigma_{i,j}$, at the interface between the two phases, i, j. If the fluids are immiscible, this tension is always positive. The interfacial tension, $\sigma_{i,j}$, is tangential to the interface and has the dimensions of Force/Length, or Work/Surface area.

In order to increase the interfacial surface, A, by an amount dA, a certain amount of work, dw, has to be permormed, with

$$dw = \sigma_{i,j} dA.$$

This work must be supplied by an external force. It increases the free energy of the system. A system is in equilibrium when its free energy is at a minimum. Left to itself, the interface between two immiscible phases will take on the minimum level. A small falling drop of a liquid will, therefore, have a nearly spherical shape. The shape of the interface depends on the geometrical form of the boundary of the interface. If this boundary is a spatial closed surface, the interface will have a saddle point.

When two immiscible liquids, or a liquid and a gas, are in contact with a solid body, the shape of the interface between the two fluids near the solid is influenced by the wetting angle at the solid. The magnitude of the wetting angle is determined by the equilibrium of the interfacial tensions at the contact point of the two fluids and the solid (Fig. 2.1.1):

$$\sigma_{f_1,f_2} \cos \phi = \sigma_{s,f_1} - \sigma_{s,f_2}. \qquad (2.1.1)$$

When $\sigma_{s,f_1} > \sigma_{s,f_2}$, $\phi < 90^o$, as

$$\cos\phi = \frac{\sigma_{s,f_1} - \sigma_{s,f_2}}{\sigma_{f_1,f_2}}.$$

The cosine of the wetting angle, ϕ, is equal to the ratio of the energy becoming free when liquid f_1 replaces f_2, to the energy gained by the system through the increase of the interface between the two fluids.

When $\sigma_{s,f_1} - \sigma_{s,f_2} = \sigma_{f_1,f_2}$, $\phi = 0^o$.

When $\sigma_{s,f_1} - \sigma_{s,f_2} > \sigma_{f_1,f_2}$, no equilibrium can be established, and the f_2-fluid will tend to replace the f_1-fluid over the entire solid surface. In this case, the f_2-fluid is the wetting fluid, or phase, and the f_1-fluid the nonwetting one. The system tends again to a state of minimum free energy.

The wetting angle, ϕ, is very sensitive to impurities, especially with respect to the solid surface. If a fluid is polar (water), the angle depends on whether or not the solid surface has been wetted by this fluid previously. The wetting angle shows hysteresis: the angle is greater when the fluid advances than when it recedes.

2.1.2 Interfacial curvature and capillary pressure

When the interface between two immiscible fluids, f_1, f_2, is curved, the interfacial tension generates a pressure discontinuity, p_c, at any point, M, of the interface. We refer to this difference as *capillary pressure*

$$p_c = p_{f_1} - p_{f_2} = \sigma_{f_1,f_2} \left(\frac{1}{R_1} + \frac{1}{R_2} \right). \tag{2.1.2}$$

Introducing a mean curvature, R_M, of the interface at M, defined by

$$\frac{1}{R_1} + \frac{1}{R_2} = \frac{2}{R_M},$$

equation (2.1.2) yields

$$p_c = p_{f_1} - p_{f_2} = 2\frac{\sigma_{f_1,f_2}}{R_M}, \tag{2.1.3}$$

in which σ_{f_1,f_2} is positive and the effective radius of curvature, R_M, is positive when the center of mean curvature is located in the f_1-fluid. Thus, the pressure is greater on that side of the interface on which the center of mean curvature is located.

When p_c is constant over an interface, then R_M is also constant. From this it follows that in a circular capillary tube with radius r, the interface is spherical and $R_M = r/\cos\phi$, so that

$$p_c = 2\frac{\sigma_{f_1,f_2}}{r}\cos\phi.$$

The above considerations are based on the assumption that the fluids are at rest. However, it can be shown that they also are valid for nonequilibrium conditions (moving interface) if

$$\left|\frac{(\mu_{f_1} - \mu_{f_2})\,u}{\sigma_{f_1,f_2}}\right| \ll 1, \tag{2.1.4}$$

where μ_{f_1}, μ_{f_2} are the viscosities of the fluids and u is the velocity of the moving interface, with respect to a fixed coordinate system. This condition is usually satisfied.

Example: f_1 : Water, f_2 : air at 20°C.

$$\mu_w = 10^{-3}\text{N. s}, \quad \mu_a = 1.8 \times 10^{-5}\text{N. s}/\text{m}^2, \quad \sigma_{w,a} = 7.3 \times 10^{-2}\text{N}/\text{m},$$

$$u = 10^{-2}\text{m}/\text{s}, \quad \frac{(\mu_{f_1} - \mu_{f_2})\,u}{\sigma_{f_1,f_2}} = 1.35 \times 10^{-4}.$$

2.1.3   Equilibrium between a liquid and its vapor

Kelvin's equation is:

$$p_c = \sigma_{l,v}\frac{2}{R_M} = \frac{RT}{v_l}\ln\frac{p_v}{p_{vo}}, \tag{2.1.5}$$

where
R :   universal gas constant,
T :   absolute temperature,
v_l :   specific volume of liquid $(= 1\text{g}/\rho_l)$,
p_{v_o} :   saturation vapor pressure over plane surface at T,
p_v :   saturation vapor pressure over curved surface at T.

In order for p_c to be positive, we require

$$\ln\frac{p_v}{p_{vo}} < 0, \quad \text{or} \quad \frac{p_v}{p_{v_o}} < 1.$$

Solving (2.1.4) for R_M, gives:

$$R_M = \frac{\sigma_{l,v}}{\sigma_l \sigma_l RT \ln(p_v/p_{vo})}.$$

Small changes in p_v produce large changes in R_M. If p_v is known, R_M is also determined.

Distribution of Immiscible Phases in a Porous Medium

In most practical problems of multiphase flow in porous media, we seek a continuum approach formulation. However, in order to perform the averaging over an REV correctly, the considered phenomena must first be studied at the microsopic level.

2.1.4 Microscopic domain

We consider a porous medium in which the pore space is interconnected. Furthermore, the dimensions of the pores in such a medium should be small enough, so that the orientation and form of the interfaces between two phases are mainly controlled by interfacial forces, and by the wetting properties of the solid phase. If different immiscible fluid phases are present, simultanously, in the pore space of such a porous medium, interfaces of complex geometry form between these phases. At every point $\mathbf{x}'$ of such an interface, a capillary pressure, $p_c(\mathbf{x}', t)$, exists according to the local curvature of the interface and the acting interfacial tension. Here, $\mathbf{x}'$ denotes the position vector in a fixed Cartesian coordinate system. This vector is determined only inside the pore space of the medium. In the microscopic domain, the interfaces are part of the boundaries of the individual phases and the values of $p_c(\mathbf{x}', t)$ are boundary ones.

2.1.5 Macroscopic space

We now introduce an REV centered at $\mathbf{x}$. The position vector $\mathbf{x}$ is defined in the entire space occupied by the porous medium in which the multiphase flow is studied. The porosity is the volume, $U_{o,v}$, of the void space in the REV divided by the volume, U_o of the REV, i.e.

$$n(\mathbf{x}) = \frac{U_{o,v}}{U_o}. \tag{2.1.1}$$

The volumetric fraction, or content, θ_{α_i}, of an α_i-phase in an REV, is given by

$$\theta_{\alpha_i}(\mathbf{x}) = \frac{U_{o,\alpha_i}}{U_o}, \tag{2.1.2}$$

where

$$U_{o,\alpha_i} = \int_{U_o} \gamma_{\alpha_i}(\mathbf{x}',t)\,dU, \qquad \gamma_{\alpha_i}(\mathbf{x}',t) = \begin{cases} 1 & \text{when } \mathbf{x}' \text{ is in the } \alpha_i - \text{phase,} \\ 0 & \text{elsewhere.} \end{cases}$$

Here the porous medium is assumed to be incompressible.

Normalizing the volumetric content of an α_i-phase with porosity, n, gives the degree of saturation, S_{α_i}, for the α_i-phase

$$S_{\alpha_i}(\mathbf{x},t) = \frac{\theta_{\alpha_i}}{n} = \frac{U_{o,\alpha_i}(\mathbf{x},t)}{U_{o,v}(\mathbf{x})}. \tag{2.1.3}$$

It is readily seen that when the number of phases is N

$$\sum_1^N \theta_{\alpha_i} = n, \qquad \sum_1^N S_{\alpha_i} = 1. \tag{2.1.4}$$

We now consider one distinct phase, α_k. At the interfaces formed within the pore space between this phase and any of the remaining phases, a capillary pressure, p_c, occurs, and the local effective radius of curvature of the interface, according to (2.1.3), is

$$R_{\mathrm{M}} = 2\frac{\sigma_{\alpha_k,\alpha_i}}{p_c}.$$

To every R_{M}, we assign a fraction $dU_o(R_{\mathrm{M}})/U_o$ of the pore space in the REV. Let the α_k-phase displace one of the other phases, say, α_i. The cumulative distribution function of the pores for this pair of phases is then given by

$$f_{R_{\mathrm{M}}} = \theta_{\alpha_i}(R_{\mathrm{M}}) = \frac{1}{U_o}\int_{R_{\mathrm{M}}}^{\infty} dU_{\alpha_i}(R_{\mathrm{M}}),$$

with

$$f(0) = n = \frac{1}{U_o}\int_0^{\infty} dU_{\alpha_i}(R_{\mathrm{M}}).$$

Normalized with n, this gives

$$S_k(R_{\mathrm{M}}) = 1 - S_{\alpha_i}(R_{\mathrm{M}}).$$

In the integrals appearing in these formulas, the lower limit, R_M, is a function of p_c:

$$\text{for } R_M \to 0, \quad p_c \to \infty, \quad \text{and} \quad \begin{aligned} \theta_{\alpha_k}(R_M) &\to \theta_{\alpha_k}(p_c), \\ \text{for } R_M \to \infty, \quad p_c \to 0, \quad\quad\quad S_{\alpha_i}(R_M) &\to S_{\alpha_k}(p_c), \end{aligned}$$

Thus, for every pair of phases, there exists a relation between the volumetric content, or the degree of saturation, and the capillary pressure. These relations are continuous functions of $\mathbf{x}$ in the macroscopic space. The same holds true for $p_c = p_c(\theta_{\alpha_k})$ and $p_c = p_c(S_{\alpha_k})$. Capillary pressure is uniquely related to the effective radius of curvature of the menisci forming at the interfaces between two immiscible fluids in the pore space of the porous medium within an REV. It follows that at a given volumetric content of an α_k-phase, the menisci formed between the pair of considered phases have the same effective radius of curvature.

2.1.6 Phase distribution in the pore space

The distribution of immiscible phases in the void space of a porous medium, depends on the wettability of its solid matrix.

Consider first two fluid phases: a wetting phase and a nonwetting one. In this case

$$p_c = p_{nw} - p_w > 0, \quad \text{or} \quad p_{nw} > p_w.$$

Figure 2.1.2 gives a schematic presentation of the distribution of water and air, as wetting and non-wetting fluids, respectively, in a porous medium. The pressure in the air is assumed constant and equal to the atmospheric one, at least as long the space occupied by air is continuous. The pressure in the water is then equal to minus the capillary pressure and, thus, smaller than atmospheric. In Fig. 2.1.2, the capillary pressure head, $h_c = p_c/\rho_w g$, is introduced. With increasing capillary pressure head, the degree of water satuation decreases. At low capillary pressure, air occurs in 'insular' form, i.e., as bubbles completly surrounded by water (entrapped air). At this stage, the pressure in the air bubbles is determined by the size of the bubbles and by the pressure in the surrounding fluid. With increasing capillary head, the air phase becomes continuous and the pressure in it is atmospheric. As long as a phase is continuous, i.e., *funicular*, pressure forces are transmitted through it. At high capillary head, the wetting fluid, here the water, ceases to be continuous and occupies only the angular spaces formed, for example,

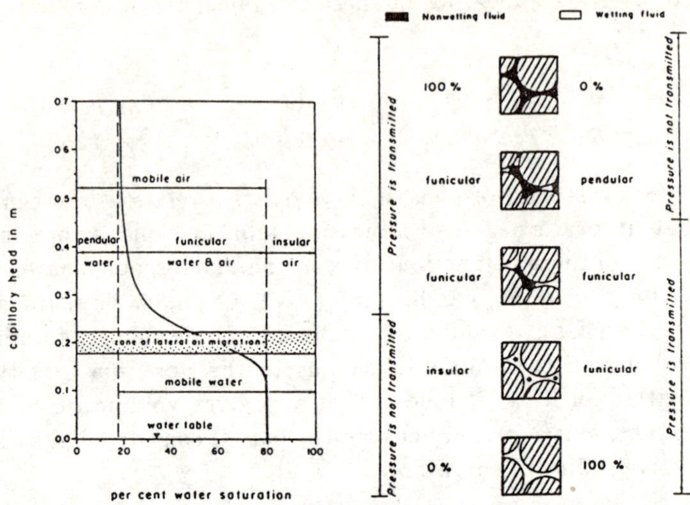

Figure 2.1.2: Distribution of a wetting (water) and a non-wetting (air) fluids in the pore space of a porous medium.

at the contact of individual grains, in a granular porous medium. We then say that the wetting fluid occurs in *pendular* form. We shall call the degree of saturation of the nonwetting fluid at which it first becomes continuous the *insular residual* degree of saturation. . The degree of saturation of the weting fluid at which it first ceases to be continuous we call *pendular residual* degree of saturation. These two residual degrees of saturation are not significantly altered when the capillary pressure is increased or decreased, respectively, beyond the values corresponding to these two limits.

A consequence of this is that when a porous medium is wetted by a liquid, without previously evacuating the pore space, the maximum saturation reached is always less then 100%, mostly near 80%, as the insular residual degree of saturation for air is close to 20%. Similarly, drainage of a porous medium cannot reduce the content of the weting fluid significantly below the pendular residual degree of saturation, i.e., below 15%.

When more than two fluids, say water, oil and air, are involved, the behaviour is somewhat more complex. If the porous matrix consists of *hydrophilic material*, water is the wetting (w) phase, both in the presence of oil and air. Oil is nonwetting in the presence of water and wetting in presence of air (wnw). Air is usually the nonwetting phase with respect to both of the other liquids (nw). If the solid matrix is *oliophilic*, the behaviour of water and oil are interchanged.

The ambivalent behaviour of one of the three fluids, has some consequences in the residual degree of saturation of this fluid. When oil is drained out of a medium in which water and air are also present, one has to distinguish two possible cases. In the part of the medium in which oil is replaced by water, only the residual degree of oil saturation is insular and close to 20%. In the parts in which oil is mainly replaced by air, but water is also present, the pendular residual degree of oil saturation can become very low, say of the order 1%, or less, because the angular spaces between the grains are occupied by water and the residual oil forms, more or less, thin films covering the water menisci.

In some cases, a continuous thin film of a wmw-phase may separate the wetting phase from the non-wetting one. As shown already in 1890 by Rayleigh, such films consisting of oil are continuous if they are not thinner then 25×10^{-7}mm. In this case, for all practical purpose, the wnw-phase can be neglected. Its influence is, however, sensed in the interfacial tension which is reduced to the value of the interfacial tension between the wnw-phase and the nw-phase. This may considerably alter the relations between $p_{c,w}$, $k_{r,w}$ and S_w, especially when this relations correspond to drainage.

If the wettability order, w–wnw–nw, is valid, it follows that no interface between phases w and nw exist as long as the wnw-phase is continuous. The residual saturations in a three phase system are presented in Fig. 2.2.1, showing the irregular behaviour of the wnw-phase. Notice that in the non-hatched region, all three phases are mobile.

2.2   Flow Equations for Immiscible Fluids

In writing the equations for flow of immiscible fluids in a porous medium, the following assumptions are made:

- There are no acting osmotic, or thermal potentials.

- All continuous phases flow simultaneously.

- Every single phase follows its own tortuous path in the pore space.

- The interfaces between phases follow the wettability order: w - wnw - nw.

- There is no transport of a vapor phase in the gaseous phase.

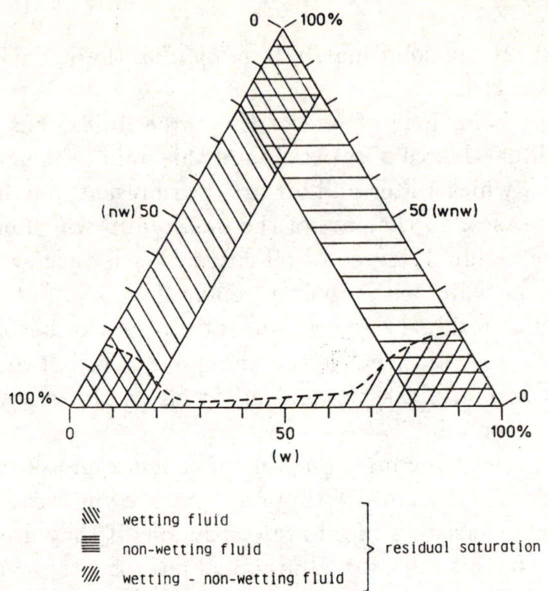

Figure 2.2.1: Tentative distribution of the residual degrees of saturation in a three fluid phase system.

Based on the macroscopic space concept, overlapping continua are introduced, one for each fluid phase involved. Darcy's law for an α_i-phase in an isotropic porous medium is

$$
\begin{aligned}
\mathbf{q}_{\alpha_i} &= -\frac{k_{\alpha_i}(S_{\alpha_i})}{\mu_{\alpha_i}}(\nabla p_{\alpha_i} + \rho_{\alpha_i} g \nabla z) \\
&= -k_{r,\alpha_i}(S_{\alpha_i})\frac{k}{\mu_{\alpha_i}}(\nabla p_{\alpha_i} + \rho_{\alpha_i} g \nabla z).
\end{aligned}
\tag{2.2.1}
$$

When ρ_{α_i} is a constant, and the porous medium is isotropic, these equations can be rewritten in the form

$$
\begin{aligned}
\mathbf{q}_{\alpha_i} &= -k\frac{k_{r,\alpha_i}(S_{\alpha_i})}{\mu_{\alpha_i}}\nabla(p_{\alpha i} + \rho_{\alpha_i} g z) \\
&= -\frac{\rho_{\alpha_i} g}{\mu_{\alpha_i}}k_{r,\alpha_i}(S_{\alpha_i})\nabla\left(z + \frac{p_{\alpha_i}}{\rho_{\alpha_i} g}\right).
\end{aligned}
\tag{2.2.2}
$$

The volumetric flux of the α_i-phase, $\mathbf{q}_{\alpha_i}$, is related to the average velocity of this phase, $\mathbf{V}_{\alpha_i}$, by

$$
\mathbf{q}_{\alpha_i} = \theta_{\alpha_i}\mathbf{V}_{\alpha_i} = n S_{\alpha_i}\mathbf{V}_{\alpha_i}.
\tag{2.2.3}
$$

In these equations, μ_{α_i}, p_{α_i} and $p_{\alpha_i}/\rho_{\alpha_i}g$ are the viscosity the pressure and the pressure head in the α_i-phase, respectively, and $k_{\alpha_i}(S_{\alpha_i})$ is the *effective permeability* of the porous medium to the α_i-phase. It depends on the degree of saturation, S_{α_i}.

For an anisotropic porous medium, the effective permeability is a second order tensor. Each component of this tensor may have a different functional relationship to saturation.

In (2.2.2) we have introduced the relative permeability, $k_{r,\alpha_i}(S_{\alpha_i})$. This is a scalar function of S_{α_i}, defined by

$$k_{r,\alpha_i} = \frac{k_{\alpha_i}(S_{\alpha_i})}{k|_{\text{sat}}}. \qquad (2.2.4)$$

Here, $k|_{\text{sat}}$ is the *intrinsic permeability*, or *permeability* at full saturation, of the porous medium. Note that this is a property of the porous medium, and does not depend on the properties of the fluid flowing in it. For nonhomogeneous porous media, the intrinsic permeability is space dependent.

If N phases flow simultaneously in a porous medium, we write N equations of the type of (2.1.10), or (2.2.2). The pressure in the α_k-phase is then related to the pressure in any other α_i-phase, having an interface with α_k, through the capillary pressure, $p_c = p_{\alpha_i} - p_{\alpha_k}$. The equations are,, thus, not independent of each other. If the gaseous phase is air in contact with the atmosphere, or a gas connected to a reservoir in which the pressure is constant, the pressure drop in the air, or gas, due to the flow of the liquid phases, is very small and does not affect significantly the flow of these phases. In this case, the equation for the air, or the gas, can be neglected, and the number of flow equations is reduced by one. However, the degree of air or gas saturation in (2.1.4), must still be taken into account.

2.3 Mass Balance Equations

The change per unit of time of the mass of an α_i-phase, contained in a unit volume of a porous medium, is equal to the difference of the mass flux of the phase into and out of the unit volume, plus the mass added, or extracted, per unit time by sources and sinks located inside the unit volume. Noting that the mass, m_{α_i}, of the α_i-phase contained in a unit volume is $m_{\alpha_i} = \rho_{\alpha_i} n S_{\alpha_i}$, one can express this balance in the form

$$\frac{\partial \rho_{\alpha_i} n S_{\alpha_i}}{\partial t} = -\nabla \cdot (\rho_{\alpha_i} \mathbf{q}_{\alpha_i}) + \sum \rho_{\alpha_i} s_{\alpha_i}. \qquad (2.3.1)$$

For a nondeformable solid matrix, (2.3.1) reduces to

$$n\frac{\partial \rho_{\alpha_i} S_{\alpha_i}}{\partial t} = -\nabla \cdot (\rho_{\alpha_i} \mathbf{q}_{\alpha_i}) + \sum \rho_{\alpha_i} s_{\alpha_i}. \qquad (2.3.2)$$

In these equations, s_{α_i} is the rate of volumetric flux of sources or sinks, per unit volume of soil, e.g., due to uptake of water by the roots of plants.

When ρ_{α_i} is constant and the solid matrix nondeformable, (2.3.2) becomes

$$n\frac{\partial S_{\alpha_i}}{\partial t} = -\nabla \cdot \mathbf{q}_{\alpha_i} + \sum s_{\alpha_i}. \qquad (2.3.3)$$

Inserting the flow equation into the mass balance equation, one gets

$$\frac{\partial n S_{\alpha_i}}{\partial t} = \nabla \cdot \left[\frac{k_{\alpha_i}}{\mu_{\alpha_i}} (\nabla p_{\alpha_i} + \rho_{\alpha_i} g \nabla z) \right] + \sum s_{\alpha_i}, \qquad (2.3.4)$$

or for constant ρ_{α_i} and n

$$n\frac{\partial S_{\alpha_i}}{\partial t} = \nabla \cdot \left[\frac{k_{\alpha_i} \rho_{\alpha_i} g}{\mu_{\alpha_i}} \nabla \left(z + \frac{p_{\alpha_i}}{\rho_{\alpha_i} g} \right) \right] + \sum s_{\alpha_i}. \qquad (2.3.5)$$

For an isotropic porous medium, we may write (2.3.5) in the form

$$n\frac{\partial S_{\alpha_i}}{\partial t} = \nabla \cdot \left[k_{r,\alpha_i} \frac{k \rho_{\alpha_i} g}{\mu_{\alpha_i}} \nabla \left(z + \frac{p_{\alpha_i}}{\rho_{\alpha_i} g} \right) \right] + \sum s_{\alpha_i}. \qquad (2.3.6)$$

For a homogeneous isotropic porous medium and constaant ρ_{α_i}, the permeability at saturation, $k \equiv k|_{\text{sat}}$, is not a function of space, and we can write $gk\rho_{\alpha_i}/\mu_{\alpha_i}$ in front of the nabla operator in (2.3.6). In contrast, k_{r,α_i}, being a function of $S_{\alpha_i}(x,y,z,t)$, is always a function of space. In an isotropic porous medium, k is a scalar. If the medium is also homogeneous it becomes a constant. Under all circumstances, the above equations are nonlinear, because k_{r,α_i} is always a function of S_{α_i}. To solve these equations, it is necessary to know the relations $p_{c,\alpha_i} = f(S_{\alpha_i})$, and $k_{r,\alpha_i} = f(S_{\alpha_i})$, or, by virtue of the first of these relations, $k_{\alpha_i} = f(p_{c,\alpha_i})$.

If only two phases are flowing through a porous medium, with the wetting phase being a liquid (water), and the nonwetting phase a gas (air) in contact with a large reservoir of constant pressure (the atmosphere), a single equation, for the wetting fluid, is sufficient to describe the flow. For an isotropic porous medium, we then write

$$\frac{\partial}{\partial t}(\rho_w n S_w) = \nabla \cdot \left[\frac{k}{\mu_w} k_{r,w} (\nabla p_w + \rho_w g \nabla z) \right] + \sum \rho_w s_w.$$

Note that if $p_w = -p_c$ is a unique relation of S_w, then ∇p_w can be replaced by $-(\partial p_c/\partial S_w)\nabla S_w$. The subscript w refers to the wetting phase, and S_w and $k_{r,w}$ are functions of p_w. Hence, we can write

$$\left[S_w\frac{\partial(\rho_w n)}{\partial p_w} + n\rho_w\frac{\partial S_w}{\partial p_w}\right]\frac{\partial p_w}{\partial t}$$

$$= \nabla\cdot\left[\frac{k}{\mu_w}k_{r,w}(\nabla p_w + \rho_w g\nabla z)\right] + \sum\rho_w s_w. \qquad (2.3.7)$$

Usually

$$S_w\frac{(\rho_w n)}{\partial p_w} \ll n\rho_w\frac{\partial S_w}{\partial p_w}.$$

If, in addition, ρ_w is constant, this equation simplifies, in the absence of sources or sinks, to

$$n\frac{\partial S_w}{\partial p_w}\frac{\partial p_w}{\partial t} = \nabla\cdot\left[\frac{k}{\mu_w}k_{r,w}\nabla\left(z + \frac{p_w}{\rho_w g}\right)\right]. \qquad (2.3.8)$$

This equation describes saturated and unsaturated flow of the wetting phase in an isotropic porous medium. In an anisotropic porous medium we replace $k\,k_{r,w}$ by k_w.

For saturated flow: $p_w \geq 0$, and $S_w = 1$, $k_{r,w} = 1$.

For unsaturated flow: $p_w = -p_{c,w} < 0$, and $S_w = f_1(p_{c.w})$, $k_{r,w} = f_2(p_{c,w})$.

When the nonwetting fluid is air at atmospheric pressure, $p_w = -p_c$, and $p_w/\rho_w g = -p_c/\rho_w g = \psi$, which the soil scientists call the *capillary potential*.

2.4 Simultaneous Flow of Two Fluids Having a Small Density Difference

The equations describing the flow of two phases, a wetting phase and a nonwetting one, are

$$\mathbf{q}_w = -\frac{k_w}{\mu_w}\nabla(p_w + \rho_w g z),$$

$$\mathbf{q}_{nw} = -\frac{k_{nw}}{\mu_{nw}}\nabla(p_{nw} + \rho_{nw} g z),$$

with

$$\nabla p_c = \nabla p_{nw} - \nabla p_w.$$

Assuming that ρ_w and ρ_{nw} are constant, these equations may be combined to give

$$\nabla p_c = \mathbf{q}_w \frac{\mu_w}{k_w} - \mathbf{q}_{nw} \frac{\mu_{nw}}{k_{nw}} - (\rho_w - \rho_{nw})g\nabla z.$$

Introducing the total flux

$$\mathbf{q}_t = \mathbf{q}_w + \mathbf{q}_{nw},$$

we can write

$$\nabla p_c = \mathbf{q}_t \frac{\mu_w}{k_w} - \mathbf{q}_{nw} \left(\frac{\mu_w}{k_w} + \frac{\mu_{nw}}{k_{nw}} \right) - (\rho_w - \rho_{nw})g\nabla z, \qquad (2.4.1)$$

or

$$\mathbf{q}_{nw} = -f_{nw} \left(\frac{k_{nw}}{\mu_{nw}} \nabla(p_c + (\rho_w - \rho_{nw})gz) - \mathbf{q}_t \right), \qquad (2.4.2)$$

where

$$f_{nw} = \frac{1}{1 - \frac{\mu_{nw}k_w}{\mu_w k_{nw}}} \qquad (2.4.3)$$

is the fractional flow function.

Introducing (2.4.2) in the continuity equation

$$n\frac{\partial S_{nw}}{\partial t} = \nabla \cdot \mathbf{q}_{nw},$$

we finally get

$$n\frac{\partial S_{nw}}{\partial t} = -\nabla \cdot \left\{ f_{nw} \left(\frac{k_w}{\mu_w} \nabla(p_c + (\rho_w - \rho_{nw})gz) - \mathbf{q}_t \right) \right\}. \qquad (2.4.4)$$

For the wetting phase, $\mathbf{q}_w = \mathbf{q}_t - \mathbf{q}_{nw}$, or

$$\mathbf{q}_w = \mathbf{q}_t(1 + f_{nw}) - f_{nw} \left\{ \frac{k_w}{\mu_w} \nabla(p_c + (\rho_w - \rho_{nw})gz) \right\}, \qquad (2.4.5)$$

and

$$n\frac{\partial S_{nw}}{\partial t} = -\nabla \cdot \left\{ \mathbf{q}_t(1 + f_{nw}) - f_{nw} \left[\frac{k_w}{\mu_w} \nabla(p_c + (\rho_w - \rho_{nw})gz) \right] \right\}. \qquad (2.4.6)$$

If $\rho_w \gg \rho_{nw}$, $\rho_w - \rho_{nw} \approx \rho_w$.

With these equations and the appropriate initial and boundary conditions, simultaneous flow of two immiscible fluids can be computed.

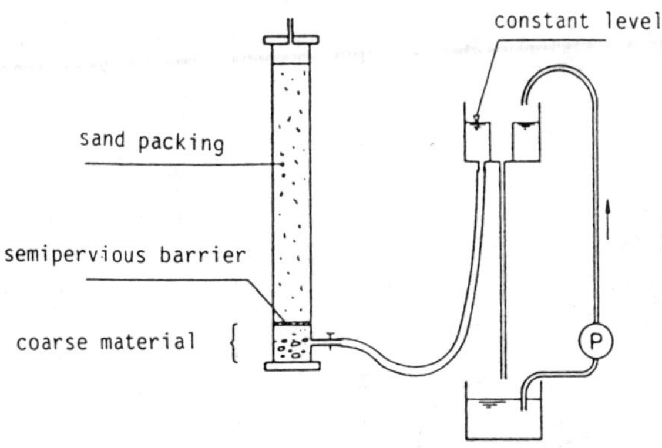

Figure 2.5.1: Schematic representation of a column experiment for the determination of $p_{c,w} = f(S_w)$.

2.5 Measurement of the relations $p_{c\alpha_i}(S_{\alpha_i})$, and $k_{r,\alpha_i}(S_{\alpha_i})$

The functions

$$p_{c,\alpha_i} = f_1(S_{\alpha_i}), \quad k_{r,\alpha_i} = f_2(S_{\alpha_i}), \quad \text{and} \quad k_{r,\alpha_i} = f_3(p_{c,\alpha_i})$$

can be measured in the laboratory by means of artifical packings in column experiments. The measurement of these functions in natural porous media under undistrubed conditions in situ are much more difficult.

A static equilibrium experiment for the determination of the capillary pressure versus degree of saturation curve for a wetting fluid, say water, and a nonwetting fluid, say air, is shown in Fig. 2.5.1.

The water content, and thus the degree of saturation, is usually measured with a γ-ray attenuation probe. The capillary pressure head is given by the elevation of the section in which the water content is measured above the free water table. Equilibrium conditions are assumed in the column. Such an experiment can be performed in the following way. First, we completly saturate the column with the wetting fluid and let it drain until equilibrium is reached (first drainage). After equilibrium is reached, we determine the $p_{c,w}$–S_w-relation. Another way to perform the experiment is to take an unsaturated column and let the wetting fluid rise in this column until equilbrium is reached by wetting, or imbibition. Then we determine the relation between $p_{c,w}$ and S_w. We further increase the capillary pressure by raising

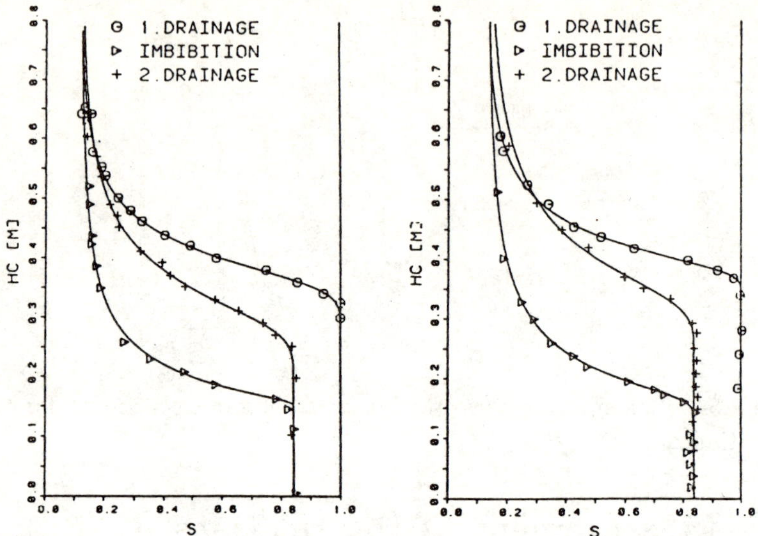

Figure 2.5.2: Capillary head as a function of degree of saturation for first drainage, wetting and second drainage (boundary drying and wetting curves)

the column. The column is then drained until equilibrium is again reached (second drainage). Figure 2.5.2 shows typical results of such measurements.

The relation between capillary pressure and degree of saturation is not unique. It shows *hysteresis*. The difference between the curves for first drainage, starting from a fully saturated column, and the second drainage, is due to the entrapment of the nonwetting phase, as was explained previously. If the entrapped phase is to some extent soluble in the wetting one, and the latter is flowing, the entrapped phase can be removed with time and complete saturation may be reached again.

The pendular residual saturation is the same for all three curves, as one would expect, and gives the *irreducible water content*. The physical causes for the hysteresis are:

- The hysteresis of the wetting angle.

- The pore-space geometry.

Individual pores are interconnected through throats. The order of change of the mean diameter from a pore to a throat is of order 2 to 1. This change occurs over a distance of the order of a pore diameter. The mean curvature

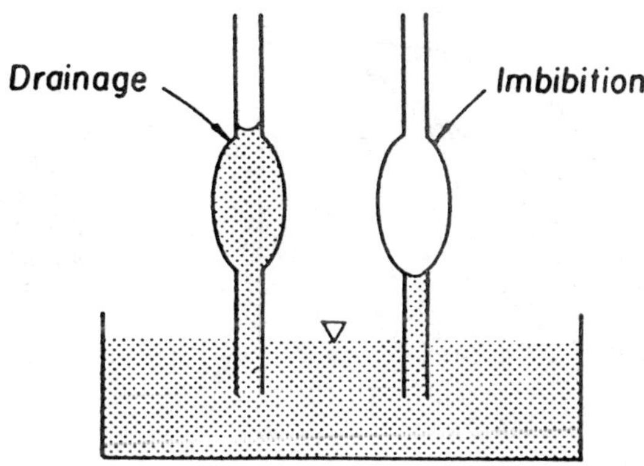

Figure 2.5.3: Capillary hysteresis due to geometry.

of the interface of two immiscible fluids is of the order of the local mean diameter of the pore space and the capillary pressure is proportional to the mean curvature of the interface. Usually the corresponding capillary pressure head, ψ, is much larger than the pore diameter, d. Thus small displacements of the interface of the order of a pore diameter, d, can produce a change in the capillary pressure, ψ, of order 1 to 2 (Fig. 2.5.3).

The hysteretic behaviour of the relation $p_c = f(S_w)$ has the consequence that the actual pressure-saturation values at a point depend on history. In fact, any combination of these values located inside the domain bound by the boundary drying and wetting curves, can occur. Reversals from a wetting process to a drying process and vice versa, occuring at a point along a boundary curve, produce primary drying or wetting curves respectively. Reversals along primary curves produce corresponding scanning curves (Fig. 2.5.4).

Brooks and Corey (1966) showed that the relation between relative permeability of a wetting phase and the capillary pressure also exhibits hysteresis (Fig. 2.5.5).

Contrary to the above reference, measurements by Riedi and Stauffer (1982) in fine sands show that the hysteresis in the degree of saturation-relative permeability relation is not significant (Fig. 2.1.4).

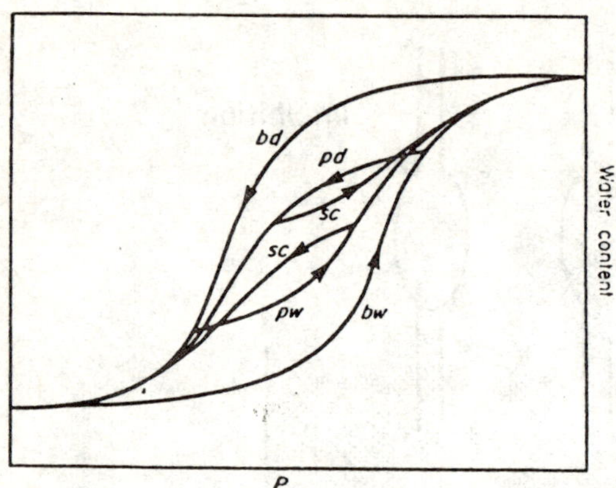

Figure 2.5.4: Hysteresis in the $p_c = f(S)$ relation after Childs.

The existance of hysteresis in the relationships between p_c and S and p_c and k_r considerably complicate the calculations of multiphase flow in porous media when reversals in the change of the capillary pressure occur.

2.6 Mathematical descripton of the relations between $p_{c,w}$, S_w, and $k_{r,w}$

When only a wetting fluid and nonwetting one are flowing through a porous medium, it is convenient for further use in computational procedures to represent the relations between $p_{c,w}$ and S_w, as well as between $k_{r,w}$ and S_w by simple mathematical functions. As only the wetting phase will be considered in the following, the subscript w will be dropped for convenience. Brooks and Corey (1966) find for the first drainage curve, the empirical relation:

$$S_e = \left(\frac{p_b}{p_c}\right)^{\lambda} \quad \text{for} \quad p_c \geq p_b. \tag{2.6.1}$$

In this relation, S_e is the *effective saturation* of the wetting fluid

$$S_e = \frac{S - S_r}{1 - S_r}, \tag{2.6.2}$$

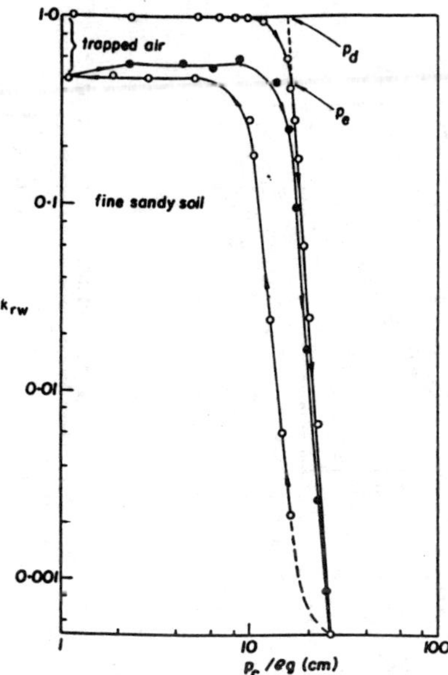

Figure 2.5.5: Hysteresis of the capillary pressure-relative permeability relation (after Brooks and Corey , 1966).

and S_r is the *pendular residual saturation* of the wetting fluid, p_b is the *bubling pressure*, that is to say, the capillary pressure at which the nonwetting fluid first enters the probe. Typically, the values of the exponent λ are around 2. Poorly graded sands have λ values greater then 2, sometimes even greater than 5.

Based on the works of Su and Brooks (1976), Riedi and Stauffer (1982) developed a relation which allows to describe the boundary curves in a more general form

$$S_e = p_b \left[\frac{S_w - S_r}{1 - S_r} \right]^{-\frac{1}{\lambda}} \cdot \left[\frac{S_m - S}{S_m - S_r} \right]^{\delta}, \qquad (2.6.3)$$

where S_m is the maximum degree of saturation of the wetting fluid. The parameters p_b, λ and δ are determined by a best fit procedure.

When hysteresis becomes important, a prediction of primary or scanning curves for drying or wetting becomes necessary. Childs (1969) introduced the concept of independent domains to solve this problem. One of the most successful approaches to this problem was developed by Mualem (1974). It enables the prediction of primary and scanning curves from known boundary drying and wetting curves only. The resulting expressions are simple. For example, a primary wetting curve which is initiated by a drop of the capillary

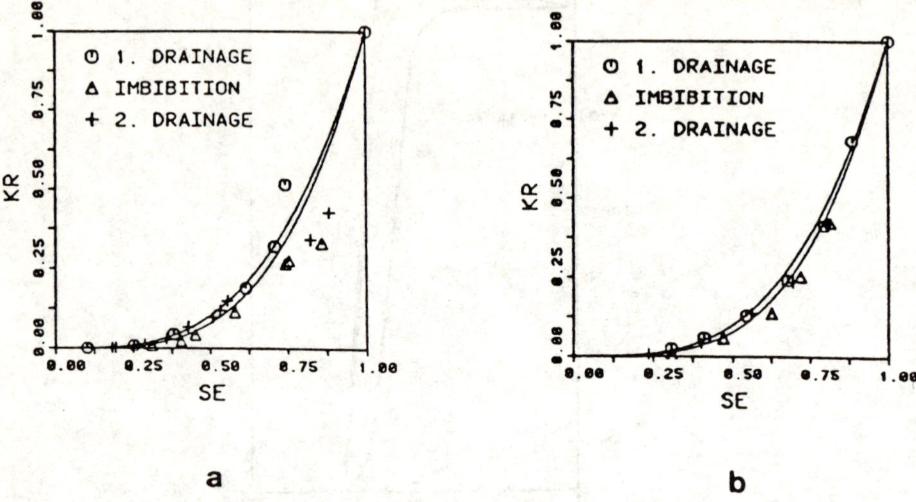

Figure 2.5.6: Measurements of the relative permeability as a function of wetting phase saturation (a) in coarse sand and (b) in fine sand (after Stauffer).

pressure at a value p_{c1}, to which corresponds a degree of saturation on the boundary drying curve $S_D(p_{c1})$, is given by

$$S_P(p_c) = S_I(p_c) + [S_m - S_I(p_c)] \cdot \frac{S_D(p_{c1}) - S_I(p_{c1})}{S_m - S_I(p_{c1})},$$

where p_c is the actual capillary pressure. The indices I, D and P indicate boundary wetting (imbibition) curve, boundary drying curve and primary wetting curve, respectivelly.

Figure 2.6.1 shows the comparison of experimental results with predictions made by Mualem's method.

Taking into account the uncertainties of the experimental determination of primary and scanning curves, the prediction is very satisfactory in this case. Most of the investigators find for the relation between relative permeability and degree of saturation an exponential law

$$k_r(S) = (S_e)^\varepsilon, \tag{2.6.4}$$

where the exponent, ε has usualy a value between 3 and 4. Brooks and Corey (1966) relate ε to λ by

$$\varepsilon = 3 + \frac{2}{\lambda}.$$

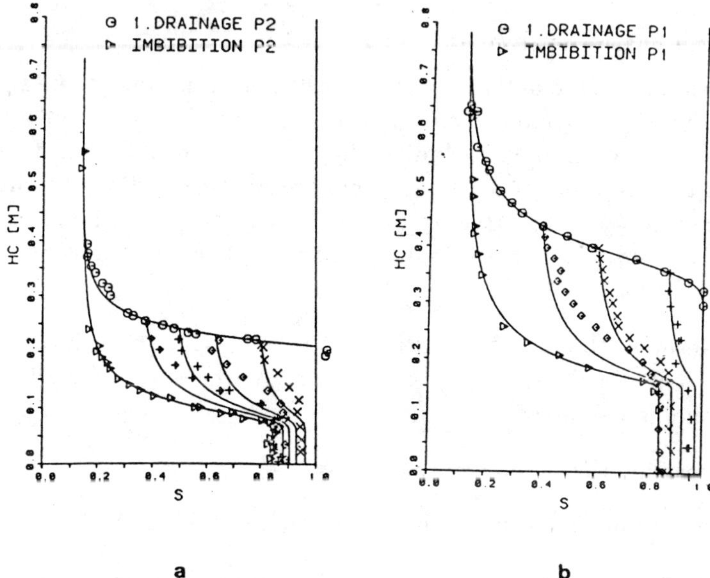

Figure 2.6.1: Primary wetting curves measured and computed by Mualem's method: (a) fine sand, (b) coarse sand.

2.7 Complete Statement of Multiphase Flow Problems

The procedure is the same as in single phase flow.

- Determine flow domain and boundaries.

- Choose independent and dependent variables. This step is important. First, one must decide if it is necessary to calculate the simultaneous flow of all phases, or of only the wetting one. According to this decision, one or more equations will be required. If only the wetting phase is considered, in a second step one has to decide if the flow of the wetting fluid must be calculated in both the saturated and unsaturated domains, simultaneously. This decision will determine the dependent variables to be employed, e.g., $S_{\alpha_i}(\mathbf{x},t)$, or $p_{\alpha_i}(\mathbf{x},t)$.

- Choose the appropriate equation.

- Determine all physical parameters needed. Here it is necessary to determine in the laboratory, or in situ, the relations

$$S_{\alpha_i} = S_{\alpha_i}(p_{\alpha_i}), \quad k_{r,\alpha_i} = k_{r,\alpha_i}(S_{\alpha_i}), \quad \text{or} \quad k_{r,\alpha_i} = k_{r,\alpha_i}(p_{\alpha_i}),$$

for all phases involved.

Then, one has to determine whether hysteresis is going to be significant in the problem under consideration. If yes, then the above relations must be determined including, hysteresis effects, that is to say, one has to determine at least the boundary wetting and drying curves and the corresponding curves in the $k_{r,\alpha_i} = k_{r,\alpha_i}(p_{\alpha_i})$ relation, if nessecary. This enables to use an approach like Mualem's in the computational procedure.

- Determine the initial conditions, if needed, i.e., the values of the dependent variables, $S_{\alpha_i}(\mathbf{x})$, $p_{\alpha_i}(\mathbf{x})$, etc. at time $t = 0$, for all α_i-phases in the flow domain.

- Determine the boundary conditions.

There are different types of boundary conditions.

(a) **Boundary conditions of the first type (Dirichlet boundary conditions).** These occur at boundaries on which p_{α_i}, or S_{α_i} is prescribed.

Example. At the surface of a porous medium, a liquid a is ponded. This liquid can be a wetting or a nonwetting one. The pressure head is given by the depth h_{α_i} of the ponded liquid, and the pressure is $p_{\alpha_i} = \rho_{\alpha_i} g h_{\alpha_i}$.

(b) **Boundary conditions of the second type (Neuman boundary conditions)** These occur at boundaries on which the flux $\mathbf{q}_{\alpha_i}$ is prescribed.

Let us consider examples.

Example 1. An impervious boundary: $q_{n,\alpha_i} = 0$, where q_{n,α_i} is the component of flux normal to the boundary.

Example 2. Infiltration of phase a at prescribed rate $\mathbf{q}_{\alpha_i}$. In case a wetting fluid is infiltrating, the rate at which this fluid infiltrates can be computed from one of the equations developed above, e.g.,

$$\mathbf{q}_w = -\frac{k_w(S_w)}{\mu_w}\nabla(p_w + \rho_w g z), \qquad (2.7.1)$$

or with

$$K_w(S_w) = \frac{k_w(S_w)\rho_w g}{\mu_w},$$

$$\mathbf{q}_w = -K_w(S_w)\nabla\left(z + \frac{p_w}{\rho_w g}\right).$$

- For $(p_w/\rho_w g) \geq (p_c/\rho_w g)$, we have $K_w = K_w(S_m)$, where S_m is the maximum degree of saturation which, as shown, is not necessarily equal to one.

- For $(p_w/\rho_w g) < (p_c/\rho_w g)$, we have $K_w = K_w(S)$, with $S < S_m$. In this case, K_w can be computed from

$$K_w = K_m\left(\frac{p}{p_c}\right)^{-(2+3\lambda)}$$

- If now the rate at which the wetting fluid infiltrates is $q_{inf} > K_w(S_m)$, $p_w/\rho_w g > 0$ at $z = 0$, and the infiltrating fluid is ponded. We then have a boundary condition of the first type.

- If $\nabla p_w = 0$, and the infiltration is vertical, $\nabla z = -1$, and $q_{inf} = K_w(S_w)$, and

$$q_{inf} > K_w(S_w), \quad \text{if} \quad \nabla\left(z + \frac{p_w}{\rho_w g}\right) < -1,$$

$$q_{inf} < K_w(S_w), \quad \text{if} \quad \nabla\left(z + \frac{p_w}{\rho_w g}\right) > -1,$$

(c) **A third type of boundary condition (Cauchy, or mixed type boundary condition)** occurs when a semipervious boundary is present.

(d) **Internal boundary conditions.** These are the conditions at the interface between two subdomains, I and II, at which a sudden change of the porous media characteristics, e.g., permeability, or porosity, occur. On such a boundary, the conditions

$$p_I = p_{II}, \quad \text{and} \quad q_{n,I} = q_{n,II},$$

must be satisfied. Here the subscripts I and II refer to the corresponding subdomains.

Example. Constant rate vertical infiltration of a wetting fluid in a layerd soil (Fig. 2.7.1).

Vertical infiltration at constant rate in homogeneous layers, implies that $\nabla p_w = 0$, $\nabla S_w = 0$ and $K_w(S_w) = $ const. At the interface between two layers

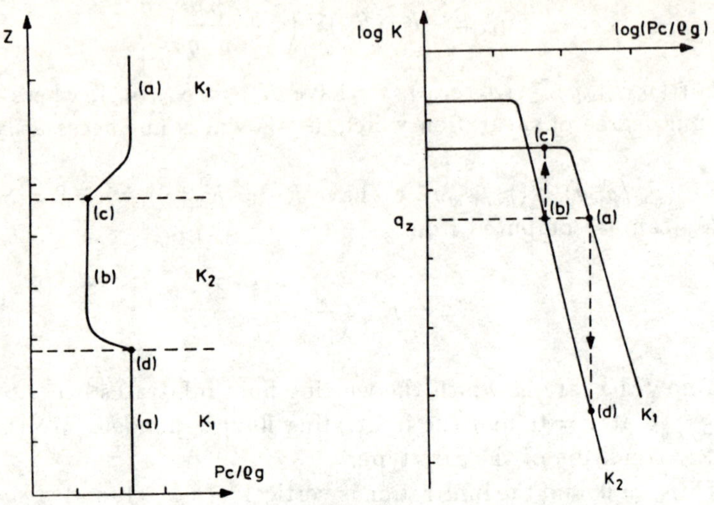

Figure 2.7.1: (a) degree of saturation– (b) Hydraulic conductivity– capillary pressure curves.

$p_I = p_{II}$. As shown in Fig. 2.7.1(b), the permeability of the two layers is not the same at the same capillary pressure, $p_c = -p_w$. From (2.7.1), we get

$$\frac{d\,(p_w/\rho_w g)}{d\,z} = 1 - \frac{q_z}{K_i}.$$

Because K_i is different in the two media, $d\,(p_w/\rho_w g)/d\,z$ is discontinous at the interface. As seen from Fig. 2.7.1(b), the permeability of the coarse material may be smaler than the permeability of the fine one for the pressure acting at the interface. The coarse material acts in this case as a less permeable barrier. The capillary pressure is reduced on the side of the fine material and the corresponding degree of saturation increases. The opposite happens when a fine layer underlies a coarser one.

2.8 Solute transport in multiphase flow through porous media

In the preceeding development, the transport of dissolved phases was not taken into account. If we accept that solution and volatility of one phase into another is weak, but not zero, transport of dissolved matter may be of interest. The balance equation for a component, γ in an α_i-phase is given by Bear in the previous chapter.

$$\frac{\partial(nS_{\alpha_i}\rho_{\alpha_i}c)}{\partial t} =$$
$$-\nabla\cdot[\rho_{\alpha_i}c\mathbf{q}_{\alpha_i} - nS_{\alpha_i}\{\mathbf{D}\cdot\nabla(\rho_{\alpha_i}c)\} - nS_{\alpha_i}\{\mathcal{D}^*\cdot\nabla(\rho_{\alpha_i}c)\}]$$
$$-f_{\alpha_i,\gamma} + nS_{\alpha_i}\rho_{\alpha_i}\Gamma_{\alpha_i,\gamma}. \tag{2.8.1}$$

In this equation c is the concentration ($=$ mass of γ per unit mass of α_i), $\mathbf{D}$ is the coefficient of mechanical dispersion, $\mathbf{D}^*$ is the coefficient of molecular diffusion, $f_{\alpha_i,\gamma}$ is the rate of transfer of γ across the boundaries of phase α_i, and $\Gamma_{\alpha_i,\gamma}$ is the rate of production (or decay) of γ in phase α_i. Each of the last two functions must be determined separatly according to the processes determing them. As flow of a phase in a multiphase system may become very slow, molecular diffusion cannot be always neglected. In fact, in some cases, it may even become dominant.

The balance equation for the α_i-phase must also be extended to include the fluxes of the α_i-phase and γ-component through the boundaries.

References

Brooks, R. H. and Corey, A.T. Properties of porous media affecting fluid flow.*J. Irr. and Drain., ASCE,* **92 (IR2)**:61-87, 1966.

Childs, E. C. *An Introduction to the Physical Basis of Soil Water Phenomena.* Wiley, London. 1969.

Mualem, Y. A conceptual model of hysteresis. *Water. Resour. Res.,* **10**:514-520, 1974.

Riedi, G. A. and Stauffer, F. *Hydraulische Charakteristika von Sandpackunqen mit Bercksichtigung des Hysteresiszyklus Drainage - Imbibition.* Report R17-82 IHW, ETH Zuerich, 1982.

Su, C. and Brooks, R. H. *Hydraulic Functions of Soils from Physical Experiments and Their Applications.* WRRI-41, Oregon State Univ., Corvallis, Oreg., 1976.

List of Main Symbols

A	Area.
c	Concentration of a component in a fluid phase (mass per unit mass).
$\mathbf{D}$	Coefficient of mechanical dispersion.
$\mathcal{D}^*$	Coefficient of molecular diffusion of a component in a porous medium.
g	Gravity acceleration.
$\mathbf{k}$	Permeability.
$\mathbf{k}_\alpha$	Effective permeability of an α-phase.
$\mathbf{K}$	Hydraulic conductivity.
n	Porosity.
p_α	Pressure in an α-phase.
p_c	Capillary pressure.
$\mathbf{q}_\alpha$	Specific discharge of an α-phase.
R	Universal gas constant.
s	As subscript, symbol denoting solid.
s_α	Source of α-phase.
s	As subscript, symbol denoting solid.
S_α	Saturation of an α-phase (e.g., $\alpha = a, w$).
t	Time.
T	Temperature.
U	Volume.
$\mathbf{V}$	Velocity.
w	As subscript, symbol denoting water, or wetting fluid.
x	Horizontal coordinate.
z	Vertical coordinate (positive upward).

Greek Letters

α	As subscript, symbol for an α-phase.
$\Gamma_{\alpha,\gamma}$	Rate of production of γ, per unit mass of α-phase.
θ_α	Volumetric fraction of an α-phase.
μ_α	Dynamic viscosity of an α-phase.
ρ_α	Mass density of an α-phase.
$\sigma_{i,j}$	Interfacial tension between i and j phases.
ϕ	Wetting angle.

Special Symbols

$\nabla \cdot \mathbf{A}$	Divergence of a vector $\mathbf{A}$ ($=$ div $\mathbf{A}$).
∇A	Gradient of a scalar, A ($\equiv$ grad A).

Chapter 3

PHASE CHANGE PHENOMENA AT LIQUID SATURATED SELF HEATED PARTICULATE BEDS

J M. BUCHLIN AND A. STUBOS
von Karman Institute for Fluid Dynamics
Rhode Saint Genèse B–1640, Belgium.

3.1   Introduction

A wide variety of industrial, agricultural and energy production processes are related to the thermohydraulics of porous media saturated with multiple fluid phases. Examples include the drying of porous solids, the freezing of soils, the geothermal application, the thermally enhanced oil recovery, the heat transfer from buried pipelines, the design of heat pipes, the underground high level nuclear energy waste disposal and the Post Accident Heat Removal PAHR. This last application, addressing the nuclear safety analysis of Liquid Metal Fast Breeder and Light Water Reactors, is the main topic of the present chapter.

The PAHR scenario, describing the sequel of a hypothetical core melt-down accident, forecasts the formation of a particulate debris bed in the reactor vessel. This porous medium, saturated by stagnant *liquid coolant*, is subjected to the decay heat of the fission products. The long range *coolability* of this heat generating particle bed is an important safety issue (Joly and Le Rigoleur, 1979). Heat removal can be achieved by single phase natural convection and by two-phase flow. Boiling is a very efficient internally cooling process, which, unfortunately, is also self-limiting in the case of the PAHR situation: the generation of high vapour flow rate prevents the replenishment of the porous matrix by liquid and subsequently *dryout* occurs.

In-pile experiments with prototypical materials have been carried out at Sandia laboratories (Gronager *et al.*, 1981; Mitchell and Ottinger, 1982; Ottinger *et al.*, 1982; Mitchell *et al.*, 1984) and in the frame of a European programme at the CEN of Grenoble and at the CEN/SCK of Mol. They are discussed in Chap. 4 (also, Mehr and Würtz, 1987). To support the preparation and the interpretation of the in-pile tests and to validate theoretical models, a large number of out-of-pile investigations have been undertaken in various laboratories. The purpose of this chapter is to review the main results obtained from out-of-pile tests and to identify the important milestones in the improvement of the modeling of two-phase flow and dryout occurrence in debris beds.

3.2   Preboiling Phenomenology

In the most general case, heat generated in the *particulate bed* is transferred downward through the supporting structure, and upward to the overlying liquid layer. As sketched in the Fig. 3.2.1, the porous medium may be divided into two layers separated by the plane of maximum temperature, called *adiabatic plane*, since through out this plane no heat flux is allowed. The bottom region of thickness H_{dow} experiences a stabilizing vertical temperature gradient which prevents the onset of *free convection*. Accordingly, conduction is the only mode of downward heat transfer, as long as T_{max} is smaller than T_{sat}. On the contrary, in the upper region of height H_{up}, the negative vertical temperature gradient can trigger the free convection motion of the liquid.

Due to the very small diameter of the solid particles that form the porous matrix, ranging from 10^{-4} to $10^{-3}m$, the homogeneous phase approximation characterized by an effective heat conductivity k_e can be applied (Buchlin *et al.*, 1982). It depends upon the packing porosity, ϵ, and the fluid and solid heat conductivities k_f and k_s, respectively. Some empirical and semi-empirical relations for the *thermal conductivity* in two-component porous bodies are presented in Table 3.2. The upper and lower bounds of k_e correspond to the classical in-parallel and in-series arrangement of the composite material. The problem of predicting k_e for the PAHR debris bed is not

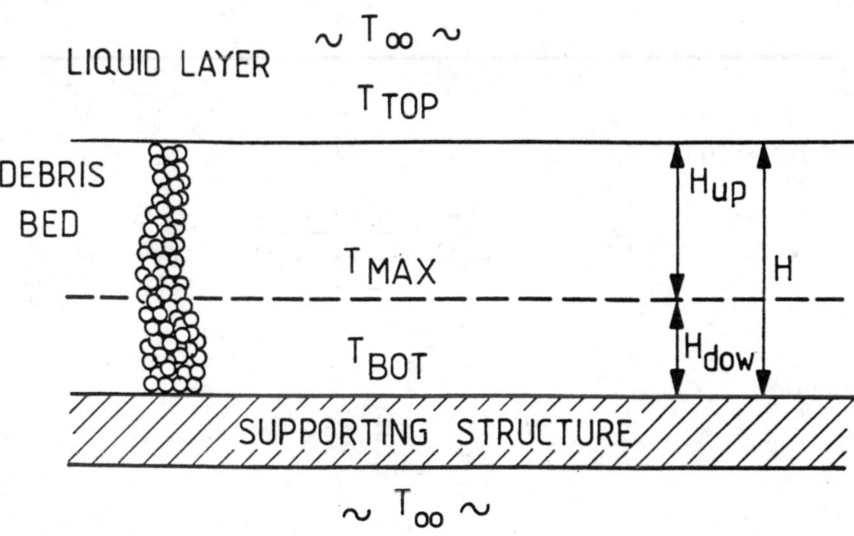

Figure 3.2.1: PAHR preboiling situation

yet considered entirely solved because the packing is composed of two types of solid components (fuel and stainless steel) and possesses an anisotropic character.

The onset of the free convective currents in the porous layer is related to the critical value of the internal *Rayleigh number*, Ra, defined as

$$\text{Ra} = \frac{g\beta\rho_\ell C_{p\ell} Q H^3 \kappa}{2k_e^2 \nu_\ell},\tag{3.2.1}$$

where κ is the bed permeability (see (3.4.3)).

Single phase heat transfer in heated beds has been investigated mostly for the special case of an adiabatic bottom. Data are available from out-of-pile experiments by means of Joule heating of the liquid phase (Sun, 1973; Buretta and Berman, 1976 ; Hardee and Nilson, 1977), or inductive heating of the solid packing (Rhee and Dhir, 1978; Barleon and Werle, 1980). The results are expressed in terms of the *Nusselt number* of the porous layer which is, Nu, defined as

$$\text{Nu} = \frac{Q H_{up}^2}{2k_e(T_{max} - T_{top})}.\tag{3.2.1}$$

Table 3.2.2 presents a summary of the single phase heat transfer correlations obtained in this field.

Kämpf-Karsten (1970):

$$k_e = k_f \left[1 - \frac{(1-\epsilon)(k_f - k_s)}{k_s + (1-\epsilon)^{\frac{1}{3}}(k_f - k_s)} \right].$$

Schulz (1981):

$$\epsilon = \frac{k_s - k_e}{k_s - k_f} \cdot \left[\frac{k_f}{k_e} \right]^{\frac{1}{3}}.$$

Zehner-Schlunder:

$$k_e = k_f \left\{ 1 - (1-\epsilon)^{0.5} + 2(1-\epsilon)^{0.5} \frac{1}{Y} \left[\frac{Y - 1 + f}{Y^2} ln\frac{k_s}{fk_f} - \frac{f+1}{2} - \frac{f-1}{Y} \right] \right\},$$

with $Y = 1 - \frac{k_f}{k_s} f$, where f is the particle shape factor.

Lower and upper bounds after Cook and Peckover (1983):

$$k_{M2} < k_e < k_{M1}, \qquad k_f > k_e.$$

$$\frac{1}{k_{M1} + 2k_f} = \frac{\epsilon}{3k_f} + \frac{1-\epsilon}{k_s + 2k_f}$$

$$\frac{1}{k_{M2} + 2k_s} = \frac{\epsilon}{k_f + 2k_s} + \frac{1-\epsilon}{3k_s}$$

Table 3.2.1: Effective bed thermal conductivity correlations (from Muller and Schulenberg, 1984).

Method of heating	Boundary conditions	Critical Rayleigh numbers	Heat transfer correlation	Authors
Joule heating	Adiabatic lower and isothermal upper boundary	31.8 experimental 32.8 theoretical	$Nu = 0.440Ra^{0.237}, \quad 0 < Ra < 200$ $Nu = 0.135Ra^{0.553}, 200 < Ra < 900$	Buretta, Berman
		33 experimental	$Nu = 0.116Ra^{0.573}, \ 30 < Ra < 800$	Sun
		32 experimental	$Nu = 0.158Ra^{0.5} \ , \ 30 < Ra < 10^4$	Hardee, Nilson
		36.1 experimental	$Nu = 0.570Ra^{0.35} \ , \ 40 < Ra < 1400$	Kulacki, Freeman
Inductive heating	Adiabatic lower boundary and overlying liquid layer	12 experimental	$Nu_p = 0.190Ra^{0.690},$ $Nu_f = 0.234Ra_f^{0.307}, \ 12 < Ra < 3000$	Rhee, Dhir, Catton
		15 extrapolated from experiments	$Nu = 0.267Ra^{0.5} \ , 100 < Ra < 2*10^4$	Barleon, Werle
		30–40 experimental $\xi = 2.8$	$Nu = 0.0786Ra^{0.707}, 30 < Ra < 7500$	Schulenberg, Muller
		15 theoretical $\xi = 1.0$	$Nu = 0.388Ra^{0.533}, \ 30 < Ra < 3000$	Schulenberg, Muller
Nuclear heating		0.8 experimental	$Nu = 0.911Ra^{0.34}, \ 0.8 < Ra < 14$ $0.57 < \xi < 1.56$	Mitchell and al.

Table. 3.2.2: Single phase heat transfer correlations in debris beds (from Muller and Schulenberg, 1984).

Figure 3.2.2 shows that the critical value of Ra, regarded as the threshold of the free motion in the bed, decreases with increasing ratio, ξ, of the depth of the overlying liquid pool to the bed height. As ξ changes, different convective regimes can be expected, both in the liquid layer and in the upper part of the bed. Figure 3.2.3 displays typical results of Sandia in-pile-tests and out-of-pile experiments. The discrepancy observed between the laboratory data and those obtained from nuclear heating of prototypical reactor material can be attributed to the strong effect of the nature of the coolant, as indicated by the numerical simulations plotted in Fig. 3.2.3 (Buchlin *et al.*, 1983), and makes questionable the straightforward application of out-of-pile correlations to reactor situations.

However, it is worth noting that the most interesting thermohydraulic behaviour of the debris bed includes the boiling regime and the dryout development during which the single phase convection, being generally of minor relevance, can be disregarded.

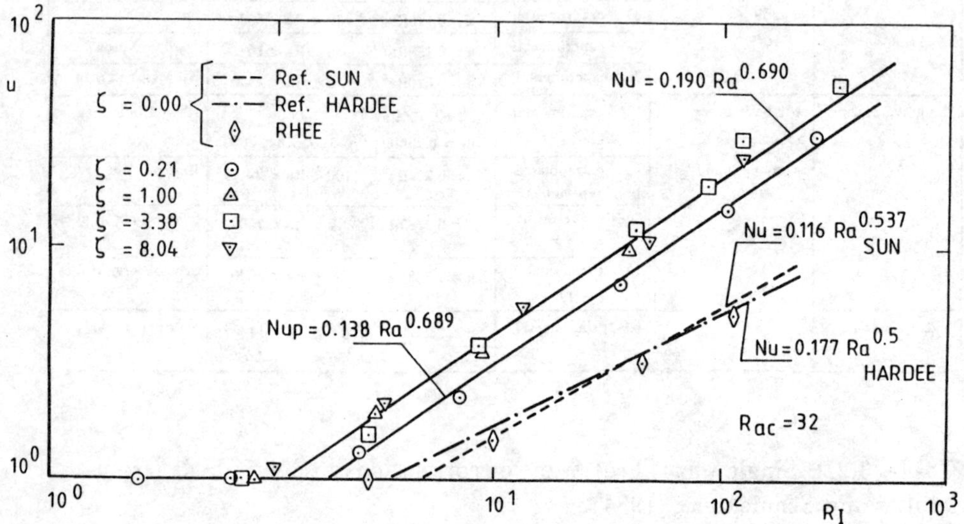

Figure 3.2.2: Effect of the overlying liquid layer.

3.3   Boiling regime and dryout heat flux

As the heat deposition increases, the temperature of the debris bed reaches the saturation condition and boiling of the liquid coolant occurs.

The heat released locally is stored as latent heat and removed advectively by the vapour phase. The latter is highly buoyant because of the large liquid-vapour density ratio (of the order of 10^3).

Several experimental investigations have been conducted in simulating facilities to analyze the boiling behaviour of the bed. All this experimental information reveals the existence of two likely types of two-phase flow in debris stacks, as illustrated in Fig. 3.3.

During packed boiling, the bed remains in a compact state, with a vapour

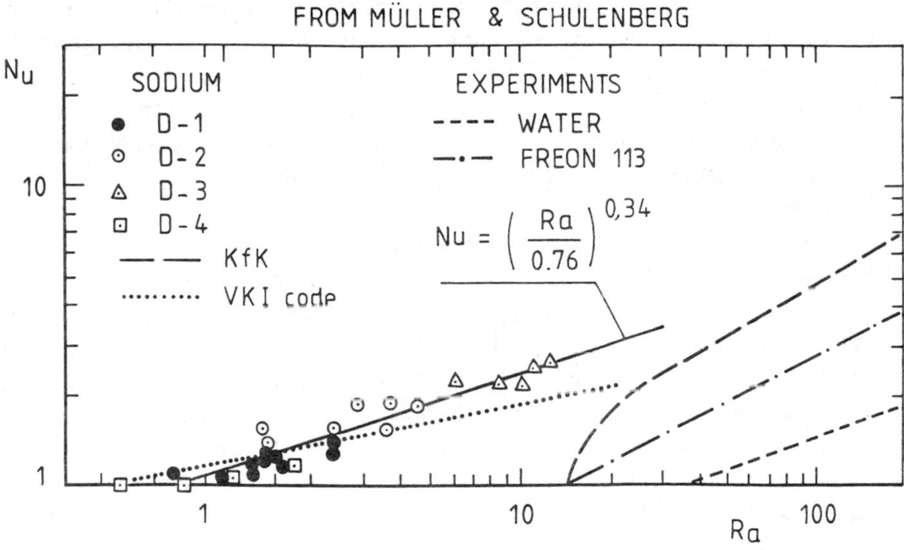

$$Nu = \left(\frac{Ra}{0.76} \right)^{0,34}$$

Figure 3.2.3: Effect of the cooling liquid.

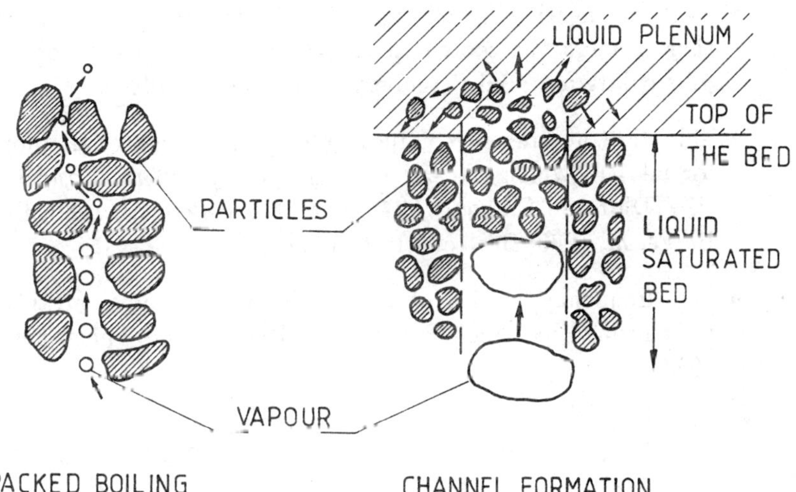

Figure 3.3.1: Packed and channeled boiling in debris beds.

bubbly flow. Under certain conditions, which are discussed in detail in a later section, the upper porous layer undergoes disturbances leading to irreversible bed changes. The pressure of the vapour phase rises enough to lift and expel the particles which are pushed aside. Usually, this event is characterized by the formation at the top of the bed of *channels*, or chimneys through which vapour escapes easily. Both packed and channeled boiling can be present at the same time. This is seen in Fig. 3.3.2, where a typical visualization of the thermohydraulic behaviour of a debris bed is shown (Buchlin and Vankoninckxloo, 1986).

Boiling as an internally cooling mechanism is self-limiting, because at high vapour flow rates the porous body is no more replenished by inflowing liquid and a portion of the bed dries. In this zone, the temperature exceeds the saturation temperature of the coolant and may reach the melting temperature of the solid material. Dryout of the porous structure is considered as a limiting condition for the coolability of the debris beds. Therefore, great attention has been given to the determination of the dryout heat flux.

Several out-of-pile investigations have been designed to identify the parameters affecting the dryout heat flux.

The most appropriate experimental simulation methods for studying the coolability of PAHR debris beds depend on the volumetric heating of the porous body. Different techniques have been developed to internally energize the particle beds:

- electrical heating by the passage of alternating current laterally, either through the coolant phase (Gabor *et al.*, 1972), or through the particulate material (Trenberth and Stevens, 1980),

- induction heating of conducting particulate material, more adapted for beds in which the particles are nearly spherical and of the same size (Dhir and Catton, 1977; Squarer and Peoples, 1980; El Genk and Bergeron, 1983; Barleon *et al.*, 1984),

- high frequency heating of a dielectric porous body, suitable for debris beds formed by irregular particles (Buchlin and Thiry, 1982; Stevens, 1985).

Figure 3.3.2: Channel visualization in a heated bed.

In all out-of-pile experiments, the dryout occurrence has been investigated by varying the bed depth, the bed width and the particle size, in the range $0.03 - 0.50$ m, $0.05 - 0.15$ m and $0.2 \times 10^{-3} - 15 \times 10^{-3}$ m, respectively.

The effective dryout heat flux, Φ_d, is defined as the integral heat generated in the porous medium divided by the superficial cross section area of the bed.

The effects of the main parameters, e.g., bed height, mean particle size, presence of bottom cooling and degree of top subcooling, on the dryout occurrence in the debris bed, are reviewed below.

Dependence of dryout heat flux on bed depth

The effect of the bed height, H, on the dryout heat flux, Φ_d, is shown in Fig. 3.3.3, where experimental data of different out-of- pile investigations are plotted. Two characteristic regions can be identified, namely the shallow-bed region and the deep-bed region.

The shallow bed

In the shallow bed region, characterized by a large scatter of data, Φ_d tends to rise as H decreases. It is generally observed that in shallow beds made of small particles, *vapour channels* penetrate a significant way into the depth of the porous packing, forming a lengthy channeled zone which provides a ready escape for the vapour and permits a high power to be generated before dryout occurs.

The deep bed

In the deep-bed region, the data are more reproducible. In this case, the dryout power depends weakly upon the height of the bed. Since Φ_d can be regarded as a constant, this means that the removable power density decreases in inverse proportion of the bed height.

In deep beds the vapour channel penetration is not prominent, compared to the bed height. Dryout occurrence is mainly governed by packed boiling.

Dependence of dryout heat flux on particle size

Figure 3.3.4 shows dryout data plotted versus the mean particle diameter, d_p.

The power to reach dryout increases as the particle diameter increases. For small-particle beds, Φ_d tends to vary as d_p^2, whereas it varies as $d_p^{0.5}$ for coarse-particle beds.

To discuss these data, the analytical zero-dimensional *Lipinski model*, described in App. A, is used.

Case of fine-particle bed

As the particles are small, typically $d_p < 10^{-3}$ m, the two-phase flow in the porous medium is governed mainly by fluid drag and the capillary force. Then the predicted *limiting heat flux* can be expressed by

$$\Phi_d = \frac{g(\rho_\ell - \rho_v)\kappa L}{\left(\nu_\ell^{1/4} + \nu_v^{1/4}\right)^4}\left(1 + \frac{\lambda}{H}\right),\qquad(3.3.1)$$

where λ is the capillary head of the porous body:

$$\lambda = \frac{\sigma\cos\theta}{g(\rho_\ell - \rho_v)}\sqrt{\frac{\epsilon}{5\kappa}}$$

The derivation of this relation is similar to that made by Shires and Stevens, 1980).

Equation (3.3.1) indicates that in deep beds, $H > \lambda$, Φ_d becomes less dependent on bed height and more proportional to κ, thus to d_p^2. This is in agreement with the trend exhibited by the experimental data in Figs. 3.3.3 and 3.3.4 The same qualitative results are also obtained by Dhir and Catton (1977), or Hardee and Nilson (1977).

As the depth of the bed decreases, the zero-dimensional model predicts an increase in Φ_d, due to a more pronounced effect of the capillary forces. However, it is worth recalling that the possible formation of vapour channels, or other disturbances, as superheat flashing, can enhance the coolability of the debris packing more easily in shallow beds.

Case of coarse-particle bed

With large particles, typically $d > 10^{-3}m$, flow in a particulate bed is dominated by inertia. The capillarity effect is now negligible (see expression for λ). In this limiting case, the Φ_d-relation, derived from the the zero-dimensional model, looks like the one of the 'flooding' model (Ostensen and Lipinski, 1981), or the 'drift flux' model (Chang and Kim, 1985), viz.

$$\Phi_d = \left(\frac{g(\rho_\ell - \rho_v)\rho_v\eta}{(1 + (\rho_v/\rho_\ell)^{0.25})^4}\right)^{0.5} L,\qquad(3.3.2)$$

where η is the *passability* of the porous medium (see (3.4.3)).

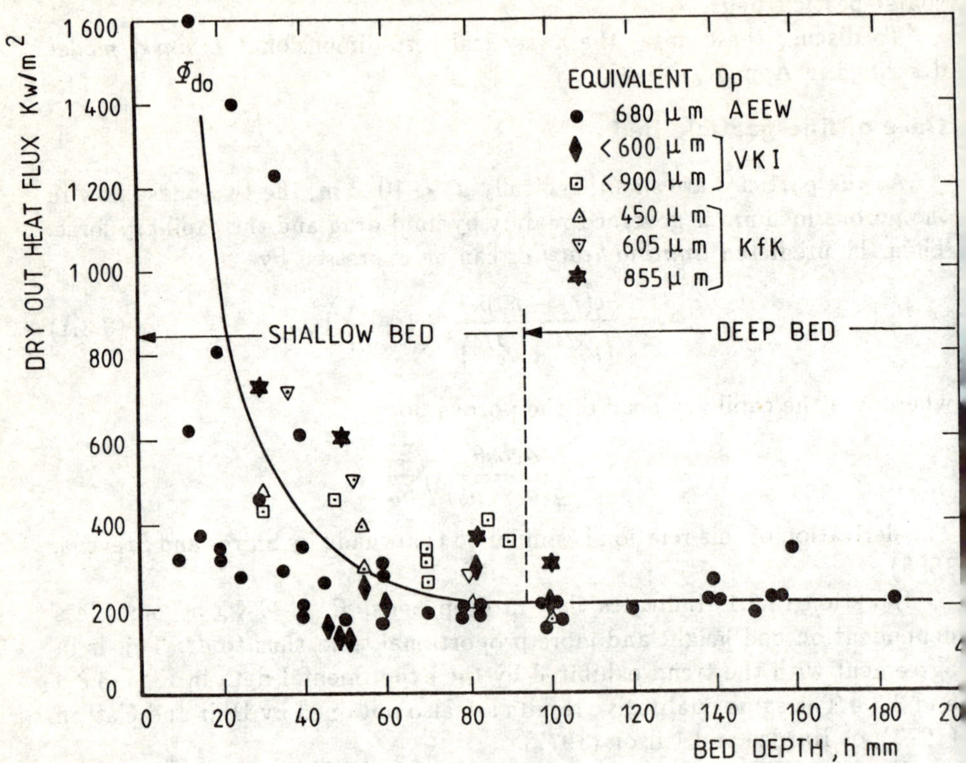

Figure 3.3.3: Effect of the bed depth on the dryout heat flux.

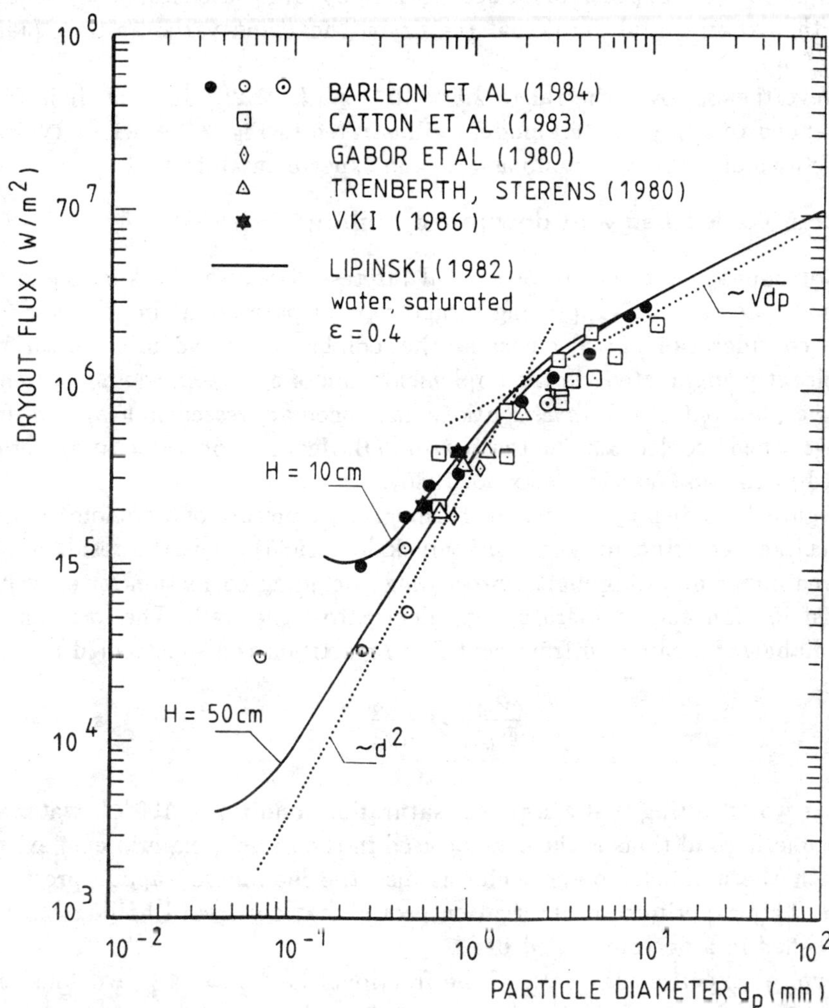

Figure 3.3.4: Effect of the particle diameter on the dryout heat flux.

Equation (3.3.2) predicts for beds formed by large particles, in agreement with the experimental data, that the dryout heat flux varies as the square root of d_p.

Nevertheless, over the range $2.5 \times 10^{-4} \le d_p \le 2 \times 10^{-3}$ m, it is recommended to apply the full model as illustrated in Fig. 3.3.4, where typical predictions of it are drawn along with the experimental data.

Bottom cooled bed and downward boiling

Bottom-cooled beds are of special interest, because the knowledge of thermal loads on the supporting structure is of paramount importance for safety consideration and also because the coolability of the debris bed can be significantly augmented via the implementation of a bottom cooling system. However, few out-of-pile investigations have been addressed to heat transfer in debris beds cooled also at the bottom (Barleon, Thomauske and Werle, 1984; Buchlin and Vankoninckxloo, 1986).

Figure 3.3.5 displays a typical thermographic picture of a bottom-cooled particulate bed (Buchlin and Vankoninckxloo, 1986). It is the result of an infrared image following digital processing, including correction for emissivity distribution and temperature gradient across the wall. The location of the adiabatic plane found from heat flux repartition is also specified by

$$\frac{\Phi_{up}}{\Phi_{dow}} = \frac{H_{up}}{H_{dow}}.$$

It is worth noting that a region at saturation conditions ($100°C$: water at atmospheric conditions is the coolant used in the present experiment) exists beneath the adiabatic plane which is also the maximum vapour pressure plane. This experimental observation proves that downward boiling can be established in a bottom cooled bed.

Typical experimental data of the fractional heat flux, Φ^*_{dow}, defined as the ratio of the absolute downward heat flux, Φ_{dow} to the total heat flux generated in the bed, Φ_{tot}, are plotted in Fig. 3.3.6 versus Φ^*_{tot}, which is the total heat flux itself normalized by its value at dryout, $\Phi_{tot,d}$.

For the sake of comparison, the values of Φ^*_{dow}, calculated with the pure heat conduction model (see App. B)

$$\Phi^*_{dow,con} = \left(\frac{2k_e(T_{sat} - T_{bot})}{H\Phi_{tot}} \right)^{0.5}, \qquad (3.3.3)$$

are also represented in Fig. 3.3.6.

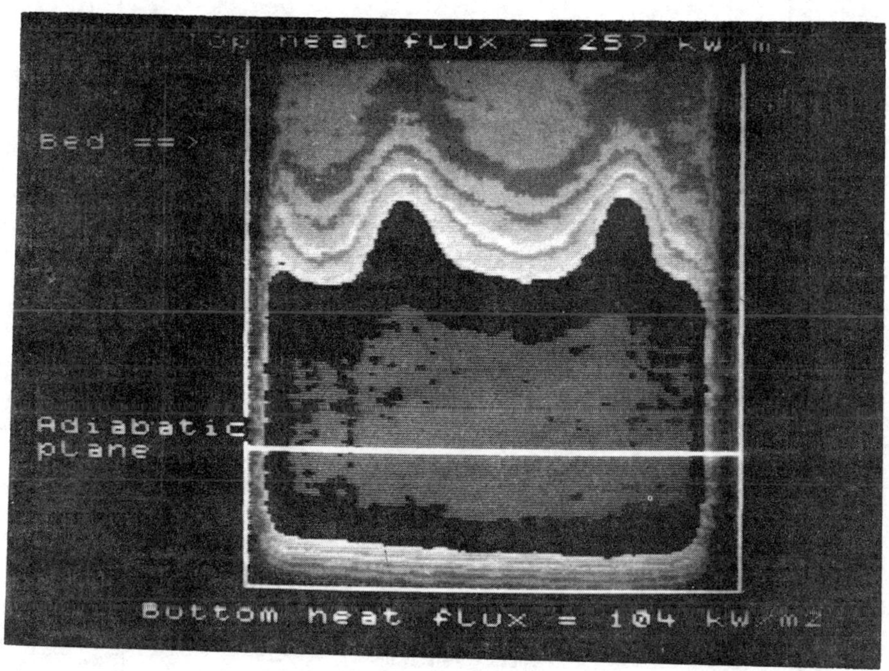

Figure 3.3.5: Digitized thermography picture of a bottom-cooled debris.

The good agreement obtained between the experimental data from coarse particle beds $(d_p > 2 \times 10^{-3}\ m)$ and the theoretical predictions from (3.3.3), confirms that only the heat generated in the subcooled layer is flowing downward through the bottom and that in this case the adiabatic plane coincides with the lower boundary of the boiling region.

On the other hand, beds of small particles $(d < 10^{-3}\ m)$ undergo different behaviour when Φ^*_{tot} exceeds the value of 0.5 and approaches the dryout limit, $\Phi^*_{tot} = 1$. Instead of decreasing, as predicted by the conduction model, the fractional downward heat flux rises to a value about twice that of $\Phi^*_{dow,con}$. Therefore, more heat is removed from the bottom than that generated in the lower subcooled layer and the adiabatic plane has to be placed inside the boiling region. This is seen in Fig. 3.3.5. In such a configuration, vapour flows downward and condenses at the lower boundary of the saturation zone

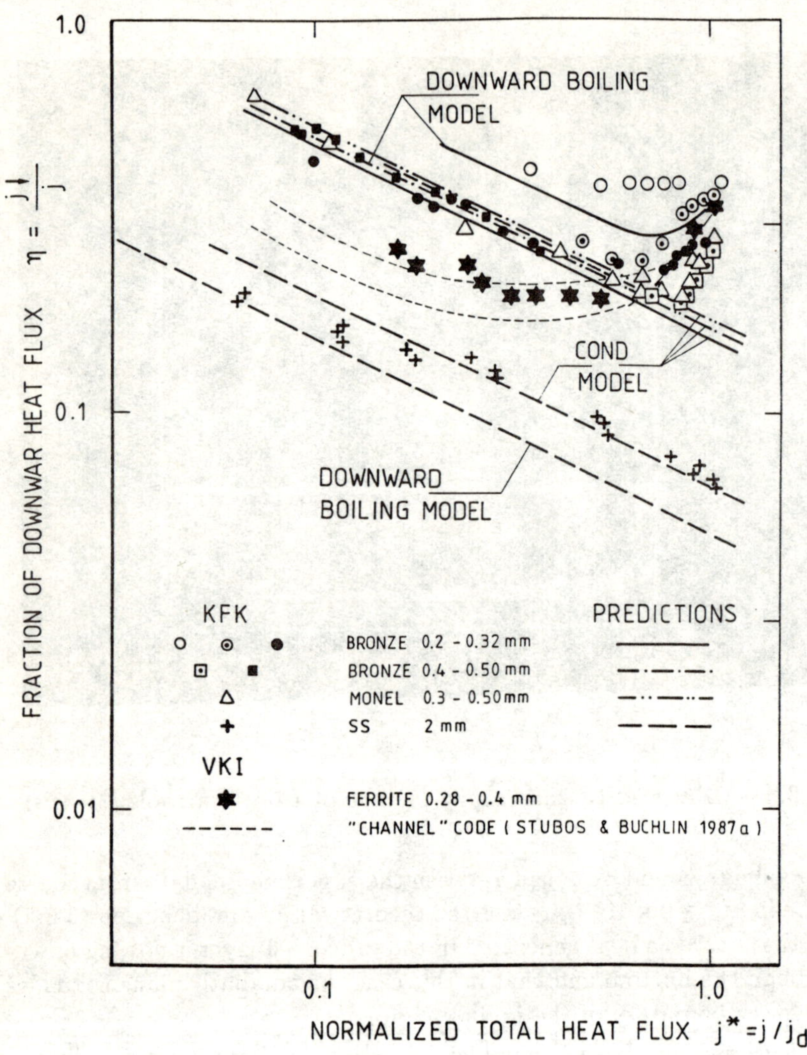

Figure 3.3.6: Dimensionless downward heat flux versus the normalized total heat flux.

from which the liquid phase is drawn up, feeding the packed boiling region by capillarity. Such a phenomenon cannot develop in a insulated-support bed because of the back pressure due to the accumulation of vapour.

The downward boiling can be predicted by the Lipinski zero-dimensional model (see App. A). Since it is expected to occur more in small particle beds, it is possible to derive an explicit expression for the downward dryout heat flux, $\Phi_{dow,d}$, in the form

$$\Phi_{dow,d} = 0.5\Phi_d \left[2\frac{\lambda}{H} + \left((2\frac{\lambda}{H})^2 + 1 \right)^{0.5} - 1 \right]. \qquad (3.3.4)$$

In (3.3.4), Φ_d is the laminar critical heat flux given by (3.3.1), without including the capillary term.

When comparing the total heat flux at dryout for the case of the bottom cooling with downward boiling, $\Phi_{tot,d} = \Phi_{up,d} + \Phi_{dow,d}$, with that characterizing the case of an adiabatic bottom, $\Phi_{d,ad}$, described by (3.3.1), it turns out that

$$\frac{\Phi_{tot,d}}{\Phi_{d,ad}} = \left(\frac{2\frac{\lambda}{H} + \left((2\frac{\lambda}{H})^2 + 1 \right)^{0.5}}{\frac{\lambda}{H} + 1} \right).$$

This leads to the conclusion that the heat removal capability of the debris bed can increase *theoretically* from 1 to 4 times as the importance of the reduced capillary head, λ/H, varies from zero to infinity.

Effect of top subcooling on dryout heat flux

Most of the out-of-pile debris beds are normally channeled. Extensive *channeling* is observed when the overlying liquid layer is at saturation conditions (Gabor and Sowa, 1974; Keowen, 1974). The common finding emerging from several out-of-pile tests carried out with volumetric heated beds is that the channel penetration is confined to a layer 0.02 to 0.05 m deep. Section 3 deals with theoretical models for a channeled bed. The typical analytical relation generally applied to estimate the length of the evolved channel can be written in the form

$$L_c - C\frac{\sigma}{g(\rho_s - \rho_\ell)\epsilon d_p}, \qquad (3.3.5)$$

where C is a constant ranging from 4.5 to 6.5 depending on the model used (Jones *et al.*, 1982; Schwalm and Nijsing, 1982; Reed, 1982).

The relation (3.3.5) is used to calculate L_c on the basis of data characterizing typical out-of-pile materials. Table 3.3.1 presents the results. These predictions correspond rather well to the experimental observations.

$$\epsilon = 0.4$$
$$\rho_\ell = 957 kg/m^3 \text{ (water)}$$
$$\rho_s = 7900 \ kg/m^3 \text{ (stainless steel)}$$

Particle d_p [m]	Channel L_c [m]
$250\text{x}10^{-6}$	$5.1\text{x}10^{-2}$
$500\text{x}10^{-6}$	$2.6\text{x}10^{-2}$
$1000\text{x}10^{-6}$	$1.3\text{x}10^{-2}$

To prevent channel penetration, the degree of subcooling of the liquid pool must be sufficiently high so that the thickness of the upper subcooled porous layer becomes comparable with the channel length. A simple analytical approach outlined in App. C, allows to estimate the size of the top subcooled zone, H_{sub}, from

$$H_{sub} = H - \left(H \left(H - \frac{2k_e(T_{sat} - T_{top})}{\Phi_{up}} \right) \right)^{0.5}. \qquad (3.3.6)$$

For a 0.20m deep bed, formed by SS particles 250×10^{-6} m in diameter, submerged in a water pool at $20°C$ and a relevant dryout heat flux of $30 \ kW.m^{-2}$ (Barleon, Thomauske and Werle, 1984), we obtain a subcooled zone thickness $H_{sub} = 0.012$ m. This is four times smaller than the expected channel penetration. This analysis explains why it is very difficult to avoid channeling in out-of-pile experiments when small particles are employed.

3.4   Constitutive Relationships–Bed Disturbances

3.4.1   Introduction

The occurrence of structural changes in the debris bed as a result of several types of disturbances during boiling, and their effect on the bed coolability are still not well understood and quantified. Therefore, more data and modeling efforts are needed in this direction. This chapter deals with the out-of-pile experimental action undertaken and the current status of progress achieved regarding the bed restructuring phenomena and the relevant parameters involved in this modeling.

First, it is necessary to provide *constitutive relationships* for the estimation of quantities like the bed absolute permeability, κ, the relative phasic permeabilities, κ_i, relative passabilities, η_i, and the capillary pressure P_c.

For the sake of simplicity, without loosing generality in the presentation, the subscripts ℓ for liquid and v for vapour or gas will denote the wetting and the nonwetting phase, respectively.

3.4.2   Bed permeability

Darcy's law for slow, unidirectional and horizontal steady flow of a Newtonian fluid moving at the superficial velocity V through a porous bed may be stated as

$$-\frac{dP}{dz} = \frac{\mu V}{\kappa}. \tag{3.4.1}$$

The permeability of the bed, κ, depends on the porosity ,ϵ, particle-shape and particle-size distribution.

When the *inertial effects* have to be accounted for, a quadratic term is introduced into (3.4.1) leading to the *Ergun equation* for the total pressure drop caused by the presence of the porous matrix

$$-\frac{dP}{dz} = \frac{\mu V}{\kappa} + \frac{\rho V^2}{\eta}. \tag{3.4.2}$$

where η is called bed passability and κ depends on ϵ and on the characteristic particle diameter, d_p.

In fact, the deviation from linearity starts to become noticeable when the particle *Reynolds number*, Re$= \rho V d_p/\mu$, reaches values in the range 1 to 5 (MacDonald *et al.*, 1979; Fand *et al.*, 1987). The following empirical relations for κ and η, respectively, have survived many experimental tests and revisions (MacDonald *et al.*, 1979):

$$\kappa = \frac{d_p^2 \epsilon^3}{A(1-\epsilon)^2} \quad \text{and} \quad \eta = \frac{d_p \epsilon^3}{B(1-\epsilon)}. \tag{3.4.3}$$

The effective particle diameter, d_p, is given by the Fair and Hatch formula

$$d_p = f \left(\sum \frac{w_i}{d_i} \right)^{-1},$$

where w_i is the weight fraction of particles with sieve diameter d_i, and f is a shape factor, the value of which is 1 for spheres and, typically, 0.78 for rough particles. Recent experiments carried out with particles smaller than $1 - 2 \times 10^{-3}$ m give $A = 181$ and $B = 1.94$ (Di Francesco, 1987). These values are in agreement with those reported for larger particles, $A = 182$ and $B = 1.92$ (Fand *et al.*, 1987).

3.4.3   Relative permeabilities and passabilities

The horizontal flow of two immiscible phases in a porous medium may be described, in general, by an extension of the Ergun equation through the use of the *phasic relative permeabilities*, κ_i, and and *passabilities*, η_i, in the form

$$-\frac{dP_i}{dz} = \frac{\mu_i V_i}{\kappa \kappa_i} + \frac{\rho_i V_i^2}{\eta \eta_i} \quad i = \ell, v. \tag{3.4.4}$$

At any given position, P_ℓ does not ordinarily equal P_v, since pressure is discontinuous across curved fluid-fluid interfaces.

Table 3.4.1 contains several forms for the functional dependence of κ_ℓ, κ_v, η_ℓ and η_v on the liquid saturation, s, or on the effective saturation, $s_e = (s - s_r)/(1 - s_r)$, for gas (vapour)-liquid flows. The quantity s_r is the *residual*, or *irreducible saturation*. .

The concept of relative permeability is practical only if κ_ℓ and κ_v are independent of pressure gradient, total flow rate, as well as of fluid properties, such as viscosity ratio and surface tension (Scheidegger, 1957).

A great number of experimental investigations indicate, subject to certain limitations, that for a given solid matrix, the relative permeabilities are functions of the wettability characteristics of the fluid pair, the fluid saturation and the saturation history only.

	κ_v	κ_ℓ	η_v	η_ℓ	ent.
Turland & Moore (1983)	$1-s$	s^3	$(1-s)^2$	s^1	1
Lipinski (1980)	$1-1.11s$	s^3	$(1-s)^3$	s^3	2
Lipinski (1982)	$(1-s_e)^3$	s^3	$(1-s_e)^3$	s_e^3	3
Lipinski (1984)	$(1-s_e)^3$	s^3	$(1-s_e)^5$	s_e^5	4
Saez & Carbonell (1985)	$(1-s)^{1.8}$	$s_e^{2.13}$	$=\kappa_v$	$=\kappa_\ell$	5
Lee & Catton (1984)	$\bar{a}^3(1.58\bar{a}^2-0.83\bar{a}+0.25)$ with $\bar{a}=\frac{1-s}{\alpha'}$	$s^{2.2}$	$=\kappa_v$	$=\kappa_\ell$	6
Schulenberg & Muller (1984)	$(1-s_e)^3$	s_e^3	$(1-s_e)^6, s_e<0.68$ $0.1(1-s_e)^4, s_e>0.68$	s_e^5	7
Chu, Dhir & Marshall (1983)	$0.2(\alpha-\alpha_0)+0.8(\alpha-\alpha_0)^2$	$0.5(1-(\alpha-\alpha_0))^4+0.5(1-(\alpha-\alpha_0))^5$	$0.4(\alpha-\alpha_0)^2+0.6(\alpha-\alpha_0)^3$	$(1-(\alpha-\alpha_0))^3$	8

Table 3.4.1: Correlations for relative permeabilities and passabilities.

The relative permeabilities are usually determined by measuring ΔP and flow rates under conditions of steady cocurrent flow along H. Typical curves are shown in Fig. 3.4.1. The features common to most relationships are

(a) The two functions are monotonic and the sum, $\sum \kappa_i$, is usually less than unity and exhibits a minimum.

(b) Hysteresis, analogous to the one found for the capillary pressure curves, occurs. Drainage and imbition relative permeability curves are generally different, as seen in Fig. 3.4.1.

(c) Both κ_ℓ and κ_v become zero for saturation values less than the phasic residual value, $s_{r\ell}$ and s_{rv}, respectively.

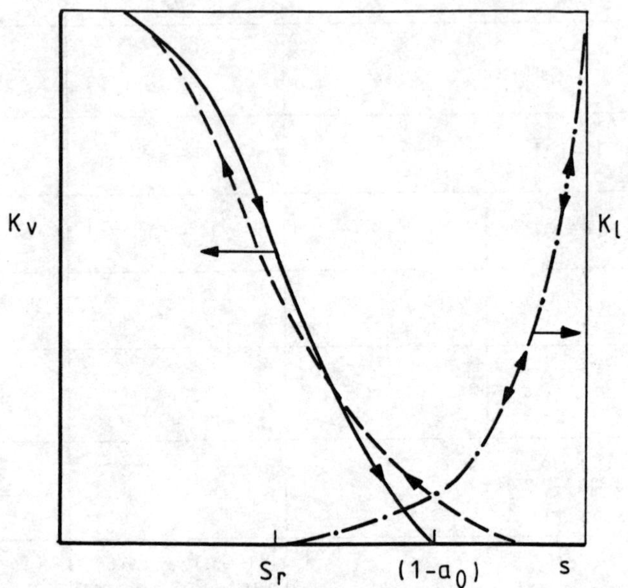

Figure 3.4.1: Imbibition and drainage relative permeabilities (Larson *et al.*, 1981).

Examining the literature, one can concluded that the relative permeabilities are independent of the viscosity ratio and pressure gradient (Larson *et al.*, 1981), except during unsteady conditions (Lin and Slattery, 1982). In addition, it can be shown that the relative permeabilities can be represented by a unique curve if plotted as functions of the effective saturation, $s_{ei} = (s_i - s_{ri})/(1 - s_{ri})$, with $i = \ell, v$, as displayed in Fig. 3.4.2. These graphs correspond to data obtained for the capillary numbers

$$N_c = \frac{\kappa \Delta P}{\sigma H},\qquad(3.4.5)$$

ranging from 0.14 to 14×10^{-5}. They show that the effect of σ is restricted only to residual saturation.

There is little consideration for the relative passabilities in the literature, mostly because the traditional studies of multiphase flow in porous media deal with low Reynolds number cases.

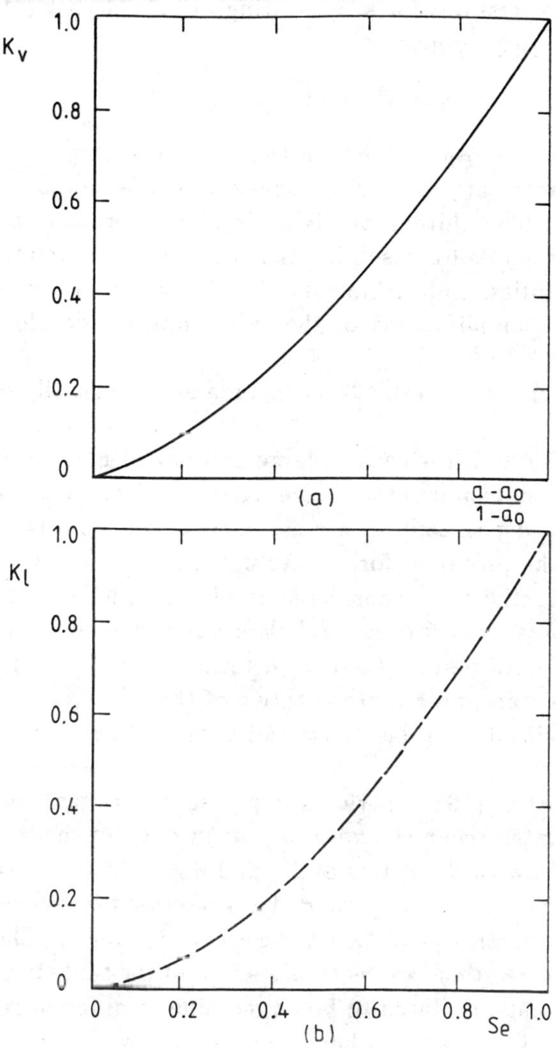

Figure 3.4.2: Phasic relative permeabilities as functions of effective saturation.

In the first entry of Table 3.4.1, Turland and Moore (1983), use a simplification of the expressions for κ_i appearing in Scheidegger (1957), while for the passabilities they propose

$$\eta_\ell = s^4 \quad \text{and} \quad \eta_v = (1 - s)^2.$$

Lipinski's initial proposals, shown in the following entry in Table 3.4.1, are based on oil-water data for various sands and may not be adequate for a liquid-gas flow. The third entry is a simplified form of the equations of Brooks and Corey (1966), assuming the equality of the relative permeabilities and passabilities. Later, Lipinski (1984) presents a revised form for η_i, resulting from a simplification of the semi-empirical development of Reed (1982).

Subsequent entries in Table 3.4.1 include experimentally derived expressions.

Saez *et al.* (1985), reviewed a large amount of measured data, mainly on air-water systems and rather large particles. The fact that the relative permeabilities and passabilities are assumed to be equal is considered as a weak point of the proposed forms. Actually, since the size of the smallest particles used in the tests considered is of the order of 2×10^{-3} m, and the range of Re is from 5 to several thousands, it is expected that for the majority of the data points the inertial term dominates and, therefore, the expressions given are more representative of the relative passabilities. This holds more specifically for gas phase and high void fraction (high flow rate) regime.

Lee and Catton (1984) performed pressure and gas fraction measurements for air-water *cocurrent upward flow* in one-dimensional beds of uniform glass beads with diameters of 2.5 and 6×10^{-3} m. In their functional forms, α' represents a constant value of gas fraction for each medium (0.9 for the 2.5×10^{-3} m particles and 1.0 for the 6×10^{-3} m ones). The disagreement between their correlations and entries 2 and 5 of the table in predicting their experimental points, is stated to be largely due to differences in the gas relative permeabilities. On the other hand, as the previously met assumption of equal permeabilities and passabilities is also made here, the comparison with entry 5 looks better, especially for values of gas fraction $\alpha = 1 - s$ larger than 0.5. For smaller α values, the present relationship appears to be less sensitive to the gas fraction.

Schulenberg and Muller (1984) conducted experiments on vertical cocurrent flow of air and water, or 48% ethanol solution, through nearly spherical glass particles of two different sizes (3 and 7×10^{-3} m). They assumed cubic dependencies for the permeabilities, and determined the passabilities as shown in the Table 3.4.1. The agreement with the improved model of Lipinski (entry 4) is rather good.

Chu *et al.* (1983) investigated cocurrent air-water flow in $1 - 6 \times 10^{-3}$ m glass beads beds. Postulating $\eta_\ell = [1 - (\alpha - \alpha_o)]^3$, the resulting rather high η_ℓ-value (compared to $\eta_\ell = s^5$ of entries 4 and 7) may be the reason for the relatively low κ_ℓ-value proposed. Furthermore it seems that in order to become equivalent to the other relations in Table 3.4.1, the expressions they propose would have to be modified, using

$$\kappa'_\ell = \frac{\kappa_\ell}{s} \quad \text{and} \quad \kappa'_v = \frac{\kappa_v}{1-s}. \tag{3.4.6a}$$

$$\eta'_\ell = \frac{\eta_\ell}{s} \quad \text{and} \quad \eta'_v = \frac{\eta_v}{1-s}, \tag{3.4.6b}$$

Tutu *et al.* (1984), in attempting to model the interfacial drag for air-water flow through beds of spheres with diameter ranging from 3.18×10^{-3} m to 12.7×10^{-3} m, produced a set of data for the gas-solid drag and the gas fraction α, that were used to test the validity of Lipinski's second form (entry 3). In doing so, they stress the effect of the residual gas fraction, α_o on the drag force which can be viewed as consisting of the sum of the dynamic gas-solid drag and the one due to the non-moving trapped gas phase. Unfortunately, they failed to account properly for the relative permeability and passability which is

$$\frac{F_{gs}}{\alpha} = \frac{\mu_v V_v}{\kappa \kappa_v} + \frac{\rho_v V_v^2}{\eta \eta_v}, \tag{3.4.7}$$

since the gas fraction was omitted from the correct form (Schulenberg and Muller, 1984; Naik and Dhir, 1982; Lee and Catton, 1984) Figure 3.4.3 presents part of their data along with the revised prediction (3.4.7), obtained by using the improved Lipinski's suggestion (entry 4) which agrees better with the Schulenberg and Muller relations (1984), shown as entry 7.

From the preceding discussion, some conclusions may be drawn concerning the PAHR problem.

- The uncertainty about the relative permeabilities and passabilities remains.

- Most of the data for packed beds were obtained with particles larger than 2×10^{-3} m. In fact, it seems that the only investigations with small unconsolidated particles involve sandpacks (Stauffer and Dracos, 1986).

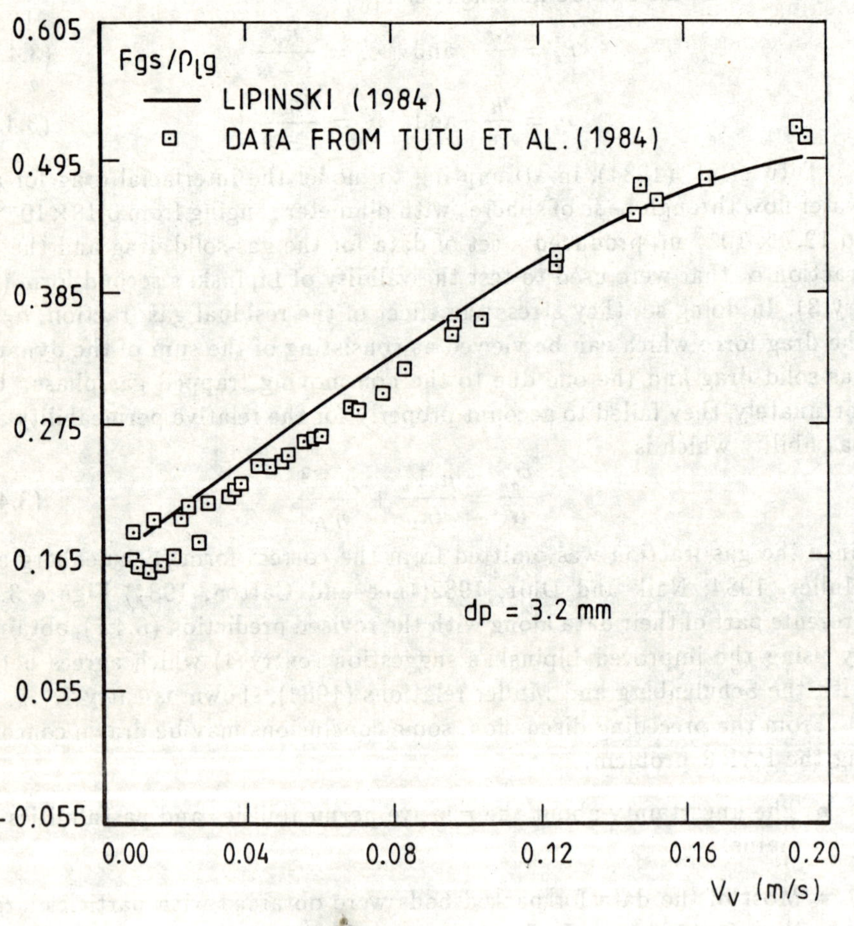

Figure 3.4.3: Gas-solid drag force as a function of superficial gas velocity.

In order to refine the situation and produce data for beds of small particles, an experimental facility, which is delineated in Fig. 3.4.4, has been built and is currently being used at the VKI. Beds of glass spheres (1 and 3×10^{-3} m), or nearly spherical steel particles (with effective diameters of 0.260, 0.440 and 0.640×10^{-3} m), with porosities ranging from 0.34 to 0.41, are formed in a cylindrical plexiglass column 11.4×10^{-2} m in diameter. Vertical cocurrent air-water flow is established through two different circuits. Liquid flow is provided by a pump from a water reservoir, and is monitored by a venturi flowmeter. The air feed line is a tube with several openings along it. The tube is located below a perforated plate that supports the particles and assures a good distribution of the air flow around the cross section. The gas flow rate is measured by two rotameters with different capacities.

Superficial velocities up to 2.0×10^{-2} m/s and up to 5.0×10^{-2} m/s, may be used for the liquid and gas phases, respectively.

An array of pressure taps are located on the wall of the test section. They allow the measurement of pressure at several locations along the bed. This is considered to be important for the small particles, since differences may show up in different parts of the bed, because of disturbances or instabilities that develop in the upper part (as will be discussed later), or because of the possible inadequacy of the working assumption of constant saturation for the packed part of the bed. The particular case of disturbances (for example channeling) seems to be the main distinguishing feature of the upward gas-liquid flow through small particles, the disturbed region being often quite appreciable. Of course, it is very probable that the relative permeability concept itself cannot account for a satisfactory description of the flow conditions in such a region.

The global gas fraction is determined by the amount of liquid displaced from the initially saturated bed. This is shown by the level in the thin tube parallel to the axis of the column when liquid is flowing. If, on the contrary, the test is done at no liquid flow, the volume of displaced liquid above the particles is measured by physically removing it by means of a syringe during the experiment.

The investigation involves tests for bed permeability, relative permeabilities and passabilities for the packed part of the bed and their eventual change in the disturbed zone. In particular a method is under development for identifying the behaviour of the different layers by analyzing wall pressure fluctuations. A typical recorded pressure signal and its frequency spectrum are shown in Fig. 3.4.5

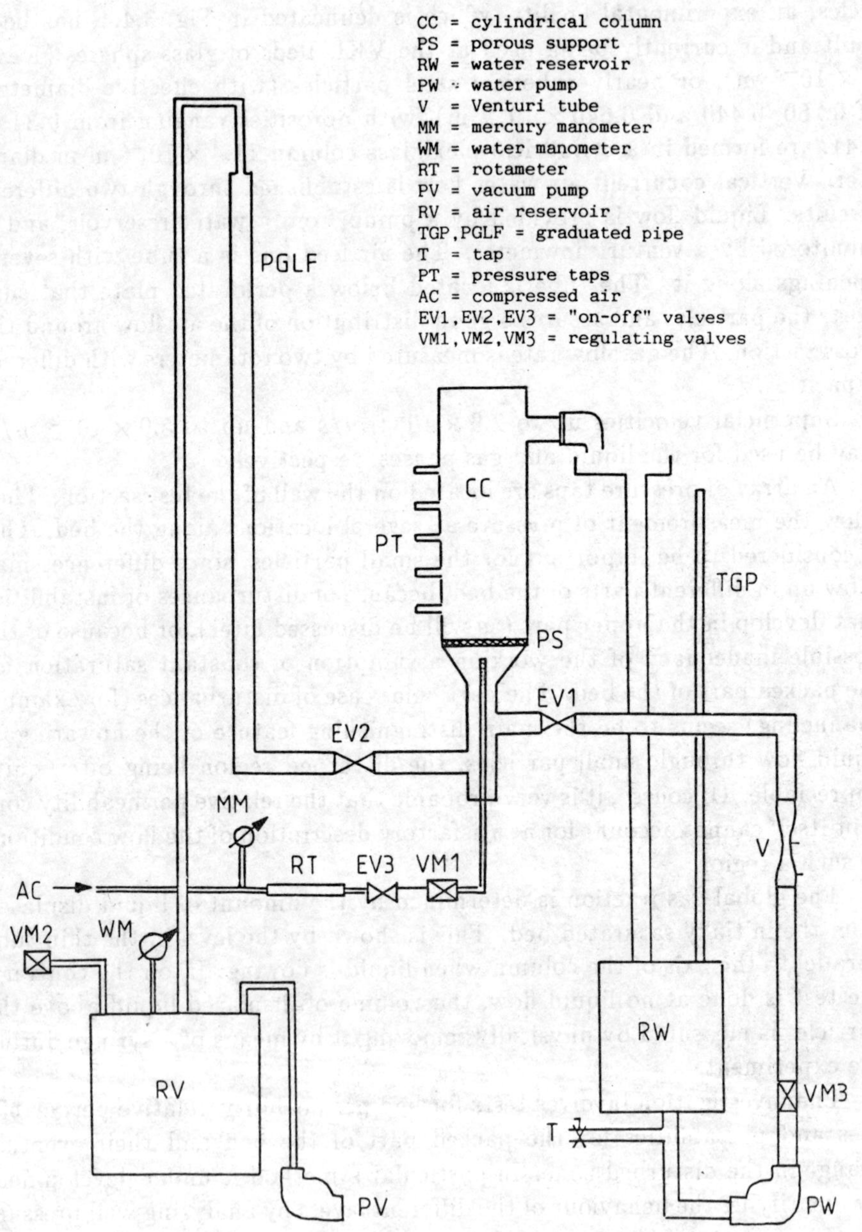

Figure 3.4.4: Sketch of the VKI-OPERA III facility.

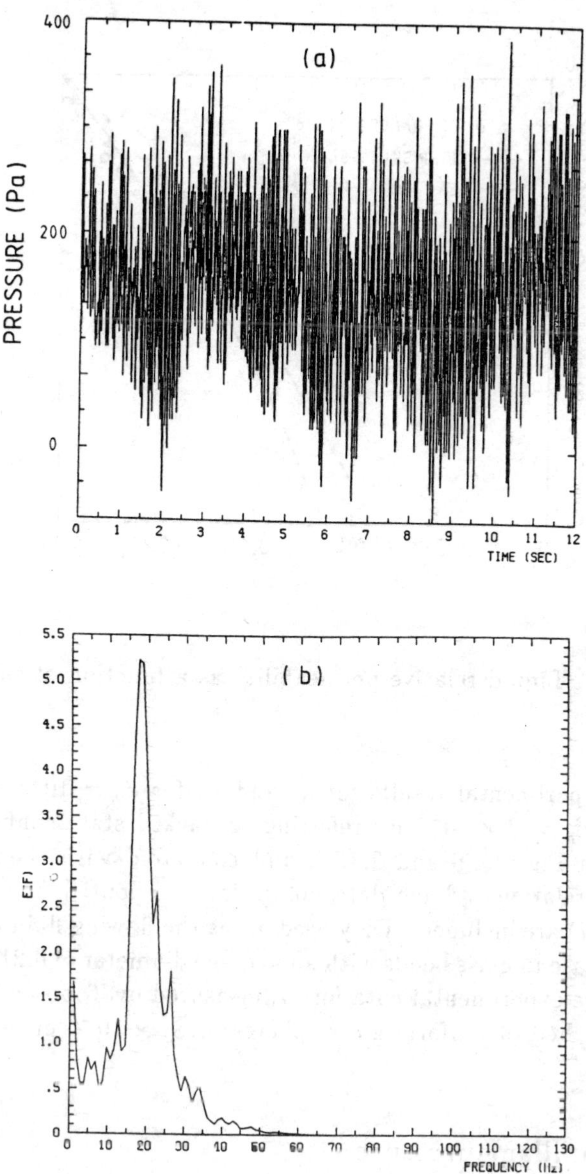

Figure 3.4.5: Wall pressure signal (a), and its frequency spectrum (b).

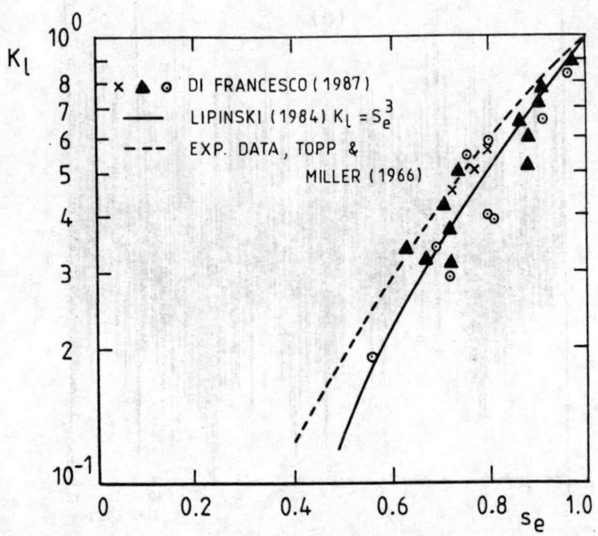

Figure 3.4.6: Liquid relative permeability as a function of effective saturation.

Some experimental results for κ_ℓ and κ_v for $d_p = 10^{-3}$ m, or smaller,
and η_v for $d_p = 3 \times 10^{-3}$ m, referring to packed state configurations, are
presented in Fig. 3.4.6 and 3.4.7, and compared with the predictions of
selected correlations. Some data points for κ_v adopted from Kaviany and
Mittal (1987) are included. They used air as the flowing fluid and water as a
stagnant phase in glass beads with an average diameter of 0.210×10^{-3} m. In
addition, the experimental data for κ_ℓ, measured by Topp and Miller (1966)
in a packed bed of uniform glass spheres 0.181×10^{-3} m in diameter, are
also shown.

3.4.4   Capillary pressure

The pressure difference which must be maintained between two fluids, e.g.,
liquid and vapour, that completely occupy the void space of a porous sample
to retain the saturation of the wetting fluid at a specified value, is defined

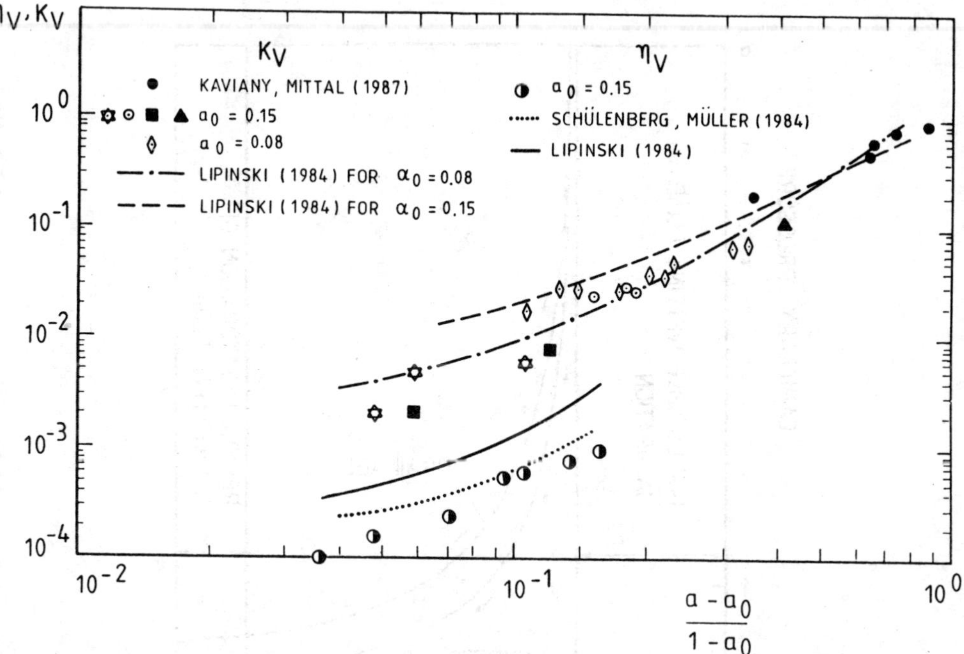

Figure 3.4.7: Gas relative permeability and passability as a function of gas effective saturation (small particle results).

as the capillary pressure, $P_c = P_v - P_\ell$.

At static equilibrium, the capillary pressure is related to the local principal radii of curvature, R and R', of each meniscus within the porous sample by the *Young-Laplace equation*

$$P_c = \sigma \left(\frac{1}{R} + \frac{1}{R'} \right). \qquad (3.4.8)$$

It has been observed that P_c depends upon the previous saturation history of the sample and is not a unique function of s. An example of a capillary pressure relationship containing *hysteresis* effects is plotted in Fig. 3.4.8.

Drainage values are, in general, larger than the ones obtained during imbibition. The role of the residual phase saturations introduced above is also clearly shown.

The capillary pressure relationships are assumed to be controlled by static fluid–fluid–solid properties, i.e., surface tension, wettability, pore

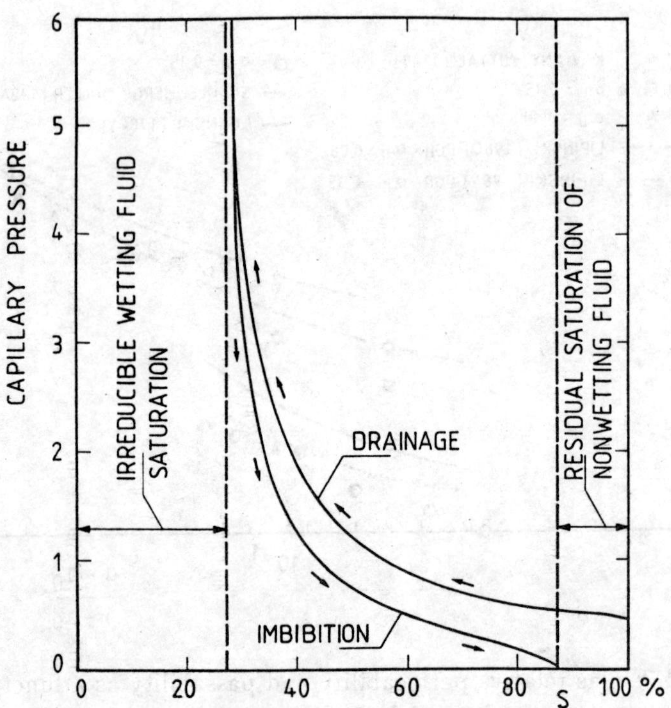

Figure 3.4.8: Typical capillary pressure-liquid saturation curve (Bear, 1972).

geometry and topology. Simple theoretical considerations and experimental evidence indicate that the dynamic capillary pressure relationship obtained from pressure measurements during steady-state two-phase flows should be in close agreement with that derived under quasi-static conditions, since in the two cases the saturation histories would be similar, (Lin and Slattery, 1982).

The following dimensionless *Leverett function* is defined as

$$J(s) = \sqrt{\frac{\kappa}{\epsilon}} \cdot \frac{P_c}{\sigma \cos \theta}.$$ (3.4.9)

For a given *contact angle*, θ, $J(s)$ should be the same for each of a set of geometrically and topologically similar porous media. The S-shape of the

Leverett curve has been well established, the actual values depending on the porous material. Measurements of this curve have been made principally for consolidated geologic media, or sands (Scheidegger, 1957; Stauffer and Dracos, 1986).

A measurement technique has been developed at VKI to obtain drainage capillary curves for particle beds (Stubos, 1985). The test rig used is shown in Fig. 3.4.9. It is part of the general facility described in Fig. 3.4.4 above. Water is evacuated out of an initially fully saturated bed with the aid of a vacuum pump. The difference between the under-pressure prevailing in the tank and the increase in the level Δh, gives the capillary pressure. This can be related to the global saturation in the bed, which, in turn, is found from the amount of liquid that was replaced by air in it.

Experimenting with shallow beds ($20 - 25 \times 10^{-3}$ m in height), to avoid the effect of a saturation profile along the bed, the capillary pressure curve is established and the residual liquid saturation is determined. Care has to be taken that the breakthrough pressure of the porous support is large enough to ensure no interference with the measurement. The tests were then repeated, using a deep bed and observing the thickness of the two-phase region during the incremental changes of the applied underpressure, ΔP. Taking advantage of the shape of the Leverett curve and the hydrostatic liquid pressure distribution in the bed, it is possible to relate the capillary pressure in the middle of the two-phase zone with the global saturation measured.

Satisfactory repeatability in the measurements can be observed in Fig. 3.4.10 (Di Francesco, 1987). For the sake of completeness, the experimental curve determined by Topp and Miller (1966) for uniform 0.181×10^{-3} m glass spheres under steady-state flow conditions, is included. The form proposed by Lipinski (1982), to fit the data given in Scheidegger (1957) on sands, as well as the approximation of the same data by Reed (1982), are drawn on the same figure. It is important to note that experiments on glass and urania beds, performed independently by Reed *et al.*, using a similar technique (with shallow beds only), provided data that compare well with the points of Fig. 3.4.10 (Reed *et al.*, 1987).

The shape of the Leverett curve deserves some further comments.

- The existence of a finite breakthrough capillary pressure at a saturation of unity in the drainage experiment is found. It corresponds to a *J*-value close to 0.4 (Scheidegger, 1957; Bear,1972 ; Stauffer and Dracos, 1986; Di Francesco,1987 ; Reed et al., 1987).

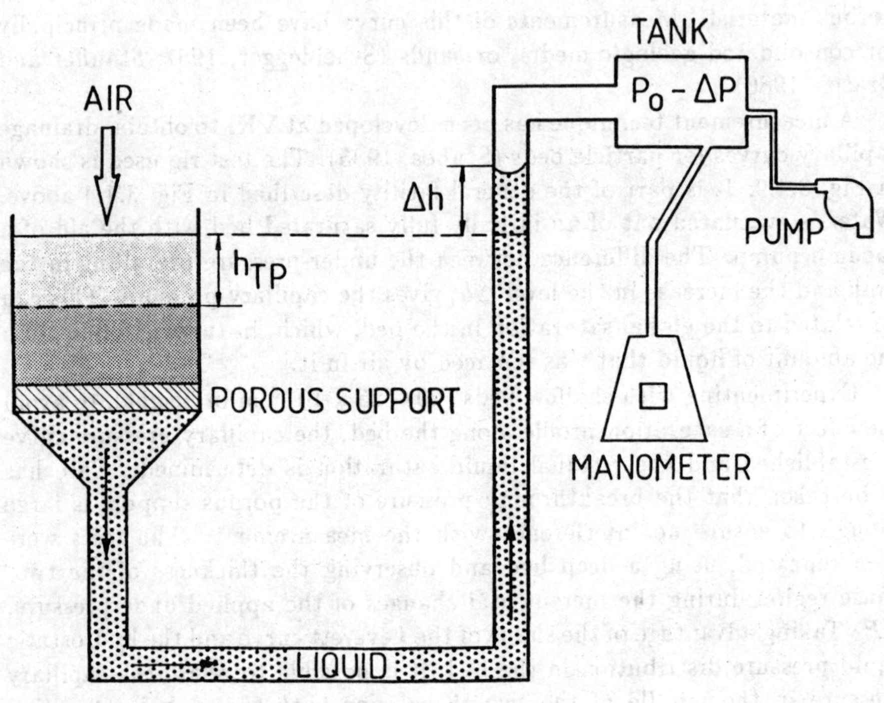

Figure 3.4.9: Quasi-static evacuation technique for J-measurement.

- By comparing the porous bed to a collection of parallel capillary tubes
  initially filled with a wetting liquid, it is observed that as the gas pressure is increased the phase interface becomes more and more concave,
  until the minimum radius of curvature in the largest tube is reached.
  At this point the gas pressure is strong enough to push the liquid from
  the largest tubes. If the gas pressure is further increased, smaller tubes
  can be evacuated and the saturation decreases. Such a representation
  implies that the Leverett function is likely to depend on the distribution of the pore sizes (i.e., on the particle size distribution for a debris
  bed). In fact the narrower is this distribution, the more horizontal
  the central part of the Leverett function should be, since a small in-

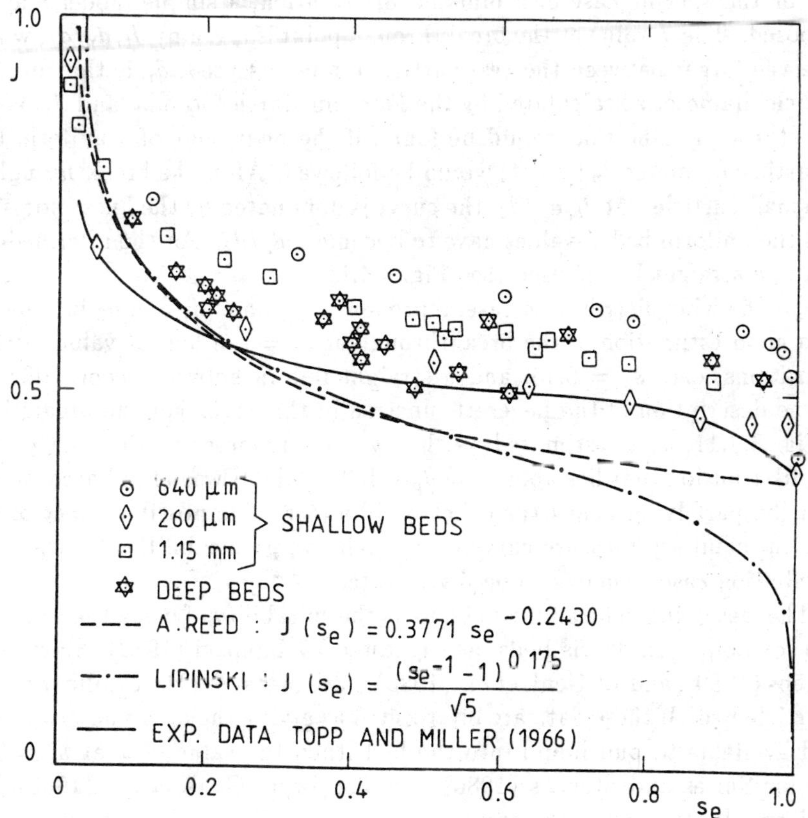

Figure 3.4.10: Leverett function as a function of effective saturation.

crease in capillary pressure above the breakthrough value is sufficient
to empty most of the uniform pores. On the other hand if the distribution is wide, the breakthrough point will be defined mainly by
the large particles, while the high capillary pressure values close to the
residual saturation will be imposed by the smaller particle sizes. The
intermediate J-values are expected to vary rather continuously with
saturation. Such a trend is emphasized in Fig. 3.4.11, where measurements for bimodal beds and beds with a wide size distribution, like the
ones used in the in-pile tests D-10 and DCC-1, are reported (Reed *et
al.*, 1987).

For the specific case of a bimodal distribution, a simple model may be proposed. The J-value at the breakthrough point is given by $J_{br} d_p/d_1$, where d_1 is the larger between the two particle diameters used, d_p is the effective particle diameter as calculated by the Fair and Hatch formula, and J_{br} is the breakthrough value that would be found if the behaviour of a uniform bed of particle diameter d_p ($= 0.4$) would be followed. After the breakthrough of the small particles, at $J_{br} d_p/d_2$, the curve is dominated by the finest pores, so that the uniform bed J-values have to become $J.d_p/d_2$. For the intermediate region, a straight line is used (see Fig. 3.4.11).

For the wide distribution case, a proper selection of d_1 and d_2 in order to get a good estimation of the breakthrough at $s_e = 1$, a high J-values at low saturations, e.g., $s_e = 0.15$, and a straight line in between seem sufficient for the description of the Leverett function in this case. For the predictions of Fig. 3.4.11, d_1 is estimated as the average diameter of the part of the size distribution that lies above the d_p-value, while d_2 was found accordingly from the part lying below the effective diameter. The possible effect of the different capillary pressure curve on the relative permeabilities for the wide distribution case, remains to be investigated.

The last point to be addressed here is the suitability of using the drainage data for boiling in debris beds, as suggested by Lipinski (1982). Shires and Stevens (1980), and El-Genk *et al.* (1983), have measured the capillary rise in a particle bed. If these data are interpreted as giving the maximum capillary head available to pull liquid into the bed, then the value of J at $s_e = 0$ is 0.35 for Shires and Stevens (1980), and 0.4 for El-Genk *et al.* (1983), i.e. much smaller than what the drainage data show. These values are consistent with the deduction of the capillary pressure at dryout in a boiling bed of particles (0.22 and 0.4×10^{-3} m) given by Macbeth and Trenberth (1984). It is the opinion of the present authors that the capillary rise in a bed and possibly the data published by Macbeth and Trenberth (1984) correspond to the breakthrough pressure rather than to the capillary pressure at dryout. In this case, the aforementioned values are consistent with the drainage data. Further support to this statement comes from recent observations in both out-of-file (Buchlin and Van Koninckxloo, 1986), and in-pile tests (PIRAMID-1), of significantly increased boiling temperature with respect to what should have been expected due to the system pressure. This increase could be explained by adding the capillary head to the system pressure. A value of 0.15 *bar* is thus obtained for a sodium-UO_2 bed, indicating the possibility of large capillary pressures close to dryout.

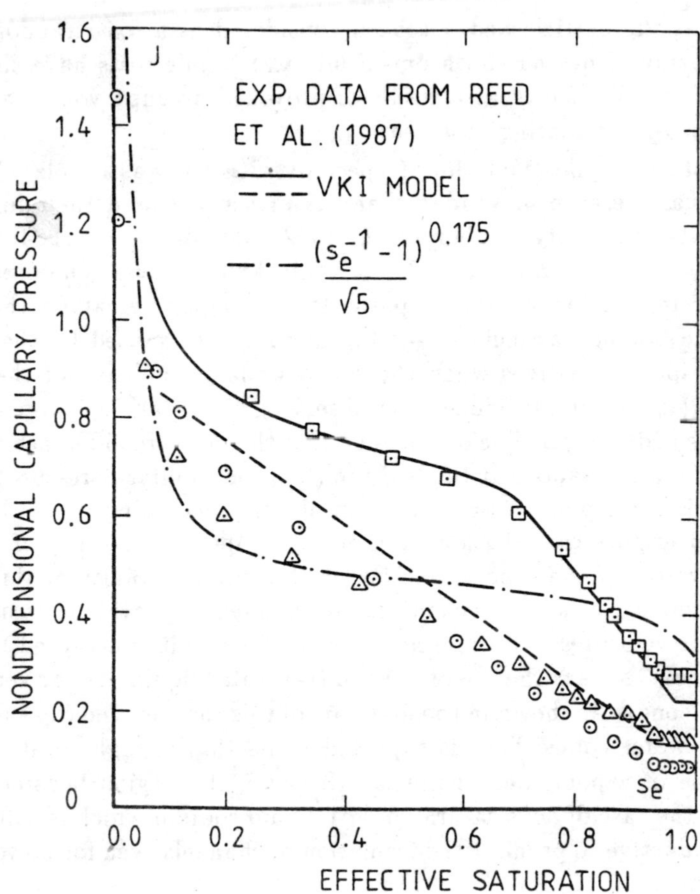

Figure 3.4.11: Leverett function for bimodal and wide particle size distributions.

3.4.5   Bed structural changes

Up to now, the particle bed has been considered as a fixed randomly distributed array. However, both out-of-pile and in-pile tests have shown an enhanced bed coolability as a result of provoked or unprovoked, stable or unstable self-restructuring processes.

Several out-of-pile simulations (especially with water as a coolant) outline the eventual presence of vapour channels in the bed and their intriguing effect on its coolability (Gabor *et al.*, 1972; Barleon *et al.*, 1984; Stevens and Trenberth, 1982; Buchlin and VanKoninckxloo, 1986). The presence of such discrete and low resistance paths for the vapour escaping the upper part of the boiling particulate (see Fig. 3.3.2), is suggested by the porous flow equations themselves when the top boundary condition of the bed is examined (Reed, 1986). Indeed, by observing the relatively large bubbles leaving the bed, it is possible to conclude that the pressure difference between the liquid and the vapour at the bed top (i.e., the capillary pressure) is very small. This corresponds to a saturation of unity according to the Leverett curve, and implies that the flow area for the vapour at the top of the bed is almost zero, i.e. the vapour velocity must approach infinity at this point. Thus, there must be some mechanism like channels to provide for reasonable vapour exit velocities. All these hold true as long as it is accepted that the vapour is able to reach the top of the bed (saturated liquid pool or very small subcooled zone). As shown in the first part of this chapter, the top subcooled region for water cooled beds is expected to be thin enough to allow early penetration of vapour, i.e., channels. However, this region becomes much thicker in the case of beds saturated with liquid sodium which is sufficiently highly conductive to prohibit the formation of channels even for power levels up to dryout.

Some modeling efforts for the prediction of the channel length have been reported. Qualitative analogy between spouting of fluidized powder and channeling, has been exploited by Buchlin *et al.* (1982), leading to the implementation of an empirical model in the TORPEDO code (Benocci *et al.*, 1981). Naik and Dhir (1982) used a pipe flow formulation for the vapour flow in the channels and obtained experimental data on the shape and number of channels. Recently, Schwalm (1985), Barleon *et al.* (1985), as well as Mehr and Wurtz (1985), suggested to model the channeled region as a porous layer of higher permeability.

In fact, the common modeling approach (Jones *et al.*, 1982; Schwalm and Nijsing, 1982; Reed, 1982) is exemplified by the Lipinski model (1982).

It is assumed that the vapour pressure at the bottom of a channel is sufficient to offset the weight of the overlying particles plus liquid, and that the flow resistances in the channeled region are negligible. However, Lipinski's model presents certain shortcomings regarding the prediction of out-of-pile (Schwalm, 1985), as well as in-pile (Mitchell *et al.*, 1984) channel data, especially when the channel length is large. For example, if the case of the D-10 in-pile test is considered, an estimation of the subcooled zone thickness close to unchanneled dryout, according to (3.3.6) gives 5.5 cm, while the predicted channel length extends to almost 11 cm, i.e., penetration should have been observed at much lower power levels. This fact along with the uncertainties concerning the influence of the vapour channel formation on the coolability of the bed (Barleon *et al.*, 1984; Schwalm, 1985) stresses the need for further refinement of the modeling of channel behaviour.

In the previous attempts,, the friction between the particles as channels are formed is neglected. Lipinski (1984) introduces the idea of a *sticking factor*, S_f, which means that the vapour pressure must be S_f times the overlying bed pressure at the bottom of a channel. The same idea, in terms of a sticking pressure, is proposed by Schwalm (1985). However, it appears that a non dimensionalization of this sticking pressure leads straight directly to the sticking factor concept. Therefore, Lipinski's model can be extended in the following way. The vapour pressure P_v at the bottom of a channel is expressed as

$$P_v = S_f \left[\rho_s(1 - \epsilon) + \rho_\ell \epsilon \right] g L_c \qquad (3.4.10),$$

where L_c is the channel length, and ρ_s and ρ_ℓ are the solid and liquid densities, respectively.

In the channeled region, liquid flows slowly and orderly downward between the channels, so that the liquid pressure P_ℓ at the channel bottom is regarded as hydrostatic

$$P_\ell = \rho_\ell g L_c. \qquad (3.4.11)$$

After introducing the capillary pressure, P_c, by means of the Leverett function, $J(s)$, given by the relation (3.4.9), the expression for channel length is finally obtained in the form

$$L_c = \frac{\sigma \sqrt{\epsilon/\kappa} J(s)}{[S_f [\rho_s(1 - \epsilon) + \rho_\ell \epsilon] - \rho_\ell] g}. \qquad (3.4.12)$$

The saturation value at the channel bottom needed for the calculation of the Leverett function, is obtained by using the continuity of the pressure gradient at this point. Equation (3.4.10) gives

$$\left.\frac{dP_v}{dz}\right|_{z=L_c} = -S_f\left[\rho_s(1-\epsilon) + \rho_\ell\epsilon\right]g, \qquad (3.4.13)$$

while from the vapour momentum equation, we obtain

$$\frac{dP_v}{dz} = -\rho_v g - \frac{\mu_v V_v}{\kappa\kappa_v} - \frac{\rho_v V_v^2}{\eta\eta_v}. \qquad (3.4.14)$$

Since the vapour velocity just below the base of a channel is a function of the heat flux at this point, which, in turn, is a function of the channel length L_c, equations (3.4.12), (3.4.13) and (3.4.14) are solved simultaneoulsy to give both the channel length and the saturation value at the top of the packed region, provided that a value of S_f is available. This fact allows for the computation of the saturation profile in the packed region and the prediction of the dryout power for a channeled bed (Lipinski, 1982).

A semi-empirical model for S_f was developed by Stubos and Buchlin (1988), using simple experiments to observe the formation and evolution of channels in glass and steel beds. It was concluded that the onset of channeling is governed by rather high S_f values, given by suitable semi-empirical expressions. Consequently the incipient channel length is small (of the order of $1-2\times10^{-2}$ m) and the occurrence of channels can be delayed for higher powers, or even suppressed completely (as in D-10). However, once the channels are formed, it is much easier for the vapour to lengthen them, as its velocity increases with power. This means that S_f decreases rapidly and tends asymptotically to a value close to unity, as demonstrated by the data obtained in the OPERA facility (ferrite particles in water) and plotted in Fig. 3.4.12. The channel length reaches its final value given by (3.4.12), with $S_f = 1$. The clear distinction between incipient ($S_f > 1$) and evolved channeling ($S_f \approx 1$) allows to account for the non occurrence of channels in D-10 and other in-pile tests even at high power levels.

It should be noted that the extended channel length is well predicted by the relation (3.3.5) which coincides with (3.4.12) when the value of the Leverett function lies between 0.37 and 0.49.

Some uncertainties exist both in experiments and predictions in the case of light particle beds, e.g., those made of small glass beads. Most of the tests performed with such beds, reveal unstable 'channel' behaviour, especially when the particles are smaller than about 0.6×10^{-3} m in diameter and the liquid is water (Jones *et al.*, 1982; Campos, 1983; Gladnick, 1985; Reed, 1986).

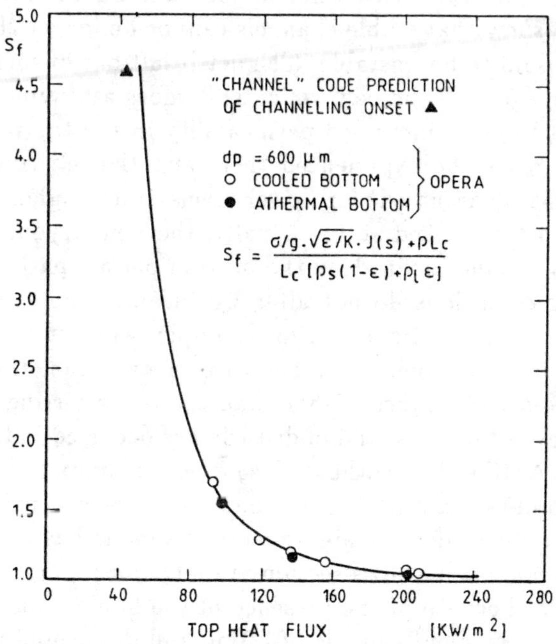

$$S_f = \frac{\sigma/g \cdot \sqrt{\varepsilon/K} \cdot J(s) + \rho L_c}{L_c [\rho_s(1-\varepsilon) + \rho_l \varepsilon]}$$

Figure 3.4.12: Sticking factor as a function of top heat flux.

There are mainly two different types of observed unstable 'channeling':

- Thin, long and twisted vapour paths characterized by a short lifetime, corresponding to the time needed for a chain of bubbles to rise through them. These occur sporadically, at random locations in the bed.

- The bed can be occasionally disturbed by the formation of large bubbles growing until the pressure is sufficient for them to rise through the solid packing. Figure 3.4.13a and b show visualizations of such instabilities performed at the VKI within bottom and volumetric heated beds, respectively.

A possible explanation for these observations is discussed by Stubos and Buchlin (1987). It seems logical to expect that there is a maximum stable channel length. Above, it the channel collapses, or a different kind of disturbances (unstable in nature) occurs. This is supported by some experimental evidence that the formal, 'well shaped' and stable channels do not extend further than 5-6 cm. The data from VKI, KfK, Winfrith, Jones at ANL, and Reed at MIT, on beds with different solid and liquid materials and particle diameters, show that stable channels cannot be longer than 6 cm. The longer ones are said to be unstable. The net result out of such disturbances in that the upper part of the bed, which undergoes a heaving motion, is (or can be simulated as) an increased permeability in the top part of the bed. Intensive boiling may be experienced there and the self coolability of the disturbed part is 'guaranteed' for power levels much higher than the ones giving dryout in the packed state. Finally, the new dryout power for the disturbed bed is mainly defined by the bottom packed part. Consequently, the top cooling conditions do not alter significantly the coolability of the disturbed bed. Its adiabatic plane, for example, will not be much affected by a change in the top cooling temperature. In addition this new dryout, power is expected to be much higher than the one referring to the packed state. It is believed that this kind of disturbance occurred in the in-pile tests with (D10, PIRAMID-1) or without (D4) a power ramp.

An effort to obtain a criterion for the occurrence of such phenomena is now briefly described. If a certain amount of vapour has to move upward, then the restriction on the cross-section available for its motion at some position in the bed, because of the presence of the liquid, could be such that a vapour pressure gradient capable of lifting and disturbing the upper part develops. In this case, the vapour chooses the easiest way to escape. For example, it prefers to gather in local cracks or cavities, forming large bubbles which then travel upward through the packing (discontinuous heaving motion of the disturbed part). If the pressure of the vapour produced during the heating of a bed at moderate power levels is at some point sufficient to lift the overlying bed, then the disturbance occurs at this point. If a power ramp is applied to provoke the disturbance, then it is likely to occur just above the position of the adiabatic plane, before the application of the steep ramp. The analysis shows that during the transient, the position of maximum vapour pressure remains very close to that of the adiabatic plane before the ramp, while the minimum saturation front moves upward (Ruel, 1986).

The present authors have developed the criterion

$$C_r = \frac{\nu_v \rho_v V_v}{\kappa \kappa_v \left((1 - \epsilon)\rho_s + \epsilon \rho_\ell \right) g}, \qquad (3.4.15)$$

occur. The form of the criterion (3.4.15) is suitable for the case of air injection experiments, where the saturation can be considered uniform. By integrating over the boiling zone above the adiabatic plane, and using an average, κ_v, we obtained for heated beds

$$C_r = \frac{\nu_v Q H_{sat}^2}{2L\kappa\kappa_v\left((1-\epsilon)\rho_s + \epsilon\rho_\ell\right)gH_{dis}}, \qquad (3.4.16)$$

where H_{sat} is the height of the top boiling zone, and H_{dis} is the height of disturbed zone. It is almost equal to the distance of the adiabatic plane from the top before the ramp. An indication for the suitability of such criteria is given in Fig. 3.4.14. The pressure drop during air-water flow through a bed of steel particles with an effective diameter of 0.440×10^{-3} m, is recorded with time. At the moment of the occurrence of the disturbance at 40 mm below the bed top, the pressure gradient is almost equal to the gravitational one ($C_r = 1$), and then is suddenly released, showing that the air escapes much easier now.

The last point to be mentioned here for the bed restructuring phenomena is their irreversibility. Disturbances in the beds do not disappear even if the vapour velocity is significantly reduced after their occurrence. Early channel formation or instabilities are then observed if the velocity is increased again.

3.5 Conclusions

The review of the various out-of-pile investigations carried out to provide support to the in-pile PAHR programme, brings out the effect of the main important parameters on the coolability of the debris bed.

- The particle size has a strong effect on the value of the dryout heat flux.

- Shallow beds and deep beds do not exhibit the same behaviour.

- Bottom cooling may induce 'downward' boiling if capillarity dominates, thus enhancing significantly the bed coolability.

- The simulation of the top-subcooling effect is difficult in out-of-pile experiments using ordinary liquids like water.

- Proper constitutive relationships for the description of two-phase flow in the PAHR debris bed have been presented.

- Disturbances of the bed structure have to be expected and further modeling is requested in this field.

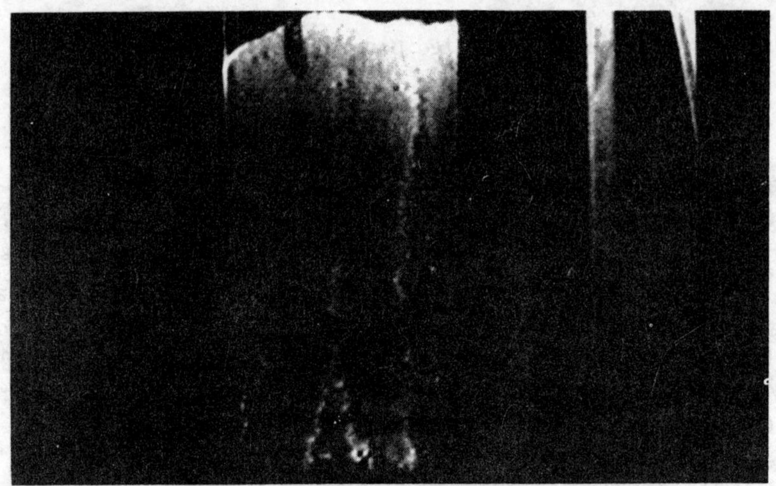

a : bottom heated bed

b : volumetrically heated bed

Figure 3.4.13: 'Channel' instabilities in particulate beds.

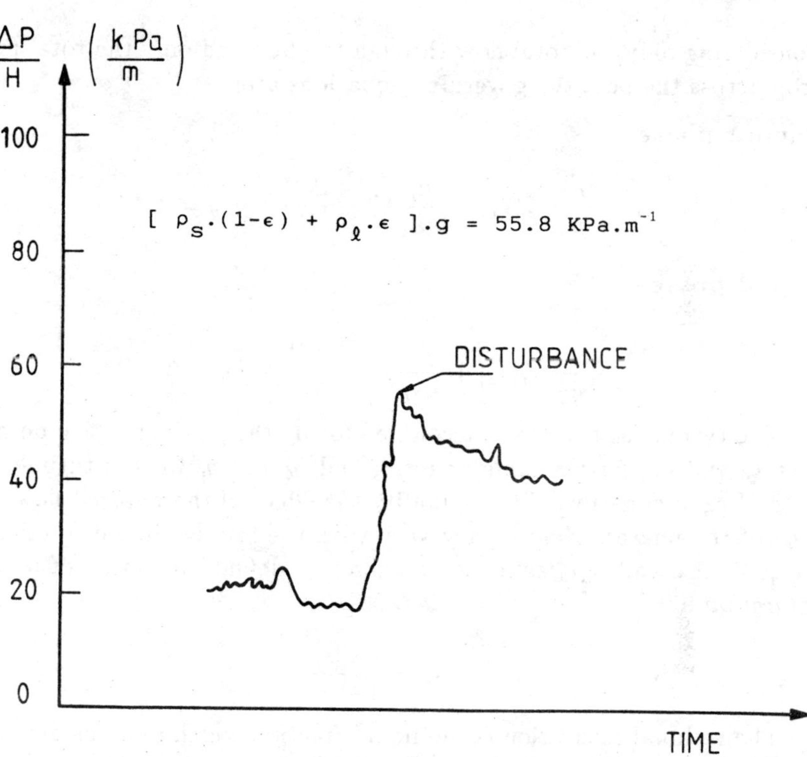

Figure 3.4.14 Out–of–pile recording of the occurrence of bed disturbance.

APPENDIX

A. Zero-Dimensional Model

Considering only the total flow through the bed and only the total pressure drop across the bed, the governing equations are:

Vapour phase

$$\frac{\rho_v}{\eta\eta_v}V_v|V_v| + \frac{\mu_v}{\kappa\kappa_v}V_v - \frac{\Delta P}{H} + g\rho_v = 0. \tag{A.1}$$

Liquid phase

$$\frac{\rho_\ell}{\eta\eta_\ell}V_\ell|V_\ell| + \frac{\mu_\ell}{\kappa\kappa_\ell}V_\ell - \frac{\Delta P}{H} + g\rho_\ell = 0. \tag{A.2}$$

The two-phase friction is accounted for by the phasic relative permeabilities κ_ℓ and κ_v, for the laminar term, and η_ℓ and η_v for the turbulent part of the Ergun equation. They quantify the effect of the reduced flow area to each of the phases. Hence, they vary with the relative liquid fraction. The latter is also called *effective saturation*, s_e, defined in terms of a *residual saturation* by

$$s_e = \frac{s - s_r}{1 - s_r}.$$

The residual saturation is the liquid fraction which remains between the particles by capillarity with a given pressure drop across the porous layer. It can be evaluated by the empirical Brown's relation

$$s_r = \frac{1}{22}\left[\frac{(1-\epsilon)^2\sigma\cos\theta}{d_p^2\epsilon^3\rho_\ell g}\right]^{0.264}, \tag{A.3}$$

where θ is the *wetting contact angle* between the liquid and the particle.

The relative permeabilities can be non-linear functions of the effective saturation. The relations of the relative permeabilities derived for laminar cocurrent flow are assumed to be applicable to counter-current flow. Thus

$$\kappa_\ell = \eta_\ell = s_e^3, \tag{A.4}$$

and

$$\kappa_v = \eta_v = (1 - s_e)^3. \tag{A.5}$$

The analytical form of the capillary pressure drop in a bed is expressed as

$$\Delta P_v - \Delta P_\ell = \sqrt{\frac{\epsilon}{5\kappa}}\,\sigma\cos\theta = \frac{6(1-\epsilon)}{\epsilon d_p}\sigma\cos\theta. \tag{A.6}$$

To be consistent with the previous substitutions, the factor of 6 in the relation $(A.6)$ ought to be replaced by $\sqrt{30}$.

The bed equations are completed with the mass and energy balance equations, with allowance for the particulate debris support to be adiabatic, or cooled, and impermeable or porous with a net mass flux. Their solution leads to a simple quadratic equation which, for an impermeable and a-thermal bottom becomes

$$A\Phi^2 + B\Phi + C = 0, \tag{A.7a}$$

with

$$A = \frac{1}{(\rho_\ell - \rho_v)\,g\eta\,L^2}\cdot\left[\frac{1}{\eta_v\rho_v} + \frac{1}{\eta_\ell\rho_\ell}\right], \tag{A.7b}$$

$$B = \frac{1}{(\rho_\ell - \rho_v)\,g\kappa L}\cdot\left[\frac{\nu_v}{\kappa_v} + \frac{\nu_\ell}{\kappa_\ell}\right], \tag{A.7c}$$

$$C = -\left[1 + \frac{\lambda}{H}\right], \tag{A.7d}$$

and

$$\lambda = \frac{6(1-\epsilon)\sigma\cos\theta}{\epsilon d_p\,(\rho_\ell - \rho_v)\,g}, \tag{A.7e}$$

where λ is the distance where the fluid would be drawn up into a dry particulate layer from below against the force of gravity. It defines the capillary head of the system. In general, the value of the liquid fraction, s_e, is not determined a-priori and the dryout heat flux is obtained by maximizing numerically the positive root of equation $(A.7)$, with respect to variations in s_e. This leads to the result

$$(\Phi_d) = \left[\frac{\Phi_t^4}{4\Phi_\ell^2} + \Phi_t^2\right]^{0.5} - \frac{\Phi_t^2}{2\Phi_\ell}, \tag{A.8a}$$

where Φ_ℓ and Φ_t are the laminar and the turbulent contributions, respectively, derived explicitly as

$$\Phi_\ell = \frac{1}{B} \cdot \left[1 + \frac{\lambda}{H} \right], \tag{A.8b}$$

$$\Phi_t^2 = \frac{1}{A} \cdot \left[1 + \frac{\lambda}{H} \right], \tag{A.8c}$$

In the case of small particles, typically $d_p < 10^{-3}$ m, the flow is regarded as laminar and the Lipinski zero model is similar to the Shires-Stevens model. It turns out that

$$\Phi_{d,\ell} = \frac{g(\rho_\ell - \rho_v)\kappa L}{\left(\nu_\ell^{1/4} + \nu_v^{1/4} \right)^4} \left(1 + \frac{\lambda}{H} \right). \tag{A.9}$$

For large particles, the dryout heat flux is given by

$$\Phi_{d,t} = L \left(\frac{g(\rho_\ell - \rho_v)\rho_v \eta}{(1 + (\rho_v/\rho_\ell)^{0.25})^4} \left(1 + \frac{\lambda}{H} \right) \right)^{0.5}. \tag{A.10}$$

As noted above, this turbulent limit of the critical heat flux does not differ much from the flooding model as proposed by Ostensen and Lipinski, except for the capillary term.

the Lipinski model predicts the possibility of a downward vapour flow when the bottom plate is cool enough to condense all the vapour. The resulting condensed liquid is then drawn up into the bed by capillary force. The process does not develop in a bed with thermally isolated support, because it is soon stopped by the back pressure, due to the accumulation of vapour.

This so-called 'downward boiling' is expected in the case of small particles. Thus, an explicit expression is derived for the laminar limit case

Upward heat flux at dryout

$$\Phi_{up,d} = 0.5\Phi_{\ell o} \left(2\frac{\lambda}{H} + \left((2\frac{\lambda}{H})^2 + 1 \right)^{0.5} + 1 \right). \tag{A.11}$$

Downward heat flux at dryout

$$\Phi_{dow,d} = \Phi_{up,d} - \Phi_{\ell o},$$

given by the relation $(A.9)$ without the capillary term. The total heat flux from the bed, $(\Phi_d)_{,tot}$, in the presence of downward boiling, is related to that obtained in the case of an adiabatic bottom ($\Phi_{d,ad}$, see $(A.9)$) by

$$\frac{(\Phi_d)_{,tot}}{\Phi_{d,ad}} = \frac{\Phi_{up,d} + \Phi_{dow,d}}{\Phi_{d,ad}} = \left(\frac{2\frac{\lambda}{H} + \left((2\frac{\lambda}{H})^2 + 1\right)^{0.5}}{\frac{\lambda}{H} + 1}\right).$$

B. Fractional Downward Heat Flux by Conduction

The bed is uniformly heated by a volumetric power, Q. It is characterized by an effective conductivity, k_e. The adiabatic plane is assumed to be located at the top of the subcooled bottom region of depth H_{sub}.

The temperature distribution in the subcooled zone is given for $0 \leq z \leq H_{sub}$ by

$$T(z) = -\frac{Q}{k_e} z \left(\frac{z}{2} - H_{sub}\right) + T_{bot}. \tag{B.1}$$

Since $T = T_{sat}$ at $z = H_{sub}$, $(B.1)$ gives

$$H_{sub} = \left[\frac{2k_e\left(T_{sat} - T_{bot}\right)}{Q}\right]^{0.5}.$$

The downward heat flux due to conduction, $\Phi_{dow,con}$, is

$$\Phi_{dow,con} = QH_{sub} = \left[2k_eQ\left(T_{sat} - T_{bot}\right)\right]^{0.5}. \tag{B.2}$$

The substitution of the total heat flux, $\Phi_{tot} = QH$, in $(B.2)$, leads to the relation of the fractional downward heat flux by conduction

$$\frac{\Phi_{dow,con}}{\Phi_{tot}} = \left[\frac{2k_e\left(T_{sat} - T_{bot}\right)}{\Phi_{tot} H}\right]^{0.5}.$$

C. Subcooled Zone Thickness At The Top Of The Bed

The power removed at the top of the bed is characterized by the upward flux, Φ_{up}, expressed by

$$\Phi_{up} = H_{up}Q, \tag{C.1}$$

where H_{up} is the upper part of the heated bed defined by the adiabatic plane.

In an insulated bottom bed $H_{up} = H$.

In the subcooled top region of thickness H_{sub}, the temperature distribution for $0 \leq z \leq H_{sub}$, is given by

$$T(z) = -\frac{Q}{2k_e}z^2 - \frac{z\Phi_{sat}}{k_e} + T_{sat}, \qquad (C.2)$$

where Φ_{sat} is the upward heat flux provided by the boiling region, with

$$\Phi_{sat} = Q\left(H_{up} - H_{sub}\right). \qquad (C.3)$$

Combining $(C.3)$ with $(C.2)$ expressed at $z = H_{sub}$, where $T = T_{top}$, gives for an adiabatic bottom bed

$$H_{sub}^2 - 2HH_{sub} + \frac{2k_e\left(T_{sat} - T_{top}\right)}{Q} = 0.$$

The solution of this equation is

$$H_{sub} = H - \left[H\left[H - \frac{2k_e\left(T_{sat} - T_{top}\right)}{\Phi_{up}}\right]\right]^{0.5}.$$

References

Barleon, L. and Werle, H. *Debris Bed Investigation with Adiabatic And Cooled Bottom*. 9th Meeting Liquid Metal Working Group, Rome 1980.

Barleon, L., Thomauske, K. and Werle, H. Cooling of debris beds. *Nuclear Technology*, **65**:67-86, 1984.

Barleon, L. Thomauske, K. and Werle, H. *Investigation On Channel Penetration At KfK*. Meeting on Debris Bed Modeling, Ispra, May 29-30, 1985.

Bear, J. *Dynamics Of Fluids In Porous Media*. American Elsevier Publishing Company Inc., New York, 1972.

Benocci, C., Buchlin, J-M. and Joly, C. *Theoretical And Numerical Modeling Of The PAHR Debris Bed Behaviour Implemented In The TORPEDO Code*. von Karman Institute, CR 1982-04/EA, November 1981.

Brooks, R. H. and Corey, A. T. Properties of porous media affecting fluid flow. *J. Irrig. Drainage Division, ASCE*, **92, IR2**:61-88, 1966.

Buchlin, J-M. and Thiry, F. *Simulation Expérimentale Du Refroidissement Post-accidentel d'Un Lit De Débris Saturé Par Un Liquide*. von Karman Institute, CR 1983-08/EA, 1982.

Buchlin, J-M., Benocci, C., Joly, C. and Siebertz, A. *A Two Dimensional Finite Difference Modeling Of The Thermohydraulic Behaviour Of The PAHR Debris Bed Up To Extended Dryout*. von Karman Institute, Preprint 1982-24, 1982.

Buchlin, J-M., Simon, G. and Barth, U. *Experimental Simulation Of The Post Accident Heat Removal Of Liquid Saturated Debris Bed*. von Karman Institute, CR 1984-05/EA, 1983.

Buchlin, J-M. and van Koninckxloo, T. *O. P. E. R. A. II– A Test Facility To Study The Thermohydraulics Of Liquid Saturated Self Heated Porous Media*. von Karman Institute, TM 41, 1986.

Buretta, R. G. and Berman, A. S. Convective heat transfer in liquid saturated porous layer. *Journal of Applied Mechanics*, June 1976.

Campos, J. M. *Theoretical And Experimental Investigation Of Channeled Boiling In Bottom Heated Particulate Beds*. von Karman Institute, PR 1983-05, June 1983.

Catton, I., Dhir, V. K., Somerton, C. W. and Squarer, D. *An Experimental Study Of Debris Bed Coolability Under Pool Boiling Conditions.* EPRI, NP-3094, May 1983.

Chang, S. H. and Kim, S. H. Derivation of a dryout model in a particle debris bed with the drift-flux approach. *Nuclear Science and Engineering*, 91:404-413, 1985.

Chu, W., Dhir, V. K. and Marshall, J. *Study Of Pressure Drop, Void Fraction And Relative Permeabilities Of Two-Phase Flow Through Porous Media.* AIChE Symposium Series, 79:224-235, 1983.

Cook, I. and Peckover, R. S. *Effective Thermal Conductivity Of Debris Beds.* In "Post Accident Debris Cooling", Proc. 5th PAHR Information Exchange Meeting, Karlsruhe, Braun, 1983.

Dhir, V. K. and Catton, I. *Study Of Dryout Heat Fluxes In Beds Of Inductively Heated Particles.* UCLA, NUREG-0262 NRC 7, 1977.

Di Francesco, M. *Etude Des Paramètres d'Écoulement En Simple Et Double Phase Dans Un Lit De Particules.* Travail de fin d'études, Univ. Libre de Bruxelles, Institut von Karman, May 1987.

El-Genk, M. S. and Bergeron, E. *Dryout Heat Flux In Particulate Beds of Steel Grit.* ANS Annual Meeting, Detroit, June 1983.

El-Genk, M. S., Louie, M., Bergeron, E. and Mitchell, D. *Dependence of Porosities And Capillary Pressures In Particle Beds On The Morphology And The Size Of The Particles.* AIChE Symposium Series 225, 79:256-267, 1983.

Fand, R. M., Kim, B. Y. K., Lam, A. C. C. and Phan, R. T. Resistance to the flow of fluids through simple and complex porous media whose matrices are composed of randomly packed spheres. *Trans. ASME, J. of Fluids Engineering*, 109(3):268, 1987.

Gabor, J. D., Epstein, M., Jones, S. W. and Cassulo, J. C. *Status Report On Limiting Heat Fluxes in Debris Beds.* ANL/RAS 80-21, Sept 1980.

Gabor, J. D., Hesson, J .C. Baker L. and Cassulo, J. C. Simulation experiments on heat transfer from fast reactor fuel debris. *Trans. ANS*, 15(2):836, 1972.

Gabor, J. D., Sowa, E. S., Baker, L. and Cassulo, J. C. *Studies And Experiments On Heat Removal From Fuel Debris In Sodium.* ANS Fast Reactor Safety Meeting, Beverly Hills, CA, 1974.

Gladnick, P. G. *Experimental Investigation Of Channeling During Boiling In Liquid Saturated Porous Media.* von Karman Institute, SR 1985-24.

Gronager, J. A., Schwarz, M. and Lipinski, R. J. *PAHR Debris Bed Experiment D4*. NUREG/CR 1809, January 1981.

Hardee, H. C. and Nilson, R. H. Natural convection in porous media with heat generation. *Nuclear Science and Engineering*, **69**:119, 1977.

Joly, C. and Le Rigoleur, C. *General And Particular Aspects Of The Particulate Bed Behaviour In The PAHR Situation For Liquid Metal Fast Breeder Reactors*. In von Karman Institute LS 1979-04 "Fluid Dynamics of Porous Media in Energy Applications", February 1979.

Jones, S. W., Baker, L., Bankoff, S. G., Epstein, M. and Pederson, D. R. A theory for prediction of channel depth in boiling particulate beds. *Trans. ASME, J. Heat Transfer*, **104**(4):806-807, 1982.

Kampf, H. and Karsten, G. Effects of different types of void volumes on the radial temperature distribution of fuel pins. *Nuclear Appl. Tech.*, **9**:288, 1970

Kaviany, M. and Mittal, M. Funicular state in drying of a porous slab. *Int. J. Heat Mass Transfer*, **30**:1407-1418, 1987.

Keowen, R. S. *Dry Out Of A Fluidized Particle Bed With Internal Heat Generation*. M.S. Thesis, University of California, 1974.

Larson, R. G., Davis, H. T. and Scriven, L. E. Displacement of residual nonwetting fluid from porous media. *Chem. Eng. Sci.*, **36**:75-85, 1981.

Lee, H. S. and Catton, I. *Two-Phase Flow In Stratified Porous Media*. 6th Information Exchange Meeting on Debris Coolability, Los Angeles, Nov. 7-9, 1984.

Lin, C-Y. and Slattery, J. C. Three-dimensional, randomized network model for two-phase flow through porous media. *AIChE Journal*, **28**:311-324, 1982.

Lipinski, R. J. A particle bed dryout model with upward and downward boiling. *Trans. ANS*, **35**:350, 1980.

Lipinski, R. J. *A Model For Boiling And Dryout In Particle Beds*. Sandia Labs Report SAND82 0765 (NUREG/CR-2646), June 1982.

Lipinski, R. J. A coolability model for post accident nuclear reactor debris. *Nuclear Technology*, **65**:53-66, 1984.

Macbeth, R. V. and Trenberth, R. *Pressure Measurements In Boiling Particle Beds With Water At 1 Bar*. AEE Winfrith Report AEEW-R 1641, 1984.

MacDonald, I. F., El-Sayad, M. S., Mow, K. and Dullien, F. A. L. Flow through porous media - The Ergun's equation revisited. *Ind. Eng. Chem. Fund.*, **18**:199, 1979.

Mehr, K. and Wurtz, J. *A Model For The Channeling Of Particle Beds Based On A Parallel Capillary Tube Approach.* Technical Note No. I.06.C1.85.166, Ispra Joint Research Center, December 1985.

Mehr, K. and Wurtz, J.: *Heat Transfer In Self-Heated Particle Beds Submerged In Liquid Coolant.* VKI Lecture Series 1988-01 "Modelling and Applications of Transport Phenomena in Porous Media", Nov.30-Dec.3, 1987.

Mitchell, G. W. and Ottinger, C. A. *The D7 Experiment Heat Removal From A Shallow Stratified UO_2-Sodium Particle Bed.* Proc. 5th Post Accident Heat Removal Information Exchange Meeting, Karlsruhe, 1982.

Mitchell, G. W., Ottinger, C. A. and Meister, H. *Coolability Of UO_2 With Downward Heat Removal - The D10 Experiment.* Proc. 6th Information Exchange Meeting on Debris Coolability, U. California, Los Angeles, November 7-9, 1984.

Muller, U. and Schulenberg, T. *Post-Accident Heat Removal Research: A State Of The Art Review.* KfK 3601.

Naik, A. S. and Dhir, V. K. Forced flow evaporative cooling of a volumetrically heated porous layer. *Int. J. Heat and Mass Transfer,* **25**:541-552, 1982.

Ostensen, R. W., and Lipinski, R. J. A particle bed dryout based on flooding. *Nuclear Science and Eng.,* **79**:110, 1981.

Ottinger, C. A., Kelly, J. E., and Lipinski, R. J. *Preliminary Report Of The D9 Debris Bed Experiment.* November 1982.

Reed, A. W. *The Effect Of Channeling On The Dryout Of Heated Particulate Beds Immersed In A Liquid Pool.* Ph.D. Thesis, MIT, Feb. 1982.

Reed, A. W. : A mechanistic explanation of channels in debris beds. *Trans. ASME, J. Heat Transfer,* **108**(1):125-131, 1986.

Reed, A. W., Meister, H. and Sasmor, D. J. Measurements of capillary pressure in urania debris beds. *Nuclear Technology,* **78**:54-61, 19871.

Rhee, S. J. and Dhir, V. K. *Natural Cconvection Heat Transfer In Beds Of Inductively Heated Particles.* UCLA, NUREG/CR-0408, 1978.

Ruel, F. *Precalculations And Sensitivity Studies For The PAHR in-pile Experiment M1.* Technical Note No I.06.C1.86.101, JRC-Ispra, October 1986.

Saez, A. E. and Carbonell, R. G. Hydrodynamic parameters for gas-liquid cocurrent flow in packed beds. *AIChE Journal,* **31**:52-62, 1985.

Scheidegger, A. E. *The Physics Of Flow Through Porous Media* University of Toronto Press, 1957.

Schulenberg, T. and Muller, U. *A Refined Model For The Coolability Of Core Debris With Flow Entry From The Bottom.* 6th Information Exchange Meeting on Debris Coolability, Los Angeles, Nov. 7-9, 1984.

Schulz, B. Thermal conductivity of porous and highly porous materials. *High Temp. High Pressures*, 13:649, 1981.

Schwalm, D. and Nijssing R. The influence of subcooling on dryout inception in sodium-saturated fuel particle beds with top cooling and adiabatic bottom. *Nuclear Engrg and Des.*, 70:201-208, 1982.

Schwalm, D. *Some Remarks On Channeling In Particle Beds.* Ispra Joint Research Center; EUR 10.005 EN, 1985.

Shires, G. L. and Stevens, G. F. *Dryout During Boiling In Heated Particulate Beds.* AEE Winfrith Report AEEW-R 1779, 1980.

Squarer, D. and Peoples, J. A. Dryout in inductively heated bed with and without forced flow. *Trans. ANS*, 34, 1980.

Stauffer, F. and Dracos, T. Experimental and numerical study of water and solute infiltration in layered porous media. *J. Hydrology*, 84:9-34, 1986.

Stevens, G. F. and Trenberth, R. *Experimental Studies Of Boiling Heat Transfer And Dryout In Heat Generating Particulate Beds In Water At 1 Bar.* AEEW-R 1545, 1982.

Stevens, G. F. *Particle-Bed Heat Transfer Studies At The Atomic Energy Establishment Winfrith (UKAEA).* Presented at the First UK National Heat Transfer Conference, Leeds, July 1985.

Stubos, A. K. *Experimental And Theoretical Investigation Of Boiling In Liquid Saturated Porous Media* VKI PR 1985-25, June 1985.

Stubos, A. K. and Buchlin, J-M. *An Attempt To Model Thermohydraulic Disturbances In Liquid Saturated Heat Dissipating Particulate Beds.* VKI Preprint, 12 February 1987. Presented at the PAHR Workshop, Ispra, Italy, February 26, 1987.

Stubos, A. K. and Buchlin, J-M. Modeling of vapour channeling behaviour in liquid saturated debris beds. *Trans. ASME, J. Heat Transfer*, 110(4A):968-975, 1988.

Sun, W. J. *Convective instability in superposed porous and free layers.* U. Minnesota, Ph.D. Thesis, 1973.

Topp, G. C. and Miller, E. E. Hysteretic moisture characteristics and hydraulic conductivities for glass-bead media. *Soil Sci. Soc. Am. Proc.*, **30**:156, 1966.

Ternberth, R. and Stevens, G. F. *An Experimental Study Of Boiling Heat Transfer And Dryout In Heated Particulate Beds.* AEEW-R 1342, 1980.

Turland, B. D. and Moore, K. *One-Dimensional Models Of Boiling And Dryout.* In Post Accident Debris Cooling, pp.192-197, Gaun, Karlsruhe, 1983.

Tutu, N. K., Ginsberg, T., and Chen, J. C. Interfacial drag for two-phase flow through high permeability porous beds. *Trans. ASME, J. Heat Transfer*, **106**:865-870, 1984.

List of symbols

A, B, C	Dimensionless coefficients	
C_p	Specific heat	$[J/kgK]$
d_i	Sieve diameter	$[m]$
d_1	Largest diameter of bimodal distribution	$[m]$
d_2	Smallest diameter of bimodal distribution	$[m]$
d_p	Effective particle diameter	$[m]$
f	Shape factor	
F_{gs}	Gas-solid drag per unit volume	$[N/m^3]$
g	Gravitational acceleration	$[m/s^2]$
H	Height of the bed	$[m]$
H_{dis}	Height of the disturbed zone	$[m]$
H_{dow}	Height of the bed bottom region	$[m]$
H_{up}	Height of the bed top region	$[m]$
J	Leverett function	
k	Thermal conductivity	$[W/m.K]$
k_e	Bed effective thermal conductivity	$[W/m.K]$
L_c	Channel length	$[m]$
L	Latent heat	$[J/kg]$
$N_c = \frac{\eta_l \Delta P}{H.\sigma}$	Capillary number	
$Nu = \frac{QH^2_{up}}{2k_e(T_{max}-T_{top})}$	Bed Nusselt number	
P	Pressure	$[Pa]$
ΔP	Pressure change	$[Pa]$
Q	Volumetric power	$[W/m^3]$
$Ra = \frac{g\beta\rho_l C_{pl} QH^3 \kappa}{2k^2_e \nu_l}$	Internal Rayleigh number	
s	Liquid fraction	
S_f	Sticking factor	
T	Temperature	$[K]$
V	Superficial velocity	$[m/s]$
w_i	Weight fraction	
z	Distance along the bed	$[m]$

Greek symbols

$\alpha = 1 - s$	Gas or vapour fraction	
α_o	Residual gas or vapour fraction	
β	Expansion coefficient	$[K^{-1}]$
ϵ	Porosity	
ζ	Liquid pool depth to bed height ratio	
η	Bed passability	$[m]$
η_l, η_v	Relative passabilities	
θ	Contact angle	
κ	Bed permeability	$[m^2]$
κ_l, κ_v	Relative permeabilities	
μ	Dynamic viscosity	$[Pa.s]$
ν	Kinematic viscosity	$[m^2/s]$
ρ	Density	$[kg/m^3]$
σ	Surface tension	$[N/m]$
Φ	Heat flux	$[W/m^2]$

Subscripts

ad	Adiabatic bed bottom
b	Boiling
bot	Bottom of the bed
c	Capillary
con	By conduction
d	Dryout
dow	Downward
e	Effective
f	Fluid
i	Phasic, $i = \ell$ or v
ℓ	Liquid
max	Maximum
p	Particle
r	Residual or irreducible
s	Solid
sat	Saturation
sub	Subcooled
v	Vapour
top	Top of the bed
tot	For the whole bed

Superscript

$*$	Normalized

Chapter 4

HEAT TRANSFER IN SELF-HEATED PARTICLE BEDS SUBMERGED IN LIQUID COOLANT

KENT MEHR AND JØRGEN WÜRTZ
Commission of the European Communities
Joint Research Centre, Ispra, Italy.

4.1   The PAHR Scenario

The PAHR (Post Accident Heat Removal) is conducted by the Joint Research Centre (JRC) of the European Communities at Ispra, Italy, within the framework of its nuclear reactor safety programme. It is one of the projects concerning *fast breeder reactors.*

The fast breeder reactor has an extremely high heat production per unit volume (typically 300 kW/liter). The heat produced at high temperature is removed at almost atmospheric pressure by a liquid metal coolant at some 400°C. Sodium or a sodium-potassium alloy is used as coolant. In view of the high power density in the reactor core, the case of a local loss of cooling is considered as a severe hypothetical accident.

Under such conditions, the non-cooled part of a reactor will heat up very fast and a certain quantity of material may melt and be ejected from the core. When molten material comes into contact with liquid sodium, it freezes and fractures into particles of many different sizes. These particles, which consist mainly of UO_2 and stainless steel, may settle on horizontal structures down in the lower plenum of the reactor.

Initially, the particles are very radioactive and the heating power due to self absorption is very high (up to some 20 kW/litre of bed). The integrity of the supporting structure can only be maintained if the major part of the heat is evacuated through the upper bed surface to the overlying sodium pool.

The scope of the PAHR project is to study the passive cooling of the self heated debris bed. The main goal is to investigate the partitioning of the total bed power into upward and downward heat flows.

4.2   Specific PAHR Phenomena

In a debris bed formed after a core *disruptive accident*, the horizontal extent will be much larger than the height of the bed. Therefore, in the major part of the bed, the heat transfer phenomena are one-dimensional. In this chapter they are treated as such.

4.2.1   Bed characteristics

The particles are irregularly shaped UO_2 particles with a broad size variation. The term *particle diameter*, often used in this chapter, refers to the sieve size equivalent for spheres. The *particle size distribution* follows is a log-normal one. The mass fraction $\Phi(d)$ of particles with

$$\Phi(d) = \int_0^d \frac{\phi(\delta)}{\delta} d\delta. \tag{4.2.1}$$

The frequency function, $\phi(\delta)$, for the log-normal distribution is given by

$$\phi(\delta) = \frac{1}{\sigma\sqrt{2\pi}} \exp\left(-\frac{(\ln\delta - \ln d_m)^2}{2\sigma^2}\right), \tag{4.2.2}$$

where d_m is the mean particle diameter, and σ is the variance of the distribution.

The space between the particles forms a continous network of pores with varying diameter. The size distribution of the pore diameters is closely related to the particle size distribution. A mean particle diameter, $\bar{d}$, for use in the hydraulic calculations is defined below.

The hydraulic diameter d_h of a body is defined by

$$d_h \equiv 6 \frac{\text{volume of body}}{\text{surface area of body}}.$$

	Sodium	Water
Saturation temperature (T_{sat})	962°C	100°C
$\dfrac{dp_{sat}}{dT}$ at $T = T_{sat}$	1560 Pa°C^{-1}	9200 Pa°C^{-1}
Liquid density at T_{sat} (ρ_l)	720 kgm^{-3}	958 kgm^{-3}
Vapour density at T_{sat} (ρ_v)	0.5 kgm^{-3}	0.6 kgm^{-3}
Latent heat of evaporation (h_{lv})	3.8×10^6 Jkg^{-1}	2.26×10^6 Jkg^{-1}
Surface tension of liquid (σ_l)	0.11 Nm^{-1}	0.059 Nm^{-1}
Dynamic viscosity of liquid (μ_l)	1.5×10^{-4} Nsm^{-2}	2.8×10^{-4} Nsm^{-2}
Dynamic viscosity of vapour (μ_v)	1.8×10^{-5} Nsm^{-2}	1.2×10^{-5} Nsm^{-2}
Themal conductivity of liquid (λ_l)	48 Wm^{-1}°C^{-1}	0.67 Wm^{-1}°C^{-1}
Specific heat of liquid ($c_{p,l}$)	1.3×10^3 Jkg^{-1}°C	4.2×10^6 Jkg^{-1}°C
Thermal diffusivity of liquid (α_l)	5.1×10^{-5} m^{2}s^{-1}	1.7×10^{-10} m^{2}s^{-1}

Table 4.2.1: Properties of sodium and water at the saturation state. The pressures for sodium and for water are 2 bars and 1 bar, respectively.

The bed consists of a number of particles with varying volume. Fair and Hatch (1933) defined the mean particle diameter as the unique diameter of a sphere which could replace every single particle maintaining the total particle volume. The mean particle diameter, $\bar{d}$, of the debris becomes

$$\bar{d} = \left(\int_0^\infty \frac{\phi(\delta)}{\delta^2} d\delta \right)^{-1}. \tag{4.2.3}$$

Initially, the entire pore space is filled with liquid sodium. Liquid sodium is a metal with properties much different from those of water (cf. Table 4.2.1). Especially, the heat conductivity and the thermal diffusivity are much higher for sodium than for water. These characteristics qualify the liquid sodium as an excellent coolant.[1]

4.2.2 Heat conduction

In the part of the bed where the temperature is below the saturation temperature of the sodium, the heat is transferred by conduction. Contrary to what is often quoted in the literature, convective single phase heat transfer can be neglected in the types of debris beds considered here.

---

[1]The cover gas in the reactor may be at atmospheric pressure. However, due to the depth of the sodium pool (about 15m), the debris may experience a pressure of ca. 2 bars.

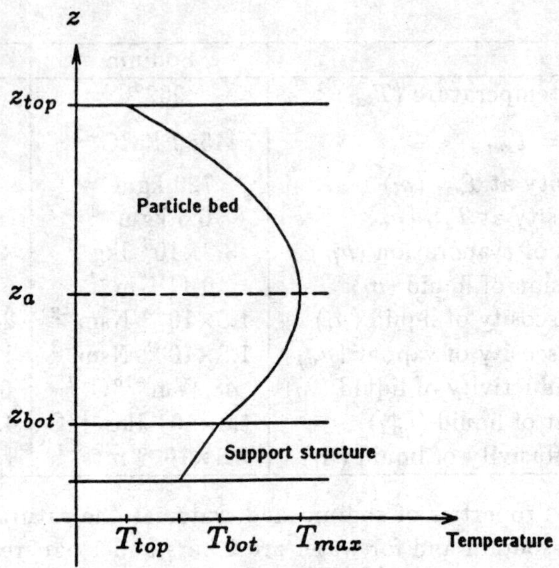

Figure 4.2.1: Temperature profile in a self-heated bed.

The general one-dimensional *heat conduction equation* for a medium with internal heat generation is

$$\frac{\partial}{\partial z}\left(\lambda\frac{\partial T}{\partial z}\right) + q''' = \rho c_P \frac{\partial T}{\partial t}, \tag{4.2.4}$$

where λ is the thermal conductivity, q''' is the rate of heat generation per unit volume, and ρc_P is the heat capacity.

For steady-state with constant conductivity, (4.2.4) becomes *Poisson's equation*

$$\frac{d^2T}{dz^2} + \frac{q'''}{\lambda} = 0. \tag{4.2.5}$$

Equation (4.2.5), integrated over the bed between z_{bot} and z_{top}, with the temperatures of the boundaries, T_{top} and T_{bot}, gives the temperature profile of the internally heated bed as illustrated in Fig. 4.2.1.

$$T(z) = T_{bot} + (z - z_{bot})\left[\frac{T_{top} - T_{bot}}{z_{top} - z_{bot}} + \frac{q'''}{2\lambda}(z_{top} - z_{bot} - z)\right]. \tag{4.2.6}$$

The height, z_a, of the *adiabatic plane* which separates the two parts of

Density (ρ_P)	10800 kgm^{-3}
Specific heat ($c_{p,P}$)	310 Jkg^{-1}°C^{-1}
Thermal conductivity (λ_P)	2.8 Wm^{-1}°C^{-1}

Table 4.2.2: Properties of UO$_2$ at 960°C.

the bed, with upward and downward heat flows, is then given by

$$z_a = \frac{\lambda(T_{top} - T_{bot})}{q'''(z_{top} - z_{bot})} + \frac{z_{top}}{2}. \tag{4.2.7}$$

For the thermal conductivity of a bed, λ_{bed}, with the porosity, n, consisting of particles with a conductivity λ_P, saturated with a liquid that has a conductivity λ_l, we use the formula of Kampf and Karsten (1970)

$$\lambda_{bed} = \lambda_l \left[1 - \frac{(\lambda_l - \lambda_P)(1 - n)}{\lambda_P + (1 - n)^{\frac{1}{3}}(\lambda_l - \lambda_P)} \right]. \tag{4.2.8}$$

Representative values for the heat conductivity of UO$_2$ and sodium are found in Tables 4.2.2 and 4.2.1, respectively.

Considering only the part of the bed above the adiabatic plane, with $z_a \equiv 0$, the temperature profile is described by

$$T(z) = T_{top} + \frac{q'''}{2\lambda_{bed}}(z_{top}^2 - z^2). \tag{4.2.9}$$

4.2.3 Boiling debris beds.

If the intrinsic bed power exceeds a certain limit, whose magnitude depends on the bed conductivity, the bed height and the boundary temperatures, the temperature in the adiabatic plane reaches the saturation temperature, T_{sat}, and the sodium in the pores starts to boil. A two phase convective heat transfer with countercurrent flows of liquid sodium and sodium vapour is established. The temperature gradient in the boiling zone is small. The heat transfer by conduction can therefore be neglected.

The boiling zone is limited to the central part of the bed as the high thermal conductivity of the sodium allows for subcooled zones between the bed boundaries and the boiling zone. A sharp boundary exists between the two subcooled regions and the boiling zone as the large thermal diffusivity of the sodium saturated bed leads to fast condensation of the vapour close to the saturation isotherm. The generated vapour travels to the cold zone boundary and condenses. Liquid sodium which is flowing back into the boiling zone maintains the evaporation process.

Again, we consider only the part of the bed above the *adiabatic plane*, so that $z_a \equiv 0$. From (4.2.9), we find the height of the boiling zone, $h_{boil,up}$, above the adiabatic plane

$$h_{boil,up} = \sqrt{z_{top}^2 - \frac{2\lambda}{q'''}(T_{sat} - T_{top})}. \qquad (4.2.10)$$

The rate of vapour generation per unit volume, Γ_v, is a function of the power density, q''', and the latent heat of evaporation, h_{lv}

$$\Gamma_v = \frac{q'''}{h_{lv}}. \qquad (4.2.11)$$

In steady state, the mass balance requires that the inflow of liquid is equal to the outflow of vapour. The mass fluxes of the vapour, $\dot{m}_v$, and of the liquid, $\dot{m}_l$, are given by

$$\dot{m}_v(z) = v_v(z)\rho_v = \Gamma_v z, \qquad (4.2.12)$$
$$\dot{m}_l(z) = v_l(z)\rho_l = -\Gamma_v z. \qquad (4.2.13)$$

The mass flows are assumed to obey *Darcy's law*

$$\dot{m}_l(z) = -\frac{\kappa_l}{\nu_l}\left(\frac{dp_l}{dz} + \rho_l g\right), \qquad (4.2.14)$$
$$\dot{m}_v(z) = -\frac{\kappa_v}{\nu_v}\left(\frac{dp_v}{dz} + \rho_v g\right), \qquad (4.2.15)$$

where κ_l and κ_v are the permeabilities, ν_l and ν_v are the kinematic viscosities, and $g(> 0)$ is the gravity.

It is assumed that no corrective term for turbulent flow in the boiling zone needs to be taken into account for the pore structure of a typical UO_2 bed.

Subtracting (4.2.15) from (4.2.15), and substituting the capillary pressure for the pressure difference ($p_c \equiv p_v - p_l$), we get

$$\frac{dp_c}{dz} - (\rho_l - \rho_v)g = \frac{\nu_l \dot{m}_l}{\kappa_l} + \frac{\nu_v \dot{m}_v}{\kappa_v}. \qquad (4.2.16)$$

From equation (4.2.16), it is seen that the driving forces are determined by buoyancy and by capillary pressure.

For the permeability of the bed, the expression of Kozeny (1927) and Carman (1941) is used

$$\kappa = \frac{\overline{d}^2 n^3}{f_s(1-n)^2},$$ (4.2.17)

where f_s is a shape factor. For irregularly shaped UO_2 particles, $f_s \approx 260$ is usually assumed. For spheres, $f_s = 150$ is recommended. The mean particle diameter $\overline{d}$, is obtained from the Fair and Hatch formula (4.2.3).

The use of the Fair-Hatch diameter in (4.2.17) is valid only when the flow takes place in the entire porous structure. In the boiling zone, the liquid sodium – which is the wetting phase – is mainly flowing in the pores of smaller diameter, whereas the vapour is flowing in the larger pores. The effective Fair-Hatch particle diameters of the parts of the porous structure through which the two phases are flowing are different from the overall mean diameter. Changes in the liquid saturation will alter the mean pore sizes for each of the phases. Therefore, the permeabilities for the individual phases are dependent on the liquid saturation. This is modelled by the relative permeabilities, κ_{rl} and κ_{rv}, of the phases, so that

$$\kappa_l = \kappa\,\kappa_{rl}, \quad \text{and} \quad \kappa_v = \kappa\,\kappa_{rv},$$

$$\kappa_{rl} = s^3,$$ (4.2.18)

$$\kappa_{rv} = (1-s)^2.$$ (4.2.19)

The saturation determines the size of the pores in which there will be liquid-vapour interfaces. The *capillary pressure* is inversely proportional to the diameters of these pores. Thus, the capillary pressure is dependent on the liquid saturation. A function $J(s)$ is used to correlate capillary pressure to liquid saturation

$$p_r = J(s)\sigma_l \cos\theta \sqrt{\frac{n}{\kappa}},$$ (4.2.20)

where σ_l is the surface tension of the liquid, and θ is the contact angle between the liquid and the particle surface.

A dimensionless expression, $J(s)$ (the *Leverett function*), was introduced by Leverett (1941) to fit his drainage curves measured in sand beds with a limited particle size range. We use

$$J(s) = \frac{1}{\sqrt{5}}\left(\frac{1-s}{s}\right)^{0.2}.$$ (4.2.21)

To keep the formalism simple in this presentation, we have excluded considerations concerning the irreducible saturation and the residual saturation.

4.2.4   Dryout

When the saturation of liquid sodium drops to zero in a certain part of the bed, dryout begins. The power density, q'''_{do}, needed in order to reach incipient dryout at the adiabatic plane is called the *dryout power density*.

In the following, we deal again only with the part of the bed which lies above the adiabatic plane, defined as $z_a \equiv 0$.

From (4.2.16), (4.2.18) and (4.2.19), we get a one-dimensional differential equation from which we can determine the liquid saturation as a function of the elevation

$$\sigma_l \cos\theta \sqrt{\frac{n}{\kappa}} \frac{dJ(s)}{ds} \frac{ds}{dz} = -\left(\frac{\nu_v}{(1-s)^2} + \frac{\nu_l}{s^3}\right)\frac{zq'''}{h_{lv}\kappa} + (\rho_l - \rho_v)g.$$

$$(4.2.22)$$

This equation is known as the one-dimensional Lipinski (1982) model for boiling in an internally heated particle bed. By integrating (4.2.22) from $z = h_{boil,up}$ to $z = 0$, with the initial value of $s(h_{boil,up}) = 1$, we can find the incipient dryout power density, q'''_{do}. The value of q''' for which the integration of (4.2.22) results in $s(0) = 0$, represents the *dryout power density*. The integration can be performed numerically by a simple Runge-Kutta method.

From (4.2.11) and (4.2.13), we obtain

$$\dot{m}_v(z) = \frac{q''' z}{h_{lv}}.$$

$$(4.2.23)$$

Inserting this equation in (4.2.22) yields

$$\dot{m}_v(z) = \left[\kappa(\rho_l - \rho_v)g - \sigma_l \cos\theta \sqrt{n\kappa}\frac{dJ(s)}{ds}\frac{ds}{dz}\right]\left(\frac{\nu_v}{\kappa_{rv}} + \frac{\nu_l}{\kappa_{rl}}\right)^{-1}.$$

$$(4.2.24)$$

We integrate (4.2.23) over the boiling zone at incipient dryout, obtaining

$$\int_0^{h_{boil,up}} \dot{m}_v(z)dz = \frac{q'''_{do} h_{boil}^2}{2h_{lv}}.$$

$$(4.2.25)$$

Integrating (4.2.24), we ignore the *buoyancy* term and set $\cos\theta = 1$. We get

$$\int_0^{h_{boil,up}} \dot{m}_v(z)dz = -\frac{\sigma_l\sqrt{n\kappa}}{\nu_l}\int_0^1 \frac{\kappa_{rl}}{1 + \frac{\kappa_{rl}\nu_v}{\kappa_{rv}\nu_l}}\frac{dJ(s)}{ds}ds$$

$$= \frac{\sigma_l\sqrt{n\kappa}}{\nu_l}\mathcal{G},$$

$$(4.2.26)$$

where $\mathcal{G}$ is a factor which reflects the bed packing. For a PAHR bed $\mathcal{G} \sim 10^{-3}$.

We now have an expression for the incipient dryout heat flux, q''_{do}, which is the heat flux out of the boiling zone

$$q''_{do} \equiv q'''_{do} h_{boil,up} = \frac{2h_{lv}\sigma_l\sqrt{n\kappa}}{\nu_l h_{boil,up}}\mathcal{G}. \qquad (4.2.27)$$

When the power density exceeds the incipient dryout power density, q'''_{do}, the dry zone expands. Since the thermal conductivity of the dry bed is very low, the maximum temperature in the dry zone may rise considerably. Melting of the particles may occur already in a relatively thin dry zone.

4.2.5 Downward boiling

Until now, we have only considered the part of the bed which is above the adiabatic plane. As the bed is also cooled from below, a significant part of the heat produced in the bed is transferred down through the bottom structure. Everything said above is also valid for the part of the bed which is below the adiabatic plane.

Equation (4.2.16) shows that the vapour and liquid flows are due not only to buoyancy, but also the capillary pressure. Thus a downward vapour flow and an upward liquid flow are to be expected below the adiabatic plane. To apply (4.2.22) to the lower part of the bed, we only have to change the sign of the buoyancy term, $(\rho_l - \rho_v)g$.

In a sodium saturated bed of UO_2 debris, the capillary forces are much larger than the buoyancy, so that in a homogeneously structured bed, the boiling zone is almost symmetric about the adiabatic plane. It means that up to incipient dryout, the partitioning of the axial heat flow into upward and a downward flows is determined by the temperatures at the top and at the bottom (cf. (4.2.7)). The heat load on the structures supporting the bed could, therefore, be estimated by the (4.2.6). However, this is only true as long as the particle packing stays homogeneous throughout the bed. This may not be the case in a dynamic situation.

4.2.6 Unsteady state

So far we have only considered a steady state, ignoring the time dependent heating. To study the transient behaviour of the bed, we have to modify the equations used above to include time as an independent variable. This is done in Sec. 4.3, which deals with the PAHR-2D model. Here we only review the physical phenomena which are intrinsic to the time variations.

When the bed resulting from a hypothetical core disruptive accident has settled, it heats up with a high power density. The central part of the bed heats up quasi adiabatically. The temperature profile which develops is rather flat in the central part of the bed. Consequently, a large part of the bed reaches the saturation temperature almost simultaneously. The boiling zone expands rapidly. Space is needed for the vapour and a high upward liquid mass flow is established. The permeability of the bed requires a certain pressure gradient to allow for the liquid flow. As the boiling zone is steadily growing, the necessary upward liquid mass flux is increasing and, therefore, also the pressure increases. After some time, the pressure in the boiling zone may become higher than the static head in the overlying bed and the upper part of the bed may be lifted.

The pressure build up in the boiling zone leads to an increase in the saturation temperature. Only liquid with temperature above the increased saturation temperature may boil. When the upper part of the bed is lifted, the overpressure of the boiling zone is released and the liquid which has a temperature between the saturation temperature at the nominal system pressure and the temporarily increased saturation temperature may flash boil and contribute to the bed disturbance.

The bed may resettle while the boiling continues and the particles might pack in a way which allows for a more separated flow of liquid and vapour. The boiling zone may not stay symmetrically about the adiabatic plane causing in effect a downward shift of the adiabatic plane. The repartition of the heat fluxes due to bed disturbance is illustrated in Fig. 4.2.2. Such a disturbed bed is commonly referred to as being *channeled*.

4.2.7 Channeling

The channeled bed is assumed to show an anisotropic flow behaviour. Figure 4.2.3 shows the influence of a small rearrangement of the particles.

A first approach to model the effects of chanelling is to increase the permeability of the upper part of the bed.

For the sake of simplicity we consider the bed as being an array of vertical parallel tubes of constant diameter, d_0. The permeability κ_a of the array of N_0 tubes, per unit area, is given by the *Hagen-Poiseuille equation*, as

$$\kappa_a = \frac{\pi}{128} N_0 d_0^4. \tag{4.2.28}$$

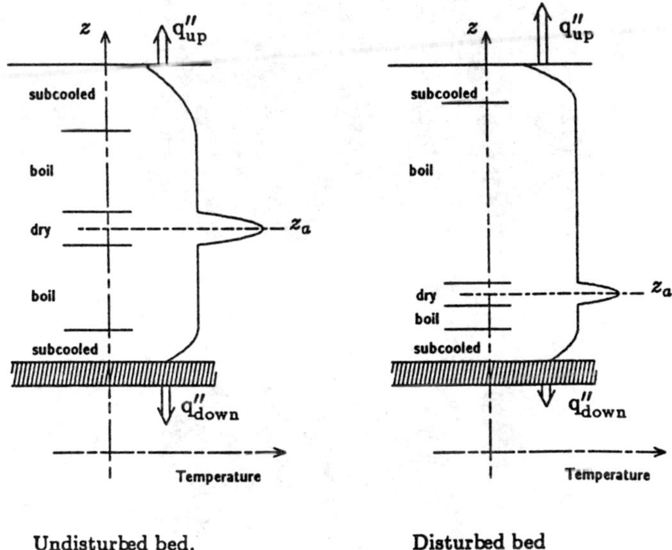

Undisturbed bed.   Disturbed bed

Figure 4.2.2: Temperature profiles in undisturbed and disturbed beds.

The number of tubes in the array, N_0, is assumed to be the same before and after the "channeling". This is a reasonable assumption as the number of pores in a debris bed is related to the number of particles. We now define the "channeling" of the tube array as being a change in the diameter of a number of tubes. Some tubes have an increased diameter, while others have a reduced one as we conservatively assume that the total cross section (porosity) would stay constant. For the sake of simplicity, we consider the case where the unchanneled bed is characterized by an array of N_0 tubes of uniform diameter, d_0, and only two new diameters appear due to "channeling". Then

$$N_0 d_0^2 = N_1 d_1^2 + N_2 d_2^2 + (N_0 - N_1 - N_2)d_0^2, \qquad (4.2.29)$$

where d_1 is the diameter of the small tubes, d_2 is the diameter of the large tubes, N_1 is the number of small tubes, and N_2 is the number of large tubes.
We shall use

$$\delta_1 = \frac{d_1}{d_0} \quad \text{and} \quad \delta_2 = \frac{d_2}{d_0}.$$

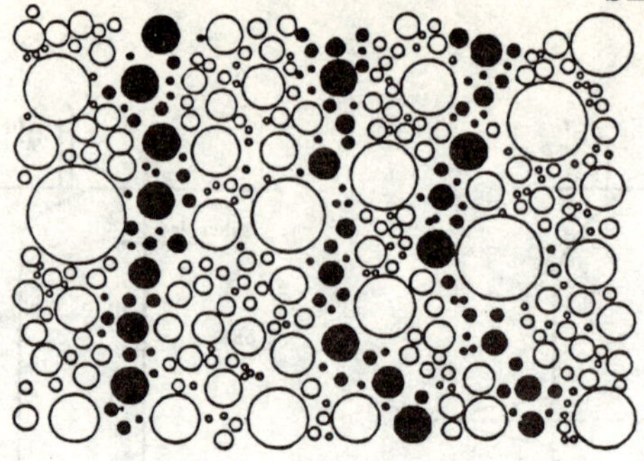

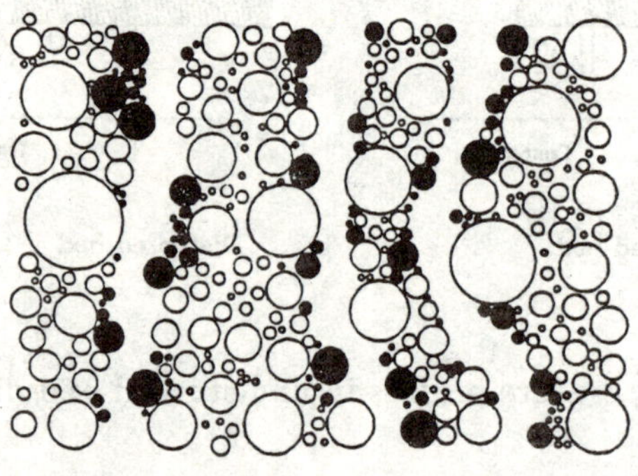

Figure 4.2.3: Channeling as a result of rearrangement of particles. The dark particles have changed position, while the wide channels have opened up for vapour flow.

The amount of 'channeling', $\mathcal{C}$, is defined as the ratio between the 'channeled' cross section and the total cross section

$$\mathcal{C} = \frac{N_1 d_1^2 + N_2 d_2^2}{N_0 d_0^2}. \tag{4.2.30}$$

Using the above definitions and some algebra, we arrive at an expression for the permeability of the 'channeled' array as function of the channeling fraction, $\mathcal{C}$

$$\kappa_c(\mathcal{C}) = \kappa_a \left[1 + \mathcal{C}(\delta_1^2 + \delta_2^2 - \delta_1^2 \delta_2^2 - 1) \right]. \tag{4.2.31}$$

From equation (4.2.31) we conclude that the channeled bed has always a higher permeability than the non-channeled bed.

Similar considerations have been made concerning the relative permeabilities. We have assumed that the liquid preferentially occupies the smallest tubes. From such an analysis, we have seen that the relative permeability of the vapour is increased, whereas the relative permeability of the liquid is decreased by 'channeling'. However, the overall flow characteristics of a channeled bed are always improved yielding an enhanced two phase convection.

4.3 PAHR-2D

In the preceding chapter we have been dealing only with heat and mass transfer phonomena in a one-dimensional space. However, in the in-pile experiments which are performed to simulate the accident scenario (cf. Sec. 4.4) the walls of the crucibles have a significant thermal conductivity and a radial heat loss has also to be considered.

Furthermore, the absorption of neutrons and γ-rays leads to a considerable variation of the power density in the experimental bed, both axially and radially.

To tackle these problems, we have to extend the model to two dimensions in space. As we want to study also dynamic bed behaviour, the time dependent terms have to be included.

The resulting system of coupled differential equations can be solved only by numerical methods. For this purpose, the computer program PAHR–2D has been developed at the Joint Research Centre, Ispra.

Here, we only outline the methods applied. The program is described in details by Ruel (1985).

4.3.1 Basic equations

We begin by defining the density, ρ, and the specific enthalphy, h, of the two-phase mixture

$$\rho = s\rho_l + (1 - s)\rho_v, \tag{4.3.1}$$

$$h = \frac{1}{\rho}\left[s\rho_l h_l + (1 - s)\rho_v h_v\right]. \tag{4.3.2}$$

The total mass flux, $\vec{m}$, and the total specific enthalpy flux, $\vec{\varphi}$, are given by

$$\vec{m} = \vec{m}_l + \vec{m}_v, \qquad\qquad (4.3.3)$$

$$\vec{\varphi} = \vec{m}_l h_l + \vec{m}_v h_v - \lambda_{bed}\,\vec{\nabla}T. \qquad\qquad (4.3.4)$$

The basic equations solved by PAHR–2D are the mass and enthalpy balance equations

$$n\frac{\partial \rho}{\partial t} + \vec{\nabla}\cdot\vec{m} = 0, \qquad\qquad (4.3.5)$$

$$n\frac{\partial \rho h}{\partial t} + (1-n)(\rho c_P)_P\frac{\partial T}{\partial t} + \vec{\nabla}\cdot\vec{\varphi} - q''' = 0, \qquad (4.3.6)$$

where $(\rho c_P)_P$ is the heat capacity per unit volume of the particles.

The mass fluxes are obtained from Darcy's law (see (4.2.15) and (4.2.15))

$$\vec{m}_l = -\frac{\kappa_l}{\nu_l}(\vec{\nabla}p_l - \rho_l \vec{g}),$$

$$\vec{m}_v = -\frac{\kappa_v}{\nu_v}(\vec{\nabla}p_v - \rho_v \vec{g}), \qquad\qquad (4.3.7)$$

where $\vec{g}\,(=-g\vec{1}_z)$ is the gravity vector, $\vec{1}_z$ is the unit vector in the vertical upwards direction.

4.3.2 Spatial discretisation

The finite volume method is applied for the spatial discretisation. Here, the balance equations are integrated over control volumes using *Green's formula*. Integration of . Integration of (4.3.6) and (4.3.6) over the volume V, with boundary A, yields

$$\int_V n\frac{\partial \rho}{\partial t}\,dV + \int_A d\vec{A}\cdot\vec{m} = 0, \qquad\qquad (4.3.8)$$

$$\int_V\left(n\frac{\partial \rho h}{\partial t} + (1-n)(\rho c_P)_P\frac{\partial T}{\partial t} - q'''\right)dV + \int_A d\vec{A}\cdot\vec{\varphi} = 0. \quad (4.3.9)$$

The above set of equations is now solved in cylindrical coordinates, (r, z). The cylinder with radius R and height H is divided into I*J control volumes, V_{ij}, defined by (see Fig. 4.3.1)

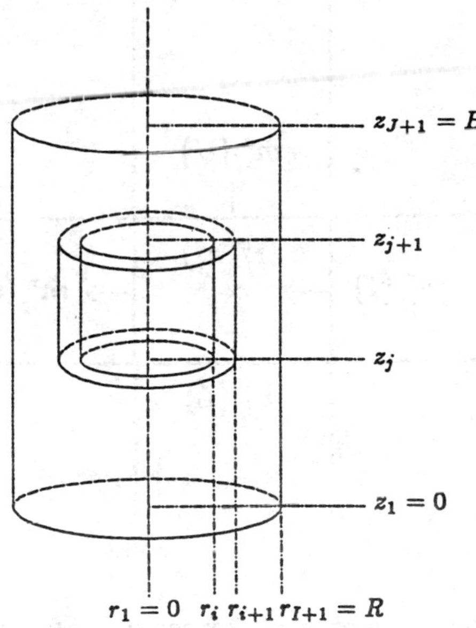

Figure 4.3.1: Definition of control volume, V_{ij}.

$$V_{ij}: \quad r_i \leq r \leq r_{i+1}, \quad 1 \leq i \leq I,$$

$$z_j \leq z \leq z_{j+1}, \quad 1 \leq j \leq J.$$

The center of the control volume, V_{ij}, is located at

$$r_i^c \equiv \frac{r_i + r_{i+1}}{2}, \qquad z_j^c \equiv \frac{z_j + z_{j+1}}{2}.$$

At each center, a set of scalar quantities

$$T_{ij}, \ s_{ij}, \ p_{l,ij}, \ p_{v,ij}, \ h_{l,ij}, \ h_{v,ij}$$

is defined.

The boundary A_{ij} of V_{ij} is divided into two horizontal sub-areas: U(pper) and L(ower), and two vertical sub-areas: (I)nner and (O)uter (cf. Fig. (4.3.2).

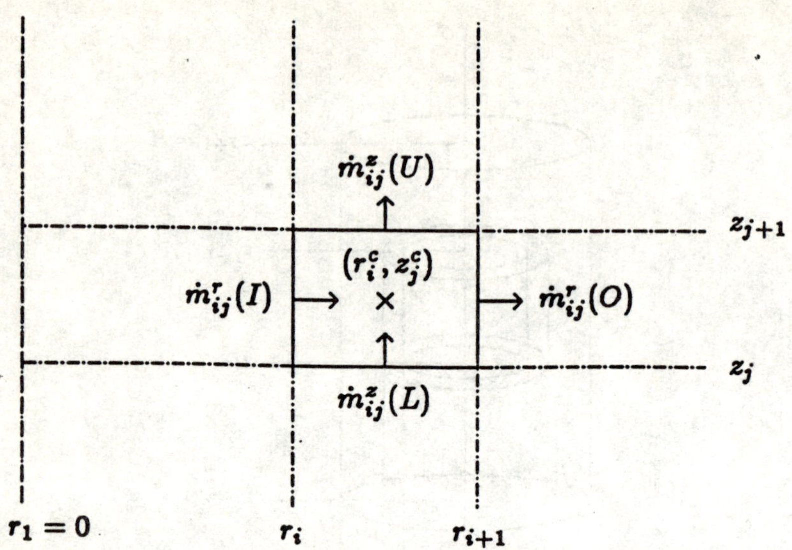

Figure 4.3.2: Definition of mass flux components for control volume, V_{ij}.

On the horizontal sub-areas, the axial components of the mass fluxes are denoted as

$$\dot{m}^z_{l,ij}(U), \ \dot{m}^z_{l,ij}(L), \ \dot{m}^z_{v,ij}(U), \ \dot{m}^z_{v,ij}(L).$$

Accordingly, the radial components of the mass fluxes on the vertical sub-areas are

$$\dot{m}^r_{l,ij}(O), \ \dot{m}^r_{l,ij}(I), \ \dot{m}^r_{v,ij}(O), \ \dot{m}^r_{v,ij}(I).$$

In view of continuity of the mass fluxes at the boundaries of the control volumes, we have

$$\begin{aligned}
\dot{m}^r_{ij}(I) &= \dot{m}^r_{i-1\,j}(O), \\
\dot{m}^z_{ij}(L) &= \dot{m}^z_{ij-1}(U).
\end{aligned} \tag{4.3.10}$$

We, therefore, have to evaluate only the mass flux components at the O(uter) and U(pper) sub-areas, and can define

$$\begin{aligned}
\dot{m}^r_{ij} &\equiv \dot{m}^r_{ij}(O), \\
\dot{m}^z_{ij} &\equiv \dot{m}^z_{ij}(U).
\end{aligned} \tag{4.3.11}$$

Also, the heat flux is continous at the boundaries. Therefore, we have to evaluate only the heat fluxes $\varphi^r_{ij}(O)$ and $\varphi^z_{ij}(U)$

$$\varphi_{ij}^r = \dot{m}_{l,ij}^r h_{l,ij}(O) + \dot{m}_{v,ij}^r h_{v,ij}(O) - (\lambda_{bed,ij}\,\vec{\nabla}T_{ij})^r, \quad (4.3.12)$$

$$\varphi_{ij}^z = \dot{m}_{l,ij}^z h_{l,ij}(U) + \dot{m}_{v,ij}^z h_{v,ij}(U) - (\lambda_{bed,ij}\,\vec{\nabla}T_{ij})^z. \quad (4.3.13)$$

The following central approximations are applied here

$$h_{l,ij}(O) \simeq \frac{h_{l,ij} + h_{l,i+1j}}{2},$$

$$h_{l,ij}(U) \simeq \frac{h_{l,ij} + h_{l,ij+1}}{2}, \quad \text{etc.,} \quad (4.3.14)$$

$$(\lambda_{bed,ij}\,\vec{\nabla}T_{ij})^r \simeq \frac{\lambda_{bed,ij} + \lambda_{bed,i+1j}}{2}\frac{T_{i+1j} - T_{ij}}{r_{i+1}^c - r_i^c},$$

$$(\lambda_{bed,ij}\,\vec{\nabla}T_{ij})^z \simeq \frac{\lambda_{bed,ij} + \lambda_{bed,ij+1}}{2}\frac{T_{ij+1} - T_{ij}}{z_{j+1}^c - z_j^c}, \quad \text{etc.,} \quad (4.3.15)$$

The flow equations (4.3.7) are also discretisized by a central approximation

$$\dot{m}_{l,ij}^r \simeq -\frac{\kappa}{\nu_l}\frac{\kappa_{rl,ij} + \kappa_{rl,i+1j}}{2}\frac{p_{l,i+1j} - p_{l,ij}}{r_{i+1}^c - r_i^c}, \quad \text{etc.,} \quad (4.3.16)$$

$$\dot{m}_{l,ij}^z \simeq -\frac{\kappa}{\nu_l}\frac{\kappa_{rl,ij} + \kappa_{rl,ij+1}}{2}\left[\frac{p_{l,ij+1} - p_{l,ij}}{z_{j+1}^c - z_j^c} + \rho_l g\right], \quad \text{etc.} \quad (4.3.17)$$

The volume integrals in (4.3.9) and (4.3.9) are evaluated in the following way:

$$\int_{V_{ij}} n\frac{\partial\rho}{\partial t}\,dV \simeq Vol(V_{ij})n\frac{d\rho_{ij}}{dt}, \quad (4.3.18)$$

$$\int_{V_{ij}}\left[n\frac{\partial\rho h}{\partial t} + (1-n)(\rho c_P)_P\frac{\partial T}{\partial t} - q'''\right]dV$$

$$\simeq Vol(V_{ij})\left[n\frac{d(\rho h)_{ij}}{dt} + (1-n)(\rho c_P)_P\frac{dT_{ij}}{dt} - q_{ij}'''\right], \quad (4.3.19)$$

where $Vol(V_{ij})$ is the volume of V_{ij}

$$Vol(V_{ij}) = 2\pi r_i^c(r_{i+1} - r_i)(z_{j+1} - z_j). \quad (4.3.20)$$

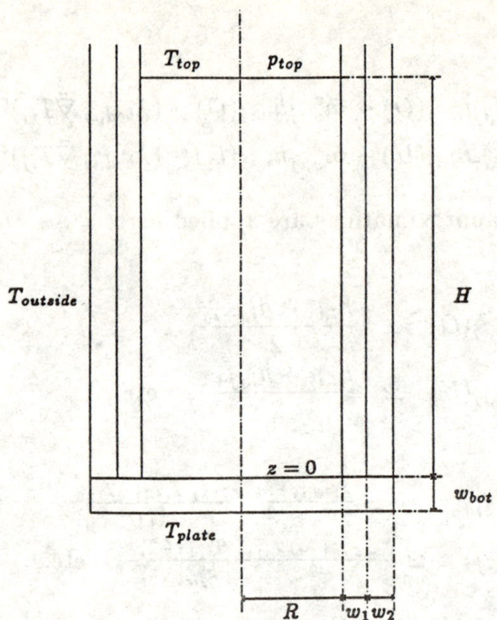

Figure 4.3.3: Crucible layout and boundary conditions.

The corresponding surface integrals are calculated as

$$\int_{A_{ij}} d\vec{A} \cdot \vec{\dot{m}} = 2\pi \Big[r_i^c (r_{i+1} - r_i)(\dot{m}_{ij}^z - \dot{m}_{ij-1}^z)$$

$$+ (z_{j+1} - z_j)(r_{i+1}\dot{m}_{ij}^r - r_i\dot{m}_{i-1j}^r) \Big], \tag{4.3.21}$$

$$\int_{A_{ij}} d\vec{A} \cdot \vec{\varphi} = 2\pi \Big[r_i^c (r_{i+1} - r_i)(\varphi_{ij}^z - \varphi_{j-1}^z)$$

$$+ (z_{j+1} - z_j)(r_{i+1}\varphi_{ij}^r - r_i\varphi_{i-1j}^r) \Big]. \tag{4.3.22}$$

4.3.3   Boundary conditions

The considered solution domain extends not only over the bed submerged in liquid sodium, but includes also the crucible, composed of two vertical walls (width w_1 and w_2) and a bottom plate (width w_{bot}). (cf. Fig. 4.3.3).

The crucible is impermeable at the inner wall and at the bottom. There-fore, the following boundary conditions are valid:

$$\dot{m}^r(r = R) = 0, \tag{4.3.23}$$

$$\dot{m}^z(z = 0) = 0. \tag{4.3.24}$$

The pressure at the top of the bed, p_{top}, is assumed to be constant:

$$p_l(z = H) - p_v(z = H) = p_{top}. \tag{4.3.25}$$

The boundary temperatures have to be specified (cf. Fig. 4.3.3)

$$\begin{aligned}
T(z = H) &= T_{top}, &\tag{4.3.26}\\
T(z = -w_{bot}) &= T_{plate}, &\tag{4.3.27}\\
T(r = R + w_1 + w_2) &= T_{outside}. &\tag{4.3.28}
\end{aligned}$$

4.3.4 Time integration

The spatial discretisation of (4.3.9) and (4.3.9), leaves us with a set of ordinary differential equations in time (cf. (4.3.18) to (4.3.22)). This set of equations is solved by means of an implicit method, the backward *Euler's scheme*. By applying this method we obtain

$$Vol(V_{ij})n\frac{\rho_{ij}^{(n+1)} - \rho_{ij}^{(n)}}{t^{(n+1)} - t^{(n)}} + 2\pi\left[r_i^c(r_{i+1} - r_i)(\dot{m}_{ij}^{z(n+1)} - \dot{m}_{ij-1}^{z(n+1)})\right.$$

$$\left. +(z_{j+1} - z_j)(r_{i+1}\dot{m}_{ij}^{r(n+1)} - r_i\dot{m}_{i-1j}^{r(n+1)})\right] = 0, \tag{4.3.29}$$

$$Vol(V_{ij})\left[n\frac{(\rho h)_{ij}^{(n+1)} - (\rho h)_{ij}^{(n)}}{t^{(n+1)} - t^{(n)}} + (1 - n)(\rho c_P)_P\frac{T_{ij}^{(n+1)} - T_{ij}^{(n)}}{t^{(n+1)} - t^{(n)}} - q_{ij}'''\right]$$

$$+2\pi\left[r_i^c(r_{i+1} - r_i)(\varphi_{ij}^{z(n+1)} - \varphi_{ij-1}^{z(n+1)})\right.$$

$$\left. +(z_{j+1} - z_j)(r_{i+1}\varphi_{ij}^{r(n+1)} - r_i\varphi_{i-1j}^{r(n+1)})\right] = 0. \tag{4.3.30}$$

where the superscript (n) indicates quantities evaluated at the time $t^{(n)}$ after the n'th timestep.

4.3.5 Solution procedure

After the integration in space and time, we finally end up with a set of 2I*J coupled highly non-linear algebraic equations. There are 6I*J unknown scalars:

$$T_{ij},\ s_{ij},\ p_{l,ij},\ p_{v,ij},\ h_{l,ij},\ h_{v,ij}.$$

When these are known, the vector quantities $\vec{m}_l, \vec{m}_v, \vec{\varphi}_l,$ and $\vec{\varphi}_v$ can be calculated by means of (4.3.13), (4.3.13), (4.3.17), and (4.3.17).

Applying the constitutive relationships

$$h_{l,ij} = h_l(T_{ij}, p_{l,ij}), \tag{4.3.31}$$

$$h_{v,ij} = h_v(T_{ij}, p_{v,ij}), \tag{4.3.32}$$

$$s_{ij} = s(p_{v,ij} - p_{l,ij}) = s(p_{c,ij}), \tag{4.3.33}$$

$$T_{sat,ij} = T_{sat}(p_{sat,ij}), \tag{4.3.34}$$

we obtain a closed system of equations. In each control volume, two unknowns are selected which exhibit significant change under the actual physical conditions in the bed:

Subcooled   (i.e., $s_{ij} = 1$; $T_{sat} > T_{ij}$): $p_{l,ij}, T_{ij}$ are chosen,

Boiling   (i.e., $0 < s_{ij} < 1$; $T_{sat} = T_{ij}$): $p_{l,ij}, p_{c,ij}$ are chosen,

Dry   (i.e., $s_{ij} = 0$; $T_{sat} < T_{ij}$): $p_{v,ij}, T_{ij}$ are chosen.

The selection of the relevant variables according to the local physical state speeds up the solution of the non-linear equations.

Only the main features of the method which was developed to solve this system of non-linear equations are listed here:

- Continuation by differentiation.

- Step-doubling technique.

- Numerical approximation of the Jacobian matrix.

- Reduction to Newton's scheme near the solution.

A detailed desciption of the method is presented by Ruel (1985).

4.3.6   D10 Post test calculation

A post-test calculation of the first dryout in the Sandia in-pile experiment D-10 has been performed.

A general description of the *in-pile experiments* follows in Sec. 4.4. In Table 4.3.1, some of the specific data are listed, which serve as input parameters to the program. A detailed report about the D-10 Experiment is given by Mitchell *et al.* (1984).

Once sodium boiling in the bed was achieved, the bed power was raised until dryout was detected at a bed power of 0.42 Wg_{bed}^{-1}. The power was then increased to 0.45 Wg_{bed}^{-1}. At this power level, which was maintained for

Bed:	Height:	0.16 m
	Radius:	0.05 m
	Particle diameter:	0.22 mm
	Porosity:	0.38 m³/m³
	Dry-out Power:	0.42 Wg_{bed}^{-1}
Top:	Pressure:	0.9 bar
	Temperature:	350°C
Bottom plate:	Temperature:	700°C
Outside crucible:	Temperature:	400°C

Table 4.3.1: Sandia D10 experiment. First dry-out.

500 seconds, the dry zone expanded and the measured temperature reached 1160°C in the center of the dry zone. Thereafter, the power was reduced to quench the dry zone.

Figure 4.3.4 shows the power history, which is used as input to PAHR–2D. In the experiment, a series of power steps was performed without waiting for a real steady-state between consecutive steps. Therefore, it is difficult to reproduce exactly the initial conditions at the start time of the calculation. The best simulation of the experimental results is obtained by assuming that initially the bed power was not stable, but rather decreasing slowly until it was stepped up to the dryout level.

Some of the experimental boundary conditions are not well known. In particular, the heat transfer coefficient of the bed bottom structure is ill defined, due to the existence of several interfaces between different materials of the bottom structure. Some tuning of input parameters was necessary in order to obtain a reasonably good agreement between the experiment and the calculations. The best results were obtained with a bottom heat transfer coefficient of 2000 $\text{Wm}^{-2}\text{°C}^{-1}$ and a bed porosity of 0.36 m³/m³. The latter was chosen to match the calculated value of the dryout power with the experimental one.

For the sake of simplicity, the Leverett function given in (4.2.21) is approximated here by a linear expression

$$J(s) = 0.6 - 0.3s. \tag{4.3.35}$$

The other input parameters (thermal conductivity and heat capacity of the structures, relative power distribution in the bed and structures, etc.) are based on the information contained given by Mitchell *et al* (1984).

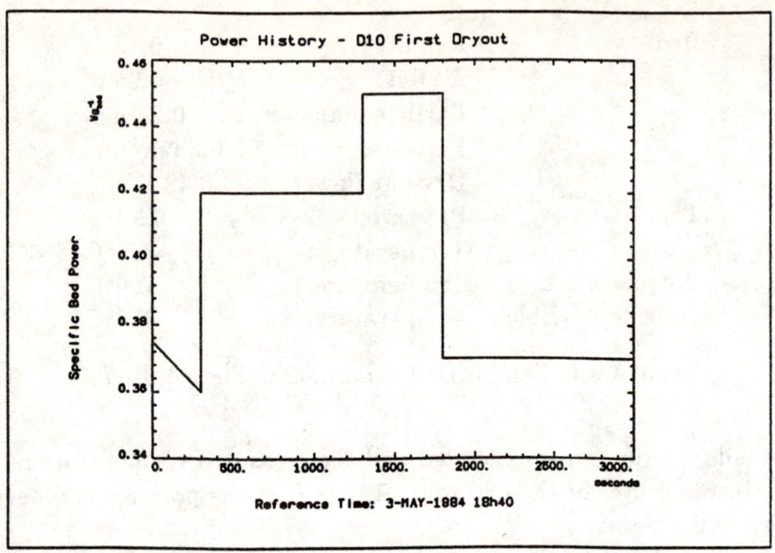

Figure 4.3.4: Power history used as input for calculation of D10 First Dry-out. Reference time: 3 May 1984 18h40.

In Fig. 4.3.5a, the experimental thermocouple readings are plotted as a function of time. (The thermocouples labelled $r=50$mm were located at the bed periphery close to the crucible wall). The corresponding calculated temperatures are shown in Fig. 4.3.5b.

Qualitatively, the calculated results agree quite well with the experimental ones. However, there are considerable differences as to the temperatures in the dry zone. But this is little surprising, since the dry zone temperatures are extremely sensitive to the bed power and to the dry bed conductivity, quantities which have a certain error margin. This margin is particularly large for the conductivity of the dry bed which is very low.

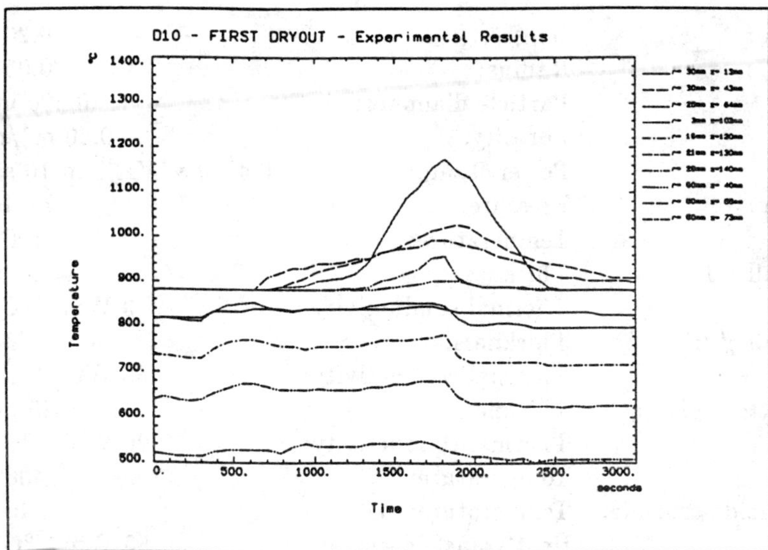

Fig. 3.5. D10 First Dryout. Thermocouple readings.

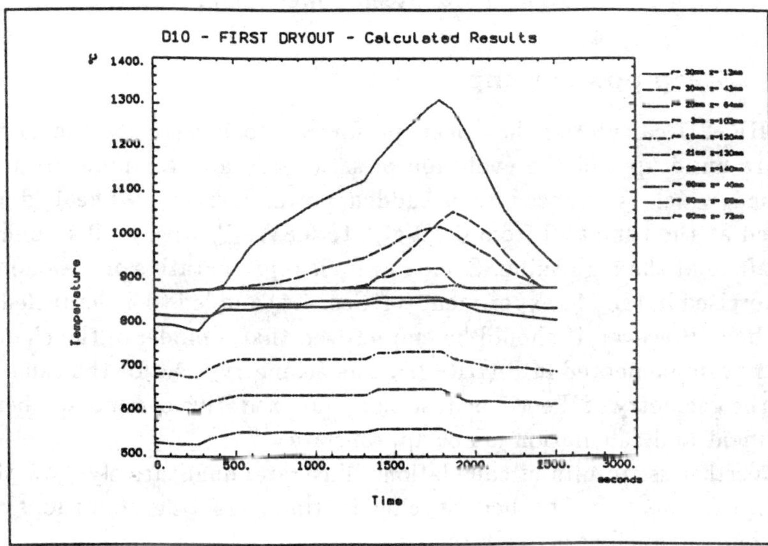

Fig. 4.3.5: D10 First Dry-out. (a) Thermocouple readings. (b) Calculated temperatures.

Bed:	Height:	0.20 m
	Radius:	0.05 m
	Particle diameter:	0.220 mm
	Porosity:	0.30 m^3/m^3
	Power Ramp:	0.4 - 0.8 Wg$_{bed}^{-1}$ in 10 sec.
Top:	Pressure:	2.0 bar
	Temperature:	400°C
Wall #1:	Thickness:	5 mm
	Thermal conductivity:	3 Wm^{-1}°C^{-1}
Wall #2:	Thickness:	5 mm
	Thermal conductivity:	50 Wm^{-1}°C^{-1}
Bottom plate:	Thickness:	20 mm
	Thermal conductivity:	20 Wm^{-1}°C^{-1}
	Temperature:	400°C
Outside crucible:	Temperature:	40°C
	Heat transf. coeff.:	50 Wm^{-2}°C^{-1}

Table 4.3.2: Steep power ramp.

4.3.7   Steep power ramp

A PAHR–2D calculation has been performed to investigate the flow and
pressure build up and the evolution of saturation and temperature profiles
in a bed which is exposed to a sudden power increase. The bed power
is raised at the time $t=0$ from 0.4Wg_{bed}^{-1} to 0.8Wg_{bed}^{-1} within 10 seconds and
maintained at the high value. Such a ramp is representative of the one which
was exercised in the M1 experiment (cf. Sec. 4.4) and led to a channeled state
of the bed. However, it should be emphasized that a model of the channeled
bed is not implemented in PAHR–2D. The geometry used for the calculation
is not the geometry of the M1 bed, so that a quantitative comparison between
experiment and calculation is not appropriate.

Nevertheless, results of calculations illustrate, qualitatively, how the selected phenomena in the bed develop in time, creating conditions under
which bed channeling is likely to occur.

The most relevant input values for the calculation are listed in Table
4.3.2.

Figure 4.3.6 shows the liquid saturation at different axial levels (radial
position 2.5mm) as a function of time. It is seen that the bed is initially
subcooled, and that incipient boiling is reached ca. 50 seconds into the

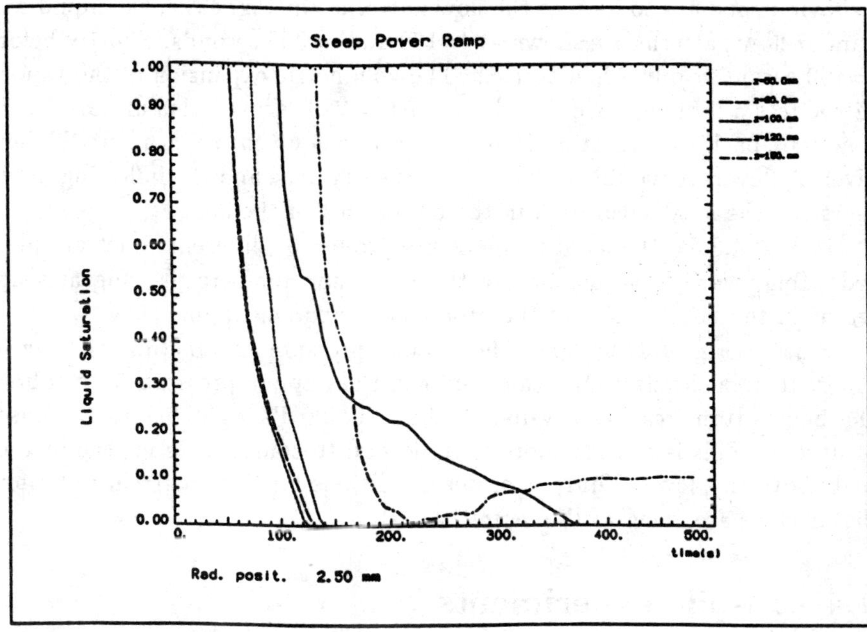

Fig. 4.3.6: Steep power ramp. Liquid saturation as a function of time.

transient. Incipient dryout occures at $t=120$ seconds close to the midplane ($z=100$mm) of the bed. Thereafter, the dry zone grows rapidl. At $t=400$ seconds it extends from ca. 5cm to ca. 15cm above the bed bottom. One observes that the liquid saturation at $z=150$mm reaches a value close to zero at $t=220$ seconds. Later it increases again. In order to interprete this behaviour one has to look at the flow patterns. In Fig. 4.3.7, the liquid and vapour flow patterns are shown at $t=100$ and $t=300$ seconds. Shortly before $t=100$ seconds boiling had started. The volumetric expansion of the vapour introduces a large upward liquid flow. At $t=220$ seconds this upward liquid flow falls back to zero at $z=150$mm as the vapour zone reaches its highest level. A flow reversal takes place above the dry zone and the inflowing liquid increases the local saturation in the upper part of the bed.

In Fig. 4.3.8a, the axial temperature profiles at different times are plotted. Temporarily, the position of the maximum temperature migrates upwards in the bed because of the pronounced upgoing vapour flow.

Finally, Fig. 4.3.8b shows the vapour pressure as function of time at different axial levels. One can see that the vapour pressure 5 cm above the bed bottom reaches a value of about 12000 Pa right before the onset of dryout. This is already more than needed to compensate for the load of bed above this level. Thus, favorable conditions for the occurrence of a bed disturbance are given at this point.

4.4   In-pile experiments

Many experiments have been carried out to study the thermo-hydraulic behaviour of internally heated particle beds. In most experiments, simulant materials, different from a real reactor debris and coolant, were used. Only in-pile tests, i.e., test performed internally in nuclear reactors, made it possible to investigate beds composed of the real reactor materials. There, the test capsule with UO_2 and sodium was placed in a test channel of a reactor, and nuclear fission in the UO_2 debris simulated the radioactive decay heat. The heat genaration rate by fission in the debris was directly related to the operating power of the reactor.

At Sandia National Laboratories, Albuquerque, New Mexico, a joint project, sponsored by the US-NRC, the JRC of the European Communities and PNC of Japan, was carried out. The project was comprised of a series of 10 in-pile experiments with sodium cooled UO_2 beds.

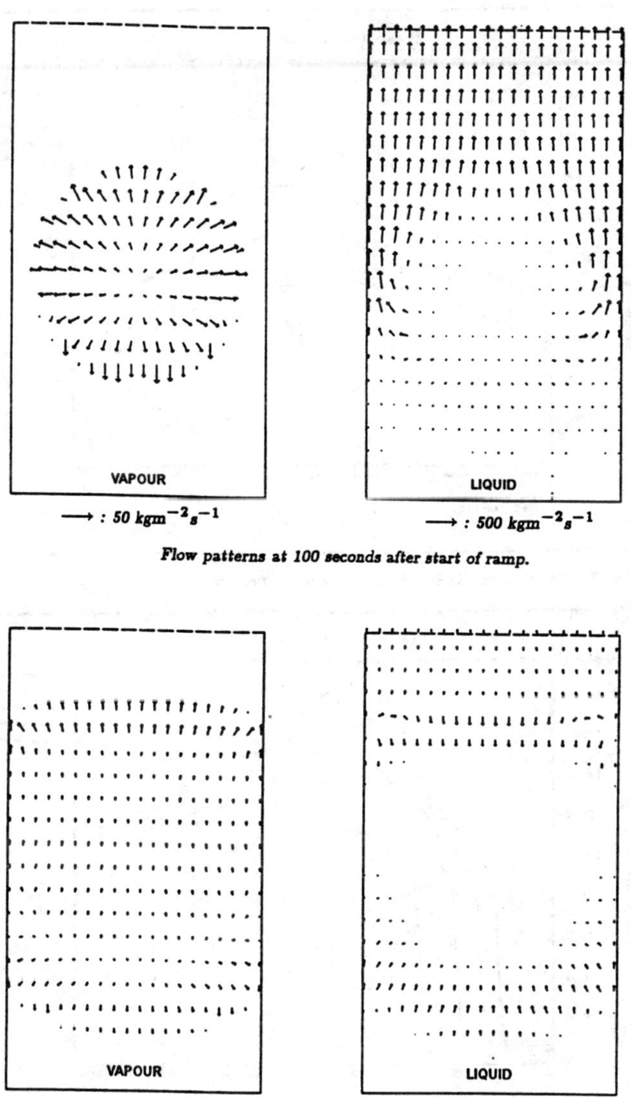

Flow patterns at 100 seconds after start of ramp.

Fig. 4.3.7: Flow patterns after a steep power ramp. (a) At 100 sec. after start of ramp. (b) At 300 sec after start of ramp.

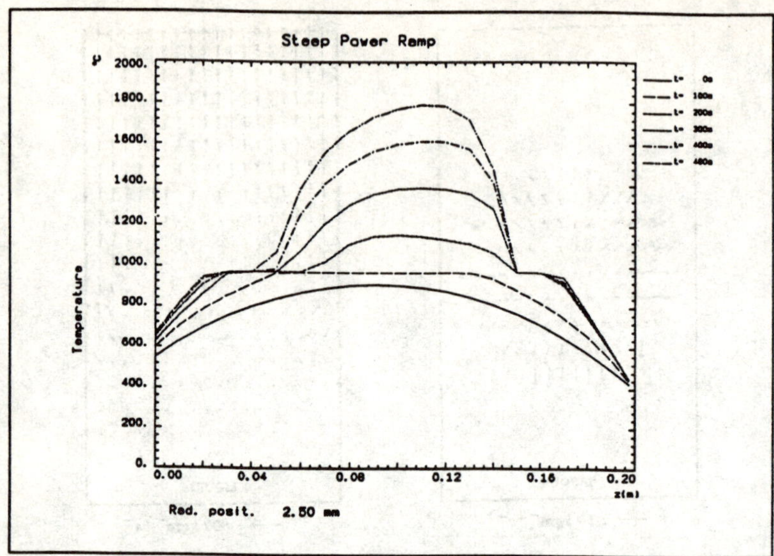

Fig. 3.9. Steep Power Ramp. Axial temperature profiles.

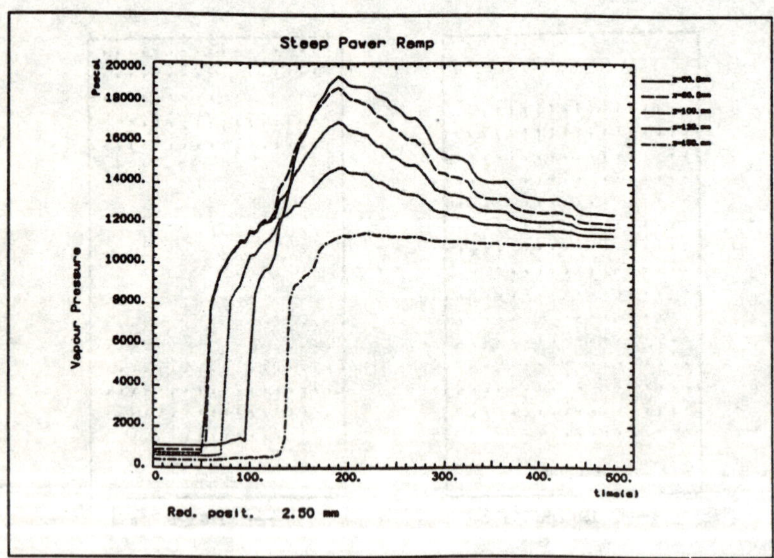

Fig. 4.3.8: Steep power ramp. (a) Axial temperature profile. (b) Vapour pressure as function of time.

The experimental programme was designed so as to quantify the influences which the height of the bed, the packing of the bed (homogeneous or stratified), the particle size range and the temperature of the overlaying sodium pool have on the heat transfer phenomena in debris beds.

The debris beds of all these experiments were contained in double-walled crucibles with radial insulation to establish next to one-dimensional conditions. In most experiments, the crucible was insulated also at the bottom, so that only phenomena occuring above the adiabatic plane could be investigated. Only in the last two experiments D10 and D13, cooling both from above and below the bed was provided. The system pressure varied during all the experiments because the containment of the crucible and the sodium pool was sealed off.

The JRC is conducting its own experimental series of which four experiments are designed for the BR-2 reactor at Mol, Belgium. The first experiment at Mol has been performed and the next three are in preparation. The European experiments aim primarily at higher temperatures than in the Sandia series. All of the European experiments are with top and bottom cooling, the heights of the beds are larger than the bed heights at Sandia and the system pressure can be varied between 1 and 5 bars.

The main instrumentation inside the bed is an array of thermometers. At Sandia, the thermometers were the only instrumentation, whereas in the European experiments the system pressure is also monitored and a sodium level meter at the pool surface indicates – for a short time scale – changes in the global liquid saturation of the bed.

4.4.1 Boiling

The coolant in a debris bed starts boiling at a temperature which is measurably higher ($\approx 10°C$) than the saturation temperature corresponding to the cover gas pressure plus the hydrostatic head of the coolant. This is due to the capillary pressure in the largest pores of the bed which has to be overcome before the liquid can be expelled to make space for the vapour.

In experiments equipped with a sodium level meter, the first indications of boiling inception come from this instrument, signalling a rising level. If a thermometer happens to be located in the boiling zone, its temperature reading will stay constant upon a slight change of bed power while thermometers outside the boiling zone will indicate varying temperatures.

Downward boiling cannot be revealed directly by temperature measurements. It has to been inferred from an analysis of the relative values of the upward and downward heat flows which yields the position of the adiabatic plane within the boiling zone. Downward boiling takes place in that part of the boiling zone which is below the adiabatic plane.

4.4.2   Dryout

Incipient dryout is detected by a sudden increase of one or more thermometer readings beyond the local saturation temperatures. Also thermometers in the subcooled zone can indicate the incipient dryout. As part of the bed dries out, the heat generated in the dry zone is temporarily used to increase the temperature of the dry zone. As a consequence, the heat flux out through the subcooled zone is reduced and this is reflected in temperature decreases.

In undisturbed beds the incipient dryout occurred with power densities of some 2-5Wcm^{-3}. When the power density was increased slightly beyond the incipient dryout power density a stable dry zone established. No sudden reentry of cold liquid quenching the dry zone was observed. When the power was reduced the liquid sodium slowly rewetted the dry zone without any vapour explosion.

4.4.3   Bed disturbance

When the boiling zone expands, liquid is expelled from the bed. The limited permeability causes a pressure build up. When the pressure within the boiling zone is superior to the static head of the overlaying bed the bed cannot stay packed any longer.

Bed disturbances were first observed in experiments in which the subcooled zones were rather shallow, typically some 1 to 3 cm thick. Here an overpressure of about 20mbars is sufficient to lift the top of the bed. From table 4.2.1 is seen that 20mbars are equivalent to a temperature increase of 1.3°C.

In deep beds (15-20cm height) with thick subcooled zones, a larger increase of bed power may be required to disturb the bed.

In the D10 experiment of the Sandia series, the power was incremented by a factor of 2.5 within 2 seconds. This caused the top part of the bed to move upwards like a piston.

In all the cases with bed disturbance, the final state of the bed showed improved boiling characteristics. In the experiments with adiabatic bottom, the dryout power density increased by up to an order of magnitude.

The first experiment at Mol also aimed at the study of bed disturbance. In all previous experiments, a balance was sought between capillary forces

Bed height:	14.6 cm
Bed diameter:	11.0 cm
Mean particle diameter ($\overline{d}$):	0.219 mm
Porosity (n):	0.32 m^3/m^3
Sodium pool temperature:	390°C
Bottom coolant temperature:	500°C
q'''_{do} for undisturbed bed:	< 3.9 Wcm^{-3}
Power step:	3.25 – 5.60 Wcm^{-3}
Ramping time:	20 seconds

Table 4.4.1: The first in-pile experiment at Mol.

in the boiling zone and the weight of the overlying part of the bed. In the experiment at Mol, we studied the influence of the fast expansion of the boiling zone. The main characteristics of the experiment are listed in Table 4.4.1.

A steep power ramp was applied to the bed when it already had a small boiling zone. This was done to avoid super heated flash boiling. The expansion of the boiling zone caused an increase in the pressure which was seen as an increase of the temperature (cf. Fig. 4.4.1). The temperature increase in the upper part of the bed was higher than in the lower part. This indicates that the heating of that part of the bed was due both to fission heating and to an inflow of hot liquid from below.

About 100 seconds after the start of the power ramp, the central temperatures dropped very fast indicating that the overpressure was released. Simultaneously the upper temperatures rose steeply which is an indication of a very large upgoing mass flow of hot material.

Finally, the bed stayed boiling at the high power density which was much higher than the dryout power density of the undisturbed bed. A heat balance based on the temperature gradient over the bottom structure reveals, that the downward heat flow at the high bed power after the bed disturbance was before the power ramping. Thus, a repartition of the axial heat flow has been obtained (cf. Fig. 4.2.2).

At a later stage, we tried to repeat the experimental sequences which lead to the bed disturbance. We applied a larger power step than the one which channeled the bed. No new disturbance was observed. Obviously, the permeability of the upper part of the bed was increased by the first disturbance (cf. Subs. 4.2.7). Therefore, a large pressure build up was not needed for the expansion of the boiling zone after the second power step.

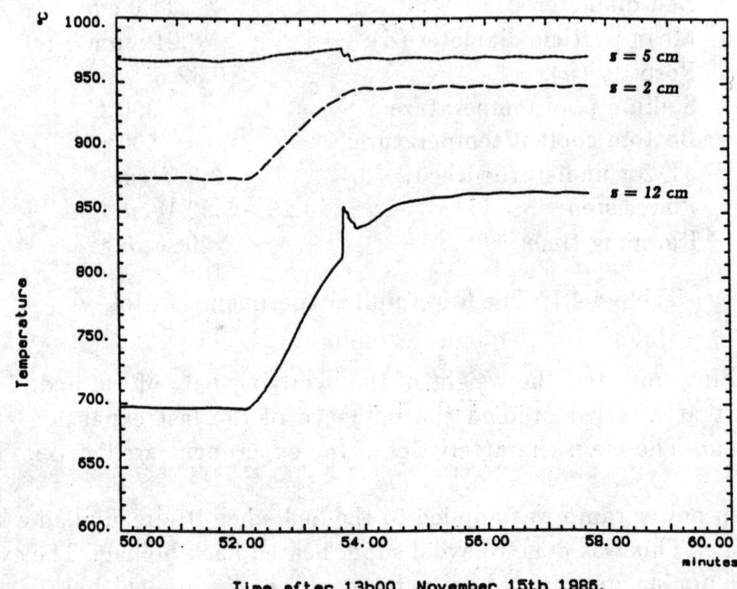

Figure 4.4.1: Temperature history following the disturbance of the bed in the first in-pile PAHR experiment at Mol. The temperatures are shown for different elevations (z) above bed bottom.

In the next experiments, we shall look for necessary and sufficient conditions for the disturbances. The intention is to find out how high the dryout heat flux could be after a bed disturbance with prototypic power densities ($\approx 15\mathrm{Wcm}^{-3}$).

References

Carman, P. C. *Soil Science*, **52**:1, 1941.

Fair, G.M., and L.P. Hatch, Fundamental Factors Governing Streamline Flow of Water Through Sand. *J. American Water Works*, **25**:1551-1565, 1933.

Kampf, H. and G. Karsten, Effects of Different Types of Void Volumes on Radial Temperature Distribution of Fuel Pins. *Nucl. Appl. Tech.*, **9**:288, 1970.

Kozeny, J. *Über kapillare Leitung des Wassers in Bodem*. Sitz Ber. Akad. Wiss., Wien, **136**:271-306, 1927.

Leverett, M. C. Capillary behaviour in porous solids. *Trans. AIME*, **142**:341-358, 1941.

Lipinski, R. *A Model for Boiling and Dryout in Particulate Beds.* Sandia Laboratories report, SAND82-0765 (NUREG/CR-2646), June 1, 1982.

Mitchell, G. W., Ottinger, C. A. and H. Meister. *The D10 Experiment: Coolability of UO$_2$ Debris in Sodium With Downward Heat Removal.* Sandia Laboratories report, SAND84-1144 (NUREG/CR-4055), December, 1982.

Ruel, F. *Outline of a Numerical Method for Solving Transient Equations for Multidimensional Two-Phase Flow Through a Porous Column.* Commission of the European Communities, EUR 10230, 1985.

List of Main Symbols

A	Surface of control volume.
c_p	Specific heat [Jkg^{-3}].
d	Pore diameter [m].
$\overline{d}$	Mean particle diameter [m].
g	Gravity [ms^{-2}].
H	Height of bed [m].
h_{boil}	Height of boiling zone [m].
h	Enthalpy [Jkg^{-1}].
h_{lv}	Latent heat of evaporation [Jkg^{-1}].
I	Number of radial subdivisions.
J	Number of axial subdivisions.
J	Leverett function.
$\dot{m}$	Mass flux [kgm^{-2}s^{-1}].
n	Porosity of bed [m 3/m 3].
p	Pressure [Pa].
p_c	Capillary pressure [Pa].
q'''	Power density of bed [Wm^{-3}].
q'''_{do}	Power density for incipient dryout [Wm^{-3}].
q''_{do}	Dryout heat flux [Wm^{-2}].
q''_{down}	Heat flux through bottom of bed [Wm^{-2}].
q''_{up}	Heat flux through top of bed [Wm^{-2}].
R	Radius of bed.
r	Radial coordinate [m].
s	Liquid saturation.
t	Time [s].
T	Temperature [°C].
T_{bot}	Temperature at bottom of bed [°C].
T_{sat}	Saturation temperature [°C].
T_{top}	Temperature at top of bed [°C].
V	Control volume.
Vol	Volume of control volume [m^3].
z	Axial coordinate [m].

z_a Coordinate for adiabatic plane [m].
z_{bot} Coordinate for bottom of bed [m].
z_{top} Coordinate for top level of bed [m].

Greek letters

θ Contact angle of liquid on particles.
λ Thermal conductivity.
Φ Distribution function.
ϕ Frequency function.
φ Enthalpy flux [Wm^{-2}].
Γ_v Vapour production rate per unit volume [$\text{kgm}^{-3}\text{s}^{-1}$].
κ Permeability [m^2].
κ_r Relative permeability.
λ_{bed} Heat conductivity of bed [$\text{Wm}^{-1\,\circ}\text{C}^{-1}$].
ν Kinematic viscosity [m^2s^{-1}].
ρ Density [kgm^{-3}].
σ Surface tension [Nm^{-1}].

Subscripts

i Radial sequence number of reference volume.
j Axial sequence number of reference volume.
l Liquid.
P Particles.
v Vapour.

Superscripts

c Center of control volume.
(n) n'th time step.
r Radial component.
z Axial component.
$\rightarrow$ Vector.

Chapter 5

PHYSICAL MECHANISMS DURING THE DRYING OF A POROUS MEDIUM

CH. MOYNE, CH. BASILICO,
J. CH. BATSALE AND A. DEGIOVANNI.
Laboratoire d'Energétique et de Mécanique
Théorique et Appliquée U.A. C.N.R.S. 875,
Ecole des Mines, Nancy, France.

The aim of this paper is to summarize the physical aspects of the drying process in order to model it.

In the first part, we describe the process from a chemical engineering point of view. We give some indications about the concept of characteristic drying curve which allows a general understanding of the phenomena. Moreover, this method is very useful for the design of dryers.

Then, a general model for the simultaneous transfer of heat and mass in unsaturated porous media is derived. The model includes: heat transfer by both conductive transport and phase change, mass transfer in the liquid phase by capillarity and filtration, and mass transfer in the gaseous phase by diffusion and convection. The model is compared with experimental results.

Finally, some simplifications of this model are discussed: the conditions under which the filtration terms are negligible, and those under which it is possible to reduce the model to one of *diffusion only*, or to a receding front one.

5.1   General Aspects of the Drying Process

In order to describe the various drying mechanisms, a set of phenomenological parameters must be determined, which depend on temperature and
moisture content (Sec. 5.2). Obviously, this task cannot be achieved for every
kind of product. Moreover, even with this set of physical properties, solving
the non-linear system of partial differential equations is a cumbersome task.

Consequently, for practical purposes (e.g., the design of dryers), a method
of analysis which gives a general understanding of the whole process is preferred. Though the operating conditions inside a dryer may vary, we shall
pay particular attention to the convective drying of solids with constant external conditions of temperature, humidity and velocity of the drying fluid.

5.1.1   The three drying periods

The drying rate, $-d\overline{X}/dt$, depends primarily on the average moisture content
of the product, $\overline{X}$, defined as the ratio of the mass of water contained in a
sample to the dry mass of this sample. Figure 5.1.1 shows a typical rate
of drying curve, $-d\overline{X}/dt$ versus $\overline{X}$, when external drying conditions are
constant.

Three periods can be distinguished during the process (Krischer, 1962):

- **First period, or constant rate period.** After a short unsteady-
  state regime, the drying rate is about constant. The capillary forces
  are strong enough to replace the evaporating water, thus maintaining the gaseous phase saturated at the medium's surface The liquid
  evaporation takes place at the interface. The resistances to heat and
  mass transfer are only within the corresponding boundary layers at
  the interface between the solid and the fluid. If there is no other heat
  source (which is the case if we disregard radiative heat transfer), the
  temperature inside the product under quasi-steady state conditions is
  uniform and identical with the *wet bulb temperature* of the fluid.

- **Second period.** It begins at a critical average moisture content, $\overline{X}_{cr}$.
  At this point, the capillary forces are no longer sufficient to transport
  water at the surface in order to maintain the vapor pressure at its saturated value. For a hygroscopic product, the surface moisture content
  falls within the hygroscopic range. For a non-hygroscopic product, the
  meniscus retreats below the surface level. Thus, internal additional
  heat and mass resistances appear and the drying rate is continuously

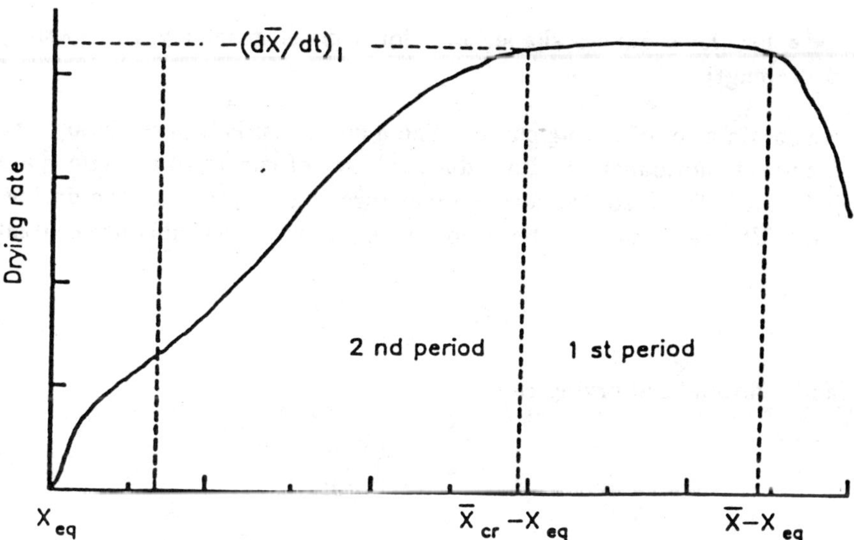

Figure 5.1.1: Typical rate of drying curve.

decreasing. The temperature inside the product is, simultaneously, increasing.

- **Third period.** Very often, the drying rate curve exhibits a second characteristic point at low moisture content. Whereas the drying rate seems to reach the equilibrium value, X_{eq} ($\simeq 0$ for a non-hygroscopic product), or given by the so-called *sorption isotherms* for a hygroscopic one, with a non-zero value, the drying rate goes to zero after an inflection point. This behaviour is sometimes attributed to the sorptive bonds of the products. Nevertheless, it can also be observed in the drying of beds of glass spheres, for which a hygroscopic effect is not expected.

5.1.2 The characteristic drying curve concept

A. Definition (Keey, 1978; Schlünder, 1983)

If we change the experimental conditions, i.e.

- for the drying fluid: temperature, humidity and velocity,

- for the product: shape, and, for a given geometry, its characteristic length,

we obtain a set of drying curves. The question is: is it possible, by means of a suitable normalization, to reduce this set of curves to a single characteristic one? To do so, the average moisture content, $\overline{X}$, and the drying rate, $-d\overline{X}/dt$, are normalized by introducing a normalized moisture content

$$W = \frac{\overline{X} - X_{eq}}{\overline{X}_{cr} - X_{eq}}, \tag{5.1.1}$$

and a normalized drying rate, F

$$F = \frac{-d\overline{X}/dt}{-(d\overline{X}/dt)_I}, \tag{5.1.2}$$

where X_{eq} is the equilibrium moisture content, $\overline{X}_{cr}$ is the critical moisture content, and $-(d\overline{X}/dt)_I$ is the constant drying rate (during the first period).

The characteristic drying curve concept leads to a unique relation, $F(W)$, which does not depend on the above-mentioned parameters. Obviously, the function F exhibits the following properties:

$$\begin{array}{lll} \text{for} \quad 1 \leq W & \quad & F = 1, \\ 0 \leq W \leq 1 & \quad & 0 \leq F \leq 1, \\ W = 0 & \quad & F = 0. \end{array} \tag{5.1.3}$$

The interest of such a concept is easily understood. During the first drying period, the drying rate, $-d\overline{X}/dt_I$, can be calculated only by using standard correlations for external heat and mass transfer phenomena, regardless of the internal phenomena. If we know the characteristic drying curve, the critical moisture content value, $\overline{X}_{cr}$, (which depends both on the external conditions and on the internal mechanisms), and the equilibrium moisture content, X_{eq}, we should be able to predict the drying kinetics.

Although the relevance of this concept can be theoretically examined, it is more realistic to consider it as an attempt to put experimental results in a convenient form. The questions are:

- Can such a process, despite its complexity, be described by this very simple analysis for practical purposes (Keey and Suzuki, 1974)?

- Does the normalized drying curve really characterize the product?

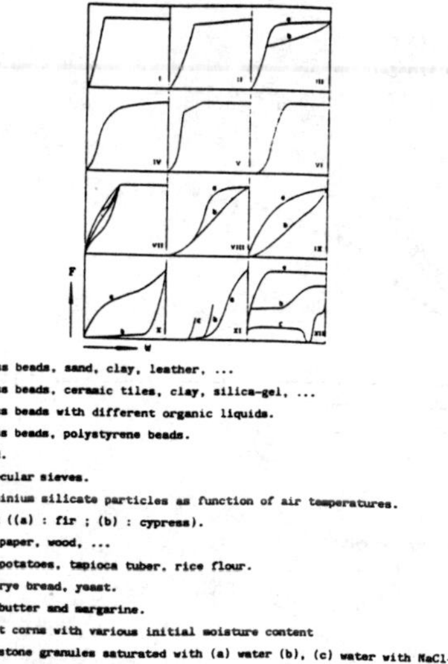

Figure 5.1.2: Characteristic drying curves (Brakel, van, 1980).

Figure 5.1.2 has been drawn by van Brakel (Brakel, van, 1980). It reproduces various forms of this curve for different classes of products which have been reported in the literature. Roughly speaking, classes I through VIII, correspond to capillary and hygroscopic porous media, classes IX-X correspond to colloidal material, while class XI corresponds to corn with different initial moisture content. The particular case of class XII corresponds to crust formation.

B. Experimental examination of the characteristic curve concept

In order to investigate experimentally the relevance of the characteristic drying curve concept, some experiments conducted in our laboratory, concerning high temperature convective drying of wood, are described below (Basilico and Martin, 1984; Moyne and Basilico, 1985, 1986) First the drying curves of beech in superheated steam for various steam temperatures are presented in Fig. 5.1.3a. If we proceed as indicated in the previous section, we obtain Fig. 5.1.3b.

As expected, the normalization of coordinates leads to a single curve.

If we change the composition of the drying fluid from superheated steam to moist air, we obtain the curves of Fig. 5.1.4a and 5.1.4b The characteristic

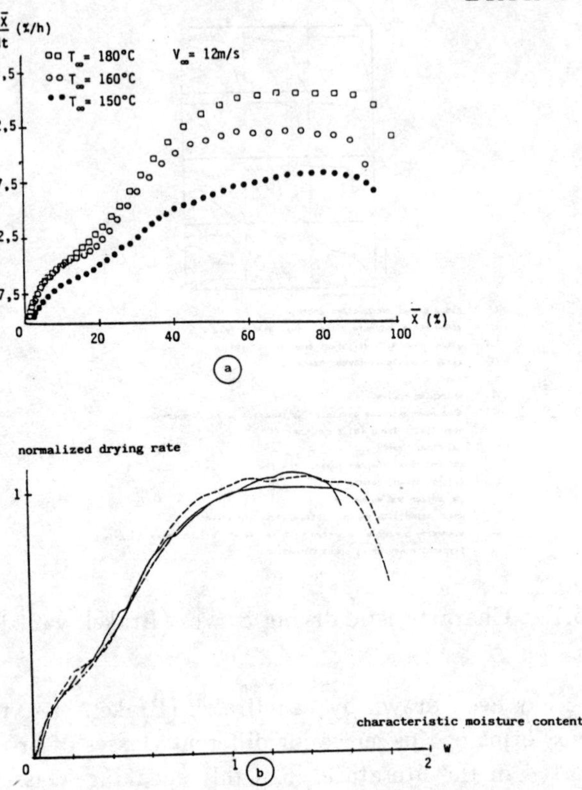

Figure 5.1.3: Drying rate curves (a) and characteristic drying rate curves (b). Convective superheated steam drying of wood (beech).

curve is no longer unique. A precise description of the internal heat and mass transfer mechanisms, facilitate the explanation of this result. On Fig. 5.1.4c, we have drawn the general aspect of a characteristic drying curve in superheated steam at high temperature (1), in moist air at high temperature (2) and at low temperature (3). The drying kinetics of a board in moist air at high temperature is characterized by two distinct parts:

- At the beginning of drying, the board temperature is lower than the boiling point of water, and drying proceeds as at low temperature.

- At the end of drying, the board temperature is above this point and drying proceeds as at high temperature in superheated steam.

The higher the fluid wet bulb temperature is, the sooner the transition between the two regimes takes place. The difference between the curves (1),

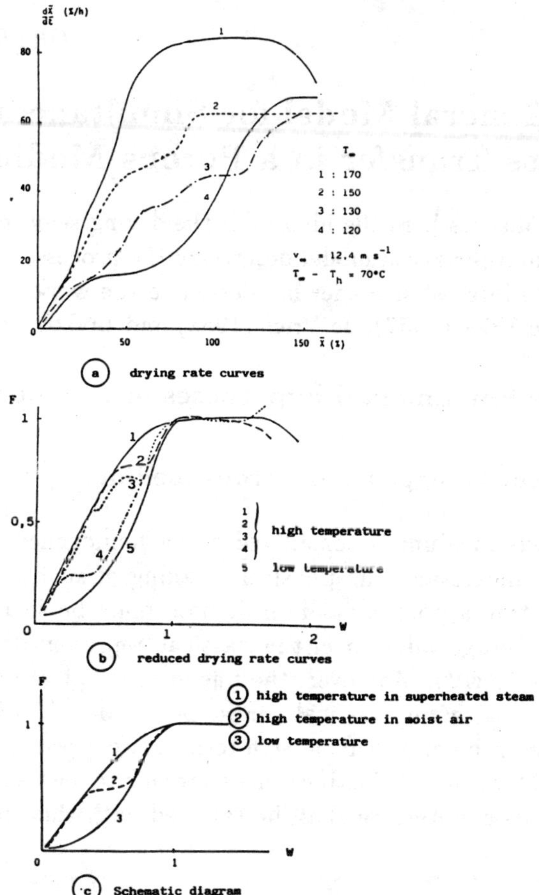

Figure 5.1.4: High temperature convective drying of wood (fir).

in superheated steam, and (3), at low temperature, can be attributed to the filtration mass transfer mechanism which only occurs in the first case.

5.1.3 Conclusion

For practical purposes, the characteristic drying curve provides an approach which leads to standard calculations of dryers and requires a minimum experimental effort. In the current state of the art, it is the most realistic and reliable method. Nevertheless, this approach, based on a macroscopic description of the drying mechanisms, is unable to provide a full understanding of the phenomena. Therefore, having these objectives in mind, a general model will be derived in the next section for the simultaneous heat and mass transfer in porous media.

5.2 A General Model for Simultaneous Heat and Mass Transfer in a Porous Medium

We shall not discuss here the history of the drying science and of the development of a mathematical model describing the process. For a survey of this last point, the interested reader is referred to the works of Krischer (1962), Philip and de Vries (1957), de Vries (1958) and Luikov (1966).

5.2.1 The fundamental hypotheses of the model

A. Equivalent homogeneous porous medium

The porous medium is considered as an homogeneous continuum. We must clearly understand the physical meaning of such an hypothesis. The properties within a porous medium domain must be seen as continuous by taking their average values over the classical 'representative elementary volume' (abbrev. REV). Moreover, the phenomenological laws are introduced by comparison with the classical continuous media theory. In other words, in the closure problem, the dispersion terms are expressed as diffusion terms. In fact, the REV analysis justifies this approach to a great extent (Whitaker, 1977). For this chapter, we shall be satisfied with this simplifying assumption.

B. Local thermodynamical equilibrium

In order to reduce the number of unknowns, we assume that the medium is in local thermodynamical equilibrium. First, the average temperature, T, is the same in the gaseous (g), liquid (l), solid (s) phases. Secondly, the vapor pressure is given by the equilibrium law

$$p_v = p_{vs}(T)\varphi(T, X), \qquad (5.2.1)$$

where p_{vs} is the saturated vapor pressure, and φ is the relative humidity (function of the local temperature T and moisture content X). Outside the hygroscopic range, $\varphi = 1$.

The validity of such a hypothesis is assured in the drying process by a long conctact time between the phases at the microscopic scale (of the pores). Thus, for characteristic observation times at the macroscopic scale of the sample, the local Fourier's numbers (for both heat and mass transfers) are always larger than unity.

5.2.2 Phenomenological laws

We consider the solid phase to be a rigid matrix.

First we have to write the phenomenological relations which express the various flux densities inside the porous medium. An attractive approach to this question is to derive the flux densities expressions from the thermodynamics of irreversible processes (Marle, 1982). However, we prefer here a simpler mechanical derivation.

A. Liquid phase

For capillary water, using a polyphasic Darcy law, the liquid mass flux density $\mathbf{n}_l$ is written in the form

$$\mathbf{n}_l = -\frac{K_l}{\nu_l}\left[\nabla\left(P - \frac{2\sigma\cos\theta}{r^*}\right) - \rho_l\mathbf{g}\right], \tag{5.2.2}$$

where P is the total pressure in the gaseous phase σ is the water surface tension, θ is the wetting angle (for water $\cos\theta \simeq 1$), r^* is the equivalent pore radius, K_l is the liquid phase permeability, ν_l is the liquid kinematic viscosity, ρ_l is the liquid density, and $\mathbf{g}$ is the gravity acceleration.

Equation (5.2.2) may be written in the form

$$\mathbf{n}_l = -\rho_o a_{ml}\left(\nabla X + \delta_l\nabla T\right) - \frac{K_l}{\nu_l}\left(\nabla P - \rho_l\mathbf{g}\right), \tag{5.2.3}$$

where ρ_o is the dry solid density.

The transport coefficient in liquid phase, a_{ml}, is defined by

$$a_{ml} = 2\frac{K_l}{\rho_o\nu_l}\frac{\sigma\cos\theta}{r^{*2}}\left(\frac{\partial r^*}{\partial X}\right)_T, \tag{5.2.4}$$

and the thermomigration coefficient, δ_l, by

$$\delta_l = -\frac{1}{\sigma}\frac{d\sigma}{dT}\Big/\left(\frac{\partial\ln r^*}{\partial X}\right)_T, \tag{5.2.5}$$

where δ_l has a positive value due to the decrease of the surface tension with temperature.

The first two terms of (5.2.3) correspond to liquid displacement under the action of capillary forces, whereas the second two terms are related to filtration movement, either under the influence of a total pressure gradient in the gaseous phase, or under the action of gravity. Although the coefficient

a_{ml} looks like a diffusion coefficient from a mathematical point of view, it must be understood that the water migration mechanism is convection inside the capillaries, due to capillary pressure gradients.

Equation (5.2.3) leads to many discussions concerning the relevance of Darcy's law and its application when the pressure gradient is primarily due to capillary forces. Moreover, the relation $r^* = f(X)$ is not unique, owing to the hysteresis of the curve of capillary pressure versus saturation of the wetting fluid. We do not intend to further discuss these points (see, for example, Dullien, 1979).

An experimental determination of the liquid transport coefficient, a_{ml}, for an unconsolidated quartz sand is drawn on Fig. 5.2.1 due to Crausse *et al.* (1984). At low moisture content, the liquid permeability and, therefore, the coefficient a_{ml}, sharply decreases for a critical moisture content corresponding to the well known irreducible saturation. Roughly speaking, this value separates the funicular state from the pendular state. In the former, at high moisture content, the liquid phase is continuous and moisture transport takes place in liquid phase. In the latter, the liquid phase is no more continuous and the liquid mobility is very weak.

In this paper, special attention is paid to capillary porous bodies for which the physical mechanisms of liquid migation are rather well understood. Nevertheless, the formalism of (5.2.3) is general enough to be applied to other cases, such as microporous material, or gels.

B. Gaseous phase

The mathematical description of the combined convection-diffusion transport in a gaseous phase in a porous medium domain is a difficult topic. According to the remarks in Subs. 2.1.1, the vapor flux density is simply given by comparison with the continuous media description using Fick's and Darcy's laws. Thus

$$\mathbf{n}_v = \omega_v \left(\mathbf{n}_a + \mathbf{n}_v \right) + \mathbf{j}_v, \tag{5.2.6}$$

with

$$\mathbf{n}_a + \mathbf{n}_v = -\frac{K_g}{\nu_g} \nabla P, \tag{5.2.7}$$

and

$$\mathbf{j}_v = -\rho_g \mathcal{D}_v f \nabla \omega_v, \tag{5.2.8}$$

where ω_v is the vapor mass fraction, $\mathcal{D}_v$ is the binary air-vapor diffusion

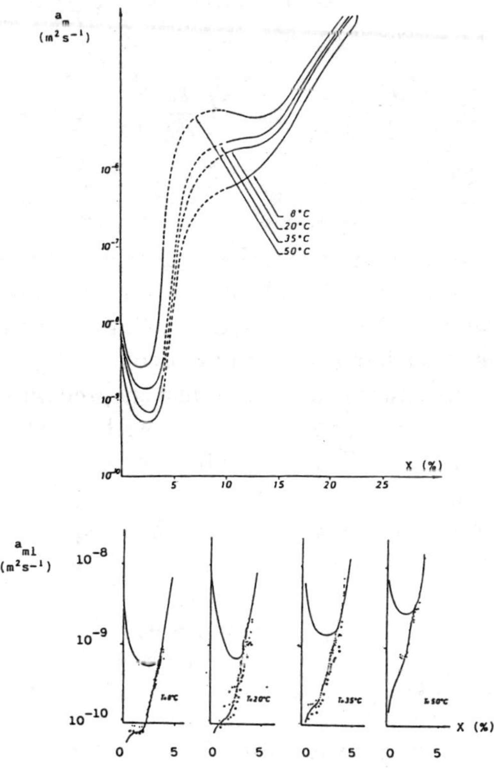

Figure 5.2.1: Liquid mass transport coefficient, $a_{m\ell}$, and total mass transport coefficient, a_m, for sand (Crausse *et al.*, 1984).

coefficient, f is the resistance factor, ρ_g is the gas density, K_g is the gas phase permeability, and ν_g is the gas kinematic viscosity.

The resistance factor, f, introduced in (5.2.8), must take into account the supplementary resistance due to the presence of the gas and liquid phases (more specifically, the surface reduction and the differences between the spatial average temperature gradient inside the medium and the gas phase). The interested reader is referred to Azizi *et al.* (1988) for a recent examination of this problem.

If we define the vapor diffusion coefficient inside the porous medium, D_v^*, as

$$D_v^* = \frac{\rho_g}{\rho_o} \mathcal{D}_v f, \qquad (5.2.9)$$

and if we choose the three driving forces as ∇T, ∇X, and ∇P, the total vapor flux, $\mathbf{n}_v$, is expressed as

$$\mathbf{n}_v = - \rho_o D_v^* \left[\left(\frac{\partial \omega_v}{\partial T}\right)_{P,X} \nabla T + \left(\frac{\partial \omega_v}{\partial X}\right)_{P,T} \nabla X + \left(\frac{\partial \omega_v}{\partial P}\right)_{T,X} \nabla P \right]$$
$$- \omega_v \frac{K_g}{\nu_g} \nabla P. \qquad (5.2.10)$$

Outside the hygroscopic range, the term related to the temperature gradient is the only significant one. The diffusive vapor movement depends only on the temperature gradient, the vapor moving from the hot region to the cold one. On the other hand, in the hygroscopic range, the transport term, depending on the moisture content gradient, is predominant, because of the variation of the equilibrium vapor pressure with moisture content. According to this fact, the total diffusion coefficient is increasing at low moisture content(Fig. 5.2.1).

In the same way, we can also derive similar equations for the transport of the inert component (air):

$$\mathbf{n}_a = \omega_a \left(\mathbf{n}_a + \mathbf{n}_v\right) + \rho_o D_v^* \nabla \omega_v. \qquad (5.2.11)$$

B. Heat flux density

Fourier's law is used to express the heat flux density,$\mathbf{q}$, through the porous medium

$$\mathbf{q} = -\lambda \nabla T. \qquad (5.2.12)$$

The thermal conductivity, λ, in (5.2.12) must be considered as a phenomenological parameter. Only an analysis at the pore scale can lead to an exact understanding of what is λ . It takes into account not only the conductive heat transfer through the three-phase material, but it is also influenced by the so-called evaporation-condensation effect (Krischer, 1962; Luikov, 1966; Whitaker, 1977; Moine and degiovanni, 1987). This interesting, but rather complex point, beyond the scope of this chapter, has recently been examined by Azizi *et al.* (1988) and Moyne *et al.* (1988).

5.2.3   Conservation laws

Next, we have to express the conservation laws for liquid, vapor, air and energy inside the porous medium.

A. Liquid phase

The liquid phase variation is due to a transport term and to a phase change term. Thus

$$\rho_o \frac{\partial X_l}{\partial t} = -\nabla \cdot \mathbf{n}_l - \dot{m},$$

(5.2.13)

where X_l is the liquid moisture content (per dry solid mass unit), and $\dot{m}$ is the rate of evaporated water mass per volume unit.

B. Vapor

In the vapor mass balance, the phase change term, $\dot{m}$, appears, of course, with an opposite sign

$$\rho_o \frac{\partial X_v}{\partial t} = -\nabla \cdot \mathbf{n}_v + \dot{m},$$

(5.2.14)

where X_v is the vapor mass fraction (per dry solid mass unit).

C. Air

For the inert component, the mass balance is written in the form

$$\rho_o \frac{\partial X_a}{\partial t} = -\nabla \cdot \mathbf{n}_a,$$

(5.2.15)

where X_a is the air mass fraction (per dry solid mass unit).

D. Energy

The application of the fundamental energy postulate, neglecting kinetic energy and pressure terms which, usually, are unimportant, yields

$$\rho_o \frac{\partial}{\partial t} (h_s + h_l X_l + h_v X_v + h_a X_a)$$
$$= -\nabla \cdot (\mathbf{q} + h_l \mathbf{n}_l + h_v \mathbf{n}_v + h_a \mathbf{n}_a),$$

(5.2.16)

where h_i is the massic enthalpy of component i, and ε is the porosity of the medium.

Using the mass balances (5.2.13) through (5.2.15), this relation becomes

$$\underset{(1)}{\rho_o c_p \frac{\partial T}{\partial t}} + \underset{(2)}{(c_{pl} \mathbf{n}_l + c_{pv} \mathbf{n}_v + c_{pa} \mathbf{n}_a) \cdot \nabla T} =$$
$$\underset{(3)}{-\nabla \cdot \mathbf{q}} \qquad \underset{(4)}{-(h_v - h_l) \dot{m}.}$$

(5.2.17)

The physical meaning of the different terms are as follows:

- (1) Sensible heat, where the specific heat of the medium, per dry solid mass unit, is given by

$$c_p = c_{ps} + c_{pl}X_l + c_{pv}X_v + c_{pa}X_a.$$

- (2) Convective transport of heat.

- (3) Conductive transport of heat (Fourier's law).

- (4) Phase change term, in which $\Delta h_v = h_v - h_l$ is the latent heat of vaporization (plus the heat of sorption, if in the hygroscopic range).

In most drying processes, the convective term is negligible compared to the phase change term, due to the high value of the latent heat compared with the sensible heat (Moyne and Degiovanni, 1987).

5.2.4   System to be solved

With the definition of the total moisture content, X, given by

$$X = X_l + X_v \simeq X_l, \tag{5.2.18}$$

adding (5.2.13) and (5.2.14), leads to the total water mass balance

$$\rho_o \frac{\partial X}{\partial t} = -\nabla \cdot (\mathbf{n}_l + \mathbf{n}_v). \tag{5.2.19}$$

The phase change term, $\dot{m}$, can be evaluated from the vapor mass balance (5.2.14), in which the 'storage' term $\rho_o \partial X_v / \partial t$ is neglected, compared to the two other ones. Thus

$$\nabla \cdot \mathbf{n}_v = \dot{m}. \tag{5.2.20}$$

Finally, we obtain the following system of three partial differential equations

$$\rho_o c_p \frac{\partial T}{\partial t} = \nabla \cdot (\lambda \nabla T) - \Delta h_v \nabla \cdot \mathbf{n}_v, \tag{5.2.21}$$

$$\rho_o \frac{\partial X_v}{\partial t} = -\nabla \cdot (\mathbf{n}_l + \mathbf{n}_v,) \tag{5.2.22}$$

$$\rho_o \frac{\partial X_a}{\partial t} = -\nabla \cdot \mathbf{n}_a. \tag{5.2.23}$$

If we use the expression of the mass flux densities, and if we choose the temperature, T, the total moisture content, X, and the total pressure in gaseous phase, P, as independent parameters, this system can be written as:

$$m_{11}\frac{\partial T}{\partial t} = \nabla \cdot (k_{11}\nabla T + k_{12}\nabla X + k_{13}\nabla P), \qquad (5.2.24)$$

$$\frac{\partial X}{\partial t} = \nabla \cdot \left(k_{21}\nabla T + k_{22}\nabla X + k_{23}\nabla P - \frac{K_l}{\nu_l}\frac{\rho_l}{\rho_o}\mathbf{g}\right), \qquad (5.2.25)$$

$$m_{31}\frac{\partial T}{\partial t} + m_{32}\frac{\partial X}{\partial t} + m_{33}\frac{\partial P}{\partial t} = \nabla \cdot (k_{31}\nabla T + k_{32}\nabla X + k_{33}\nabla P), \qquad (5.2.26)$$

where the m_{ij} values are given by:

$$m_{ij} = \begin{pmatrix} \rho_o c_p & 0 & 0 \\ 0 & 1 & 0 \\ m_{31} & m_{32} & m_{33} \end{pmatrix}, \qquad (5.2.27)$$

with

$$m_{31} = -\frac{\varepsilon M_a(1 - S_l)}{\rho_o RT}\left(\frac{\partial p_v}{\partial T} + \frac{P - p_v}{T}\right),$$

$$m_{32} = -\frac{\varepsilon M_a(1 - S_l)}{\rho_o RT}\left(\frac{\partial p_v}{\partial X} + \frac{\rho_o}{\rho_l(1 - S_l)\varepsilon}(P - p_v)\right),$$

$$m_{33} = \frac{\varepsilon M_a(1 - S_l)}{l},$$

where

$$S_l = \frac{\rho_o X}{\rho_l \varepsilon}$$

is the pore saturation in liquid phase.

The k_{ij} values are given by:

$$k_{11} = \lambda + \rho_o \Delta h_v D_v^* \frac{\partial \omega_v}{\partial T},$$

$$k_{12} = \rho_o \Delta h_v D_v^* \frac{\partial \omega_v}{\partial X},$$

$$k_{13} = \Delta h_v \left(\rho_o D_v^* \frac{\partial \omega_v}{\partial P} + \frac{K_g}{\nu_g} \omega_v \right),$$

$$k_{21} = a_{ml} \delta_l + D_v^* \frac{\partial \omega_v}{tialT},$$

$$k_{22} = a_{ml} + D_v^* \frac{\partial \omega_v}{\partial X},$$

$$k_{23} = \frac{1}{\rho_o} \left(\frac{K_g}{\nu_g} \omega_v + \frac{K_l}{\nu_l} \right) + D_v^* \frac{\partial \omega_v}{\partial P},$$

$$k_{31} = -D_v^* \frac{\partial \omega_v}{\partial T},$$

$$k_{32} = -D_v^* \frac{\partial \omega_v}{\partial X},$$

$$k_{33} = \frac{1}{\rho_o} \frac{K_g}{\nu_g} (1 - \omega_v) - D_v^* \frac{\partial \omega_v}{\partial P}. \tag{5.2.28}$$

5.2.5   Numerical solution

The system of equations, (5.2.25) through (5.2.26), with appropriate boundary conditions, has been solved numerically in one-dimensional cases. The space discretization uses GALERKIN's method. The time discretization uses CRANK-NICHOLSON's scheme. Because variations occur in the phenomenological coefficients, the problem is *non-linear* and is solved by NEWTON's method (Moyne, 1987).

5.3   Application to Drying

In this section, we want to employ the general model in order to describe typical drying situations. Our first interest will be to investigate high temperature (i.e., above the boiling point of water) convective drying. The mathematical analysis of this problem requires the use of the complete model, (5.2.25) through (5.2.26). Then, we shall examine low temperature' drying, for which a simplifying assumption allows us to disregard gas phase momentum equation by using the classical *dilute gas hypothesis*. Thus, the three partial differential equations of the model can be reduced to only two equations. This model is still rather complex. For practical purposes, drastic simplifications are often useful. Therefore, in the last part of this section, we shall discuss the reduction of drying models to the simple diffusion model and to the receding front model.

5.3.1   High temperature convective drying

In the first part of this paper, we have shown some results connected with high temperature convective drying of wood. However, wood is a rather complex material: it is inhomogeneous, anisotropic. Moreover, because of its living character, its properties are varying during the drying process. Therefore, to test the relevance of the mathematical model, we choose a light concrete ('Ytong'). The choice of such a material has been made because Krischer (1962) has used it in low temperature drying experiments. IIe has also measured some of its thermophysical properties.

A. Experimental results

An Ytong slab (0.05 m thick, 0.20 m wide, 1.0 m long) is dried in an aerodynamic return flow wind tunnel as shown in Fig. 5.3.1. The ambient pressure is atmospheric. The fluid is superheated steam (temperature T_∞ = 186°C; incipient flow velocity u_∞ 13 ms^{-1}. An electronic balance allows a continuous weighing. The temperatures of the slab are measured by means of thermocouples implanted 30 mm deep in the slab. The gas pressure inside the slab is measured by the following technique: a hollow steel needle penetrates in a 50 mm drilled hole in the thickness of the slab. An epoxy resin ensures imperviousness. A very short flexible pipe connects the needle to a pressure transducer. Rather than an absolute measurement, this device allows us to follow the pressure evolution inside the material.

The recordings of the average moisture content, temperature at various positions in the thickness, and pressure at the center of the board, are shown on Fig. 5.3.2.

The analysis of the experimental results shows a first period with a rather linear decrease in the average moisture content, $\overline{X}$. The surface temperature of the slab is close to the boiling point of water.

At the beginning of the falling rate period, the gas pressure increases and the temperatures in the outer part of the slab are above the boiling point of water, whereas a steady temperature, slightly under this point close to 95°C, is observed in the inner part.

A more precise description of the transport mechanism will be presented with the comparison between theoretical and experimental results.

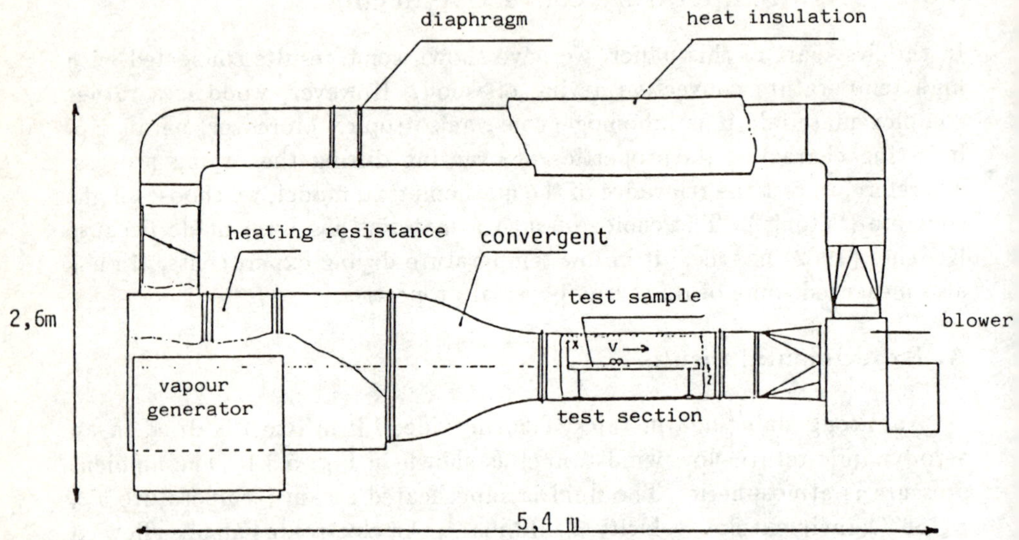

Figure 5.3.1: Return flow aerodynamic wind tunnel.

B. Mathematical model

The mathematical model is that descibed in Sec. 5.2. As initial conditions, uniform temperature, moisture content and gaseous pressure are chosen.

The boundary condition have to be prescribed. In the symmetry plane, of the slab ($x = 0$), these are symmetry conditions:

$$\text{At} \quad x = 0, \qquad \frac{\partial T}{\partial x} = \frac{\partial X}{\partial x} = \frac{\partial P}{\partial x} = 0. \tag{5.3.1}$$

At the other boundary ($x = e$), two cases must be distinguished. We first define a boiling temperature in the hygroscopic range, T_b, by:

$$p_{vs}(T_b)\varphi(T_b, X) = P_\infty. \tag{5.3.2}$$

If $T_{x=e} < T_b$, using the two heat and mass transfer convective coefficients, h and k, we can write:

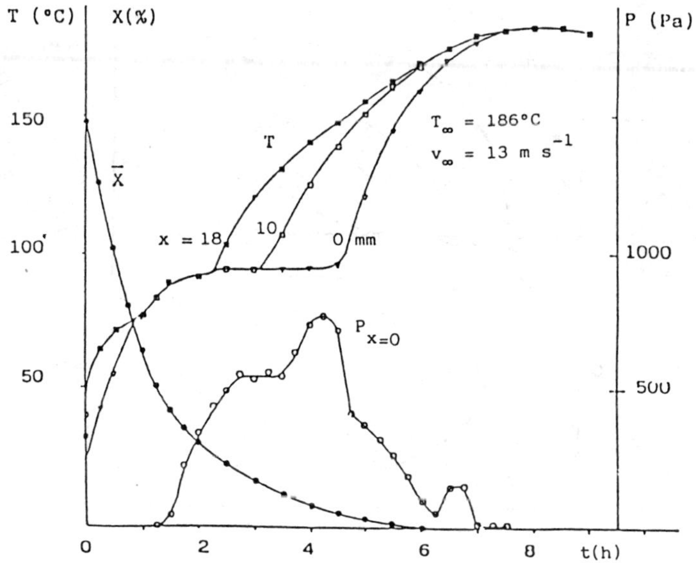

Figure 5.3.2: Drying of light concrete in superheated steam. Experimental results.

$$h\left(T_\infty - T_{x=e}\right) \;=\; k_{11}\left(\frac{\partial T}{\partial x}\right)_{x=e} + k_{12}\left(\frac{\partial X}{\partial x}\right)_{x=e} + k_{13}\left(\frac{\partial P}{\partial x}\right)_{x=e}$$
$$+\Delta h_v\left(n_l + n_v\right), \qquad (5.3.3)$$

and

$$n_l + n_v = kcM_v \ln\left(\frac{1 - x_{v\infty}}{1 - x_v|_{x=e}}\right). \qquad (5.3.4)$$

The two coefficients, h and k, are linked together by the principle of heat and mass transfer analogy.

In the case of drying in superheated steam, (5.3.4) is indeterminate. Therefore, instead of the boundary condition (5.3.4), we must use the equilibrium relation:

$$T_{x=e} = T_b. \qquad (5.3.5)$$

The third boundary condition simply states that the interface pressure, $P_{x=e}$, is equal to the atmospheric pressure, P_∞:

$$P_{x=e} = P_\infty. \qquad (5.3.6)$$

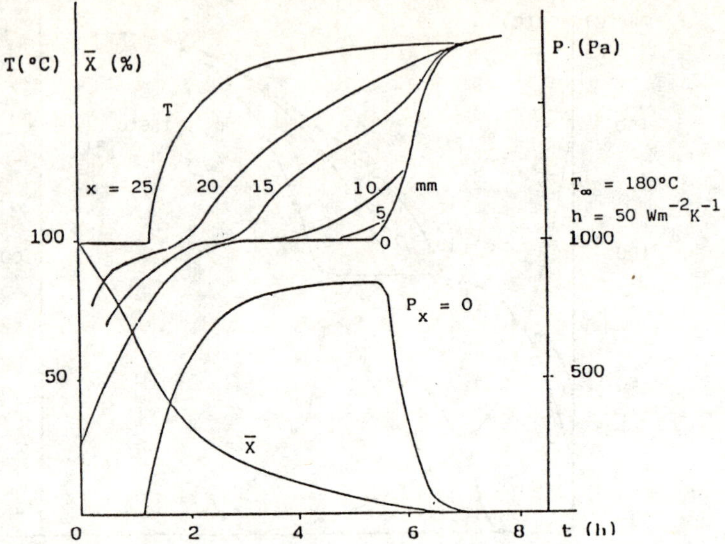

Figure 5.3.3: Drying of light concrete in superheated steam. Numerical results.

The numerical results are shown on Figs. 5.3.3 and 5.3.4.

C. Comparison between theory and experiment

The comparison of Figs. 5.2.4 and 5.2.5, shows a good qualitative agreement between theory and experiment. Thus, we can use the numerical results (Fig. 5.2.6) in order to analyze the phenomena.

During the first drying period, the essential driving force is the capillary pressure. The moisture profile is rather parabolic. The surface temperature of the slab is the boiling point of water. The gas pressure is equal to the atmospheric one.

When the capillary forces are no longer sufficient to transport water at the interface, the surface moisture content falls down quickly from the critical value X_{ir} (characteristic of the transition funicular-pendular state) to the hygroscopic range. Thus, in order to satisfy (5.3.5), the surface temperature increases. Simultaneously the total gas pressure increases. The moisture content profiles have an S-shaped form. Three zones can be distinguished inside the body:

- an inner zone, where the moisture content has a high value. The mass

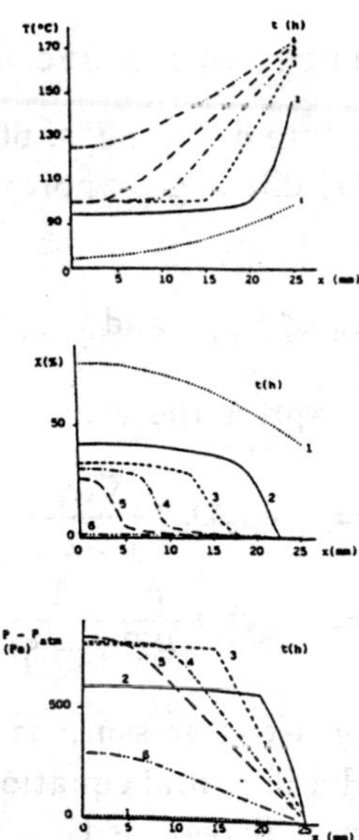

Figure 5.3.4: Temperature, moisture content and pressure profiles during light concrete drying in superheated steam. Numerical results.

transfer takes place in liquid phase by capillarity; the mass diffusivity and the apparent thermal conductivity (due to the high moisture content and temperature values) are large. Hence, temperature, moisture content and pressure are approximately constant.

- a narrow transition zone, where moisture content falls down from capillary funicular state to hygroscopic state.

- an outer zone, corresponding to the hygroscopic range. Temperature and pressure gradients are rather constant: heat transfer takes place by conduction and mass transfer by vapor filtration under the influence of a total pressure gradient in quasi steady-state.

The third drying period begins with the decrease of the internal total gas pressure. In the entire slab, moisture content lies in the hygroscopic state. The temperature, moisture content and pressure reach asymptotically their equilibrium values: external flow temperature, T_∞, equilibrium moisture content, $X_{eq} \simeq 0$, and atmospheric pressure, P_∞.

5.3.2 Low temperature convective drying

In the case of low temperature drying, $T \approx 60°C$, for the vapor transport in gaseous phase, the classical dilute gas approximation can be applied. That is:

$$n_a \ll n_v, \quad \text{and} \quad \omega_a \gg \omega_v,$$

so that we are allowed to express the vapor flux density as:

$$\begin{aligned}
n_v &= -\rho_g \mathcal{D}_v f \frac{\nabla \omega_v}{1 - \omega_v} \\
&= -\mathcal{D}_v f \frac{P}{RT} \frac{1}{1 - \frac{p_v}{P}} \nabla \left(\frac{p_v}{P} \right),
\end{aligned} \tag{5.3.7}$$

and to consider the total gaseous pressure as nearly constant.

Thus, the three partial differential equations of the model are reduced to two equations (gravity being disregarded):

$$\rho_o c_p \frac{\partial T}{\partial t} = \nabla \cdot \left(k'_{11} \nabla T + k'_{12} \nabla X \right), \tag{5.3.8}$$

and

$$\frac{\partial X}{\partial t} = \nabla \cdot \left(k'_{21} \nabla T + k'_{22} \nabla X \right), \tag{5.3.9}$$

where

$$k'_{11} \simeq k_{11}, \qquad k'_{12} \simeq k_{12}, \tag{5.3.10}$$

$$k'_{21} \simeq k_{21}, \qquad k'_{22} \simeq k_{22}. \tag{5.3.11}$$

5.3.3 The diffusion model

A drastic simplification of the system is to reduce it to the simple diffusion equation:

$$\frac{\partial X}{\partial t} = \nabla \cdot \left(a_m \nabla X \right). \tag{5.3.12}$$

This simplification requires that the temperature inside the porous medium be nearly uniform. By examining the energy equation, this hypothesis assumes that the thermal equilibrium is reached more rapidly than the mass equilibrium and that the evaporation takes place at the interface.

The first assumption is generally verified. Because thermal diffusivity is larger than mass diffusivity, we are allowed to treat the thermal problem as in steady or quasi steady-state (to take into account temperature effects on the diffusion coefficient, or in this case, an integral energy balance can be written). The second one is valid only for hygroscopic material, as indicated in the pioneer work of Ceaglske and Hougen (1937). For example, wood drying below the fiber saturation point at low temperature can be correctly described by (5.3.12).

The search for analytical solutions leads to more drastic assumptions: the diffusion coefficient is often taken as constant; if we assume that at the beginning of the falling rate period the surface moisture content is equal to its equilibrium value X_{eq}, for a plate (thickness $2e$), the problem can be put in the form:

$$\frac{\partial X}{\partial t} = a_m \frac{\partial^2 X}{\partial x^2}, \tag{5.3.13}$$

with the boundary conditions

$$x = 0, \qquad \frac{\partial X}{\partial x} = 0, \tag{5.3.14}$$

$$x = e, \qquad X = X_{eq}. \tag{5.3.15}$$

At the initial time, $t = 0$ (beginning of the falling rate period), the average moisture content value must be equal to the average critical moisture content $\overline{X}_{cr}$ (see Sec. 5.1). However, the initial profile remains to be defined.

If we choose an initial uniform repartition, the solution is

$$\frac{\overline{X} - X_{eq}}{\overline{X}_{cr} - X_{eq}} = \frac{8}{\pi^2} \sum_{n=0}^{\infty} \frac{1}{(2n+1)^2} exp\left[-(2n+1)^2 \frac{\pi^2}{4} \frac{a_m t}{e^2}\right]. \tag{5.3.16}$$

This very simple model was used by Sherwood in one of the first scientific approaches of drying science (Sherwood, 1929). The drying kinetics are represented by the curve 1 on Fig. 5.3.5. Of course, the initial drying rate is infinite. To avoid this singularity, it is more elegant to assume that at the beginning of the falling rate period, the moisture content profile has a parabolic form, corresponding to the quasi steady-state conditions. It is easy to see that the constant drying rate $-(d\overline{X}/dt)_I$, and the critical average moisture content, $\overline{X}_{cr}$, are linked together by the relation

$$-\left(\frac{d\overline{X}}{dt}\right)_I = 3\frac{\overline{X}_{cr} - X_{eq}}{(e^2/a_m)}. \tag{5.3.17}$$

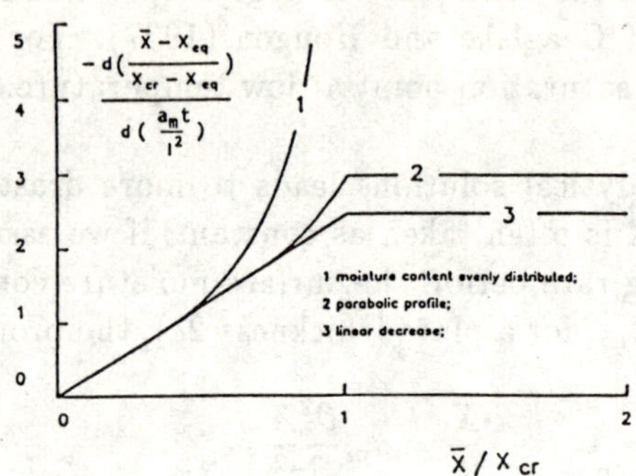

Figure 5.3.5: Drying rate curves for the diffusion model.

The analytical solution is then given by

$$\frac{\overline{X} - X_{eq}}{\overline{X}_{cr} - X_{eq}} = \frac{96}{\pi^4} \sum_{n=0}^{\infty} \frac{1}{(2n+1)^4} exp\left[-(2n+1)^2 \frac{\pi^2}{4} \frac{a_m t}{e^2}\right]. \qquad (5.3.18)$$

The drying rate deduced from (5.3.18) is represented by curve 2 on Fig. 5.2.7. Many authors assume that it is legitimate to assimilate this rate to a linear decrease (curve 3 on Fig. 5.2.7) by only retaining the first term of the series (5.3.18), with the approximation: $96/\pi^4 \simeq 1$. Thus

$$-\frac{d\overline{X}}{dt} = \frac{\pi^2 a_m}{4e^2}\overline{X}. \qquad (5.3.19)$$

In fact, these computed drying curves do not enable to describe the wide range of experimental drying curves which are shown on Fig. 5.1.2.

5.3.4   Receding drying front

Another common drying model is to assume that water evaporates from a front which migrates into the porous body during the falling rate period. In particular, this model is often applied to non-hygroscopic capillary bodies. Its interest lies in a schematic description of the sharp variation of the water transport coefficients. Above irreducible saturation, water moves in liquid phase, mostly by capillarity. In the pendular state, the liquid water transport coefficient is very low. Therefore, water moves in gaseous phase, the mass fraction vapor gradient being the driving force. This model distinguishes two zones in the body:

- An inner zone, with nearly uniform temperature and moisture content.

- An outer zone, where the heat flux and mass vapor flux are constant.

The heat and mass transfer resistances are located in the external boundary layers and across the dry zone, where the problem is treated as quasi-steady. The dry zone has a thickness Δ, and a dry thermal conductivity, λ^*. At low temperature, vapor migrates by diffusion, with a mass diffusivity $f\mathcal{D}_v$. A similar analysis could be made at high temperature, with a vapor transport by filtration in the dry zone (Moyne, 1987).

We examine this problem in plane geometry. The heat flux density, q, across the dry zone is given by

$$q = h\left(T_\infty - T_o\right) = \frac{\lambda^*}{\Delta}\left(T_o - T_f\right) = \frac{T_\infty - T_f}{\dfrac{1}{h} + \dfrac{\Delta}{\lambda^*}}, \qquad (5.3.20)$$

where the suscriptss f, o and ∞, respectively, indicate the front, the surface and the bulk of the fluid, and h is the heat transfer coefficient.

In the same way, the mass flux density, with the assumption of stagnant air, is given by

$$\begin{aligned}
n_v &= ckM_v \ln\left(\frac{1 - x_{v\infty}}{1 - x_{vo}}\right) \\
&\quad - c\frac{f\mathcal{D}_v}{\Delta}M_v \ln\left(\frac{1 - x_{vo}}{1 - x_{vf}}\right) \\
&= cM_v \frac{x_{vo} - x_{v\infty}}{\dfrac{x_{vo} - x_{v\infty}}{k\ln\left(\dfrac{1 - x_{v\infty}}{1 - x_{vo}}\right)} + \dfrac{x_{vf} - x_{vo}}{\dfrac{f\mathcal{D}_v}{\Delta}\ln\left(\dfrac{1 - x_{vo}}{1 - x_{vf}}\right)}}, \qquad (5.3.21)
\end{aligned}$$

where x_v is the vapor molar fraction, and $c = \frac{P}{RT}$ is the molar concentration.

Since the sensible heat can be disregarded, the heat balance front location leads to

$$q = \Delta h_v n_v. \qquad (5.3.22)$$

The front temperature can be calculated by

$$\frac{h(T_\infty - T_f)}{1 + \mathrm{Bi}_t} = cM_v k \Delta h_v \frac{x_{vf} - x_{v\infty}}{\dfrac{x_{vo} - x_{v\infty}}{\ln\left(\dfrac{1 - x_{v\infty}}{1 - x_{vo}}\right)} + \mathrm{Bi}_m \dfrac{x_{vf} - x_{vo}}{\ln\left(\dfrac{1 - x_{vo}}{1 - x_{vf}}\right)}},$$

$$(5.3.23)$$

where

$$\mathrm{Bi}_t = \frac{h\Delta}{\lambda^*} = \frac{(\Delta/\lambda^*)}{(1/h)}$$

is the thermal Biot number, and

$$\mathrm{Bi}_m = \frac{k\Delta}{f\mathcal{D}_v} = \frac{(\Delta/f\mathcal{D}_v)}{(1/k)}$$

is the mass Biot number.

In order to avoid cumbersome calculations, we assume that the vapor molar fraction is much lower than one ($x_v \ll 1$). Then, we can write equation (5.3.23) in the form

$$\frac{T_\infty - T_f}{x_{vf} - x_{v\infty}} = \frac{cM_v k \Delta h_v}{h} \frac{1 + \mathrm{Bi}_t}{1 + \mathrm{Bi}_m}. \qquad (5.3.24)$$

When $\mathrm{Bi}_t = \mathrm{Bi}_m = 0$, the front is located at the surface of the body.

If heating occurs only by convection, the temperature of the body is equal to the wet bulb temperature T_w. Thus

$$\frac{T_\infty - T_f}{x_{vf} - x_{v\infty}} = \frac{T_\infty - T_w}{x_{vw} - x_{v\infty}} \frac{1 + \mathrm{Bi}_t}{1 + \mathrm{Bi}_m}. \qquad (5.3.25)$$

When the internal resistances are larger than the external ones, $\mathrm{Bi} \gg 1$, the front temperature (which is also the temperature of the inner zone) is equal to a pseudo wet bulb temperature $T_{w'}$ by

$$\frac{T_\infty - T_{w'}}{x_{vw'} - x_{v\infty}} = \frac{T_\infty - T_w}{x_{vw} - x_{v\infty}} \frac{h}{\lambda^*} \frac{f\mathcal{D}_v}{k}. \qquad (5.3.26)$$

With this model, the drying rate is equal to

$$\frac{-(d\overline{X}/dt)}{-(d\overline{X}/dt)_{\Delta=0}} = \frac{T_\infty - T_f}{T_\infty - T_w} \frac{1}{1 + \mathrm{Bi}_t}. \qquad (5.3.27)$$

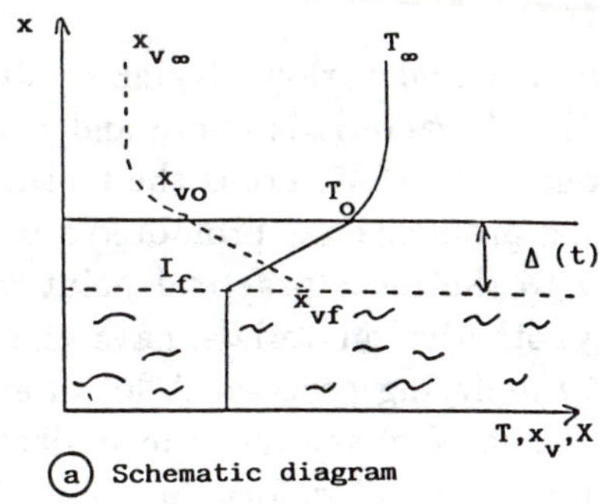

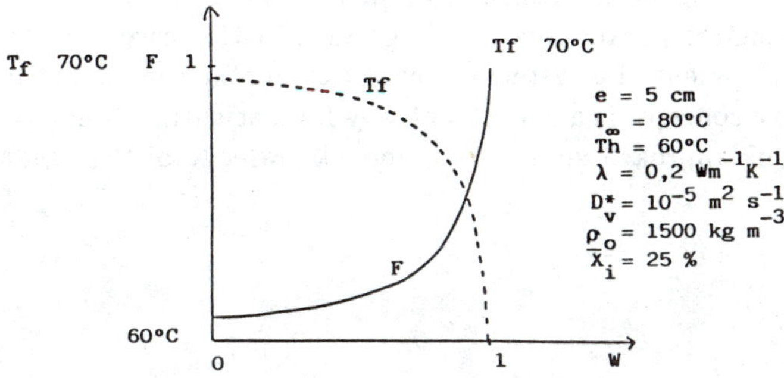

Figure 5.3.6: Receding drying front model.

If we assume that at the beginning of the falling rate period, the front is located at the surface of the body, $\Delta = 0$, and that the average moisture content is $\overline{X}_i$, we have

$$\overline{X} = \overline{X}_i \left(1 - \frac{\Delta}{e} \right), \tag{5.3.28}$$

where e is the half-thickness of the plate.

Figure 5.3.6 represents the drying kinetics for the receding front model.

5.4   Conclusions

In this paper, we have summarized various drying models, including both the empirical method of the characteristic curve and a sophisticated three partial differential equations system. Where is the truth?

The modernity of drying problems has promoted a wide range of works, from the most theoretical to the most practical point of view. Academic studies, obviously, with hypothetical materials, have succeeded in providing a better understanding of the drying process. A lot of energy has been devoted to capillary porous media. A renewal of the subject may be to extend such an approach to microporous and deformable materials. Nevertheless, for non-academic materials (those which have to be industrially dried...), it is unrealistic at the present time to hope for a complete physically reliable description of drying mechanisms. The measurement of the necessary set of phenomenological parameters is a huge task in this case. On the other hand, the development of a systematic empirical method, as the characteristic drying curve concept, is a convenient way for a scientific design of dryers, especially if this approach goes with a good knowledge of the fundamental phenomena.

References

Azizi, S., Moyne, C. and A. Degiovanni. Approche théorique et expérimentale de la conductivité thermique des milieux poreux humides. I. Expérimentation, *Int. J. Heat Mass Transfer*, **31**:2305-2317, 1988.

Basilico, C. and M. Martin, Approche expérimentale des mécanismes de transfert au cours du séchage convectif à haute température d'un bois résineux, *Int. J. Heat Mass Transfer*, **27**:657-668, 1984.

Brakel, J. van, Mass transfer in convective drying, *Advances in Drying*, **1**:217-267, 1980.

Ceaglske, N. H. and O. A. Hougen, *Drying of granular solids, Ind. Eng. Chem.*, **29**:805-813, 1937.

Crausse, P., G. Bacon, and S. Bories, Etude fondamentale des transferts couplés chaleur-masse en milieu poreux, *Int. J. Heat Mass Transfer*, **24**:991-1004, 1984.

De Vries, D. A. Simultaneous transfer of heat and moisture in porous media, *Trans. Amer. Geophys. Union*, **39**:909-916, 1958.

Dullien, F. A. L. *Porous Media: Fluid Transport and Pore Structure*, Academic Press, 1979.

Keey, R. B. *Introduction to Industrial Drying Operations*, Pergamon Press, 1978.

Keey, R. B. and M. Suzuki, On the characteristic drying curve, *Int. J. Heat Mass Transfer*,**17**:1455-1464, 1974.

Krischer, O. *Die Wissenchaftlichen Grundlagen der Trocknungstechnik*, Springer-Verlag, 1962.

Luikov, A. V. *Heat and Mass Transfer in Capillary Porous Bodies*, Pergamon Press, 1966.

Marle, C. M. On macroscopic equations governing multiphase flow with diffusion and chemical reactions in porous media, *Int. J. Engng. Sci.*, **20**: 643-662, 1982. Moyne, C. *Transferts Couplés Chaleur-Masse lors du Séchage: Prise en Compte du Mouvement de la Phase Gazeuse*, Thèse de Doctorat d'Etat, Institut National Polytechnique de Lorraine, Nancy, 1987.

Moyne, C. and C. Basilico, High temperature convective drying of softwood and hardwood. Drying, pp. 376-381, 1985.

Moyne, C. and C. Basilico, Moisture transport analysis during high temperature drying of wood, in *Drying of Solids, Recent International Developments*, pp. 317-327, Wiley Eastern Publications, 1986.

Moyne, C., Batsale, J. C. and A. Degiovanni, Approche théorique et expérimentale de la conductivité thermique des milieux poreux humides. II. Théorie, *Int. J. Heat Mass Transfer*, **31**:2319-2330, 1988.

Moyne, C. and A. Degiovanni, Conductivité thermique des matériaux poreux humides: évaluation théorique et possibilité de mesure, *Int. J. Heat Mass Transfer*, 30:2225-2245, 1987.

J.R. Philip, J. R. and D. A. de Vries, Moisture movement in porous material under temperature gradients, *Trans. AmerGeophys. Union*, 38:222-232, 1957.

Schlünder, E. U. *Handbook of Heat Transfer*, Sec. 3.13, Dryers, Hemisphere Publishing Corporation, 1983.

Sherwood, T. K. The drying of solids, *Ind. Eng. Chem.*, **21**:12-16, 1929.

Whitaker, S. Simultaneous heat, mass and momentum transfer in porous media: a theory of drying, *Adv. Heat Transfer*, 13:119-203, 1977.

List of Main Symbols

a_m	Mass diffusivity.
c	Molar concentration.
c_p	Specific heat at constant pressure.
D_v^*	Vapor diffusion coefficient in porous media.
$\mathcal{D}_v$	Binary air-water diffusion coefficient.
e	Thickness.
f	Vapor diffusion resistance coefficient in porous media.
F	Normalized drying rate.
g	Intensity of gravity.
h	Heat transfer coefficient.
h_i	Mass enthalphy of component i.
j	Diffusive mass flux density.
k	Mass transfer coefficient.
K	Permeability.
$\dot{m}$	Rate of phase change.
M	Molar mass.
n	Mass flux density.
p	Partial pressure.
P	Total gaseous pressure.
r^*	Equivalent radius in a porous medium.

q	Heat flux density.
S	Saturation.
R	Universal gas constant.
t	Time.
T	Temperature.
u	Fluid velocity.
W	Characteristic moisture content.
x	Space coordinate.
x_v	Vapor molar fraction.
X	Moisture content (dry basis).

Greek symbols

δ	Thermomigration coefficient.
Δ	Dry zone thickness.
Δh_v	Enthalpy of vaporization.
ε	Porosity.
φ	Relative humidity.
θ	Wetting angle.
λ	Thermal conductivity.
λ^*	Thermal conductivity of the dry solid.
ν	Kinematic viscosity.
ρ	Density.
σ	Superficial tension.
ω	Mass fraction in gaseous phase.

Dimensionless number

Bi	Biot number.

Subscripts

a	Air.
b	Boiling point.
cr	Critical.
eq	Equilibrium.
f	Front.
g	Gaseous phase.
l	Liquid phase.

m Mass.
s Solid phase.
t Thermal.
v Vapor.
vs Saturated vapor.
w Wet bulb temperature.
w' Pseudo-wet bulb temperature.
o Dry solid or at the interface.
∞ Fluid bulk.

Superscript

$\overline{(..)}$ Average spatial value of $(..)$.

Chapter 6

STOCHASTIC DESCRIPTION
OF POROUS MEDIA

G. DE MARSILY
Ecole des Mines de Paris
l'Université Pierre et Marie Curie,
Paris, France.

The definition of the properties of porous media in space can be made using the concept of *random functions*. This stochastic approach has two major advantages:

- It conceptually defines the properties in space at a given point, without having to define a volume over which these properties must be integrated.

- It provides means for studying the inherent heterogeneity and variability of these properties in space, and for evaluating the uncertainty of any method of estimation of their values.

We will first define these concepts, then provide a few examples of their use in solving flow and transport problems, and finally address the problem of estimation.

6.1   Definition of Properties of Porous Media: The Example of Porosity

The *porosity* of a porous medium is defined as:

$$n = \frac{\text{volume of the voids}}{\text{volume of the porous medium}}.$$

This notion is intuitively easy to understand. However, on reflection, it poses some problems if we want to define it with precision. We shall discuss them here while keeping in mind that the following applies to other properties of the porous medium as well, e.g.,s permeability.

There are two accepted ways of defining the local properties of a porous medium: the notion of the *representative elementary volume* (REV) *random functions* (RF, which is also expressed as *ensemble average*). We shall see that these two notions implicitly influence any description of the spatial variations of the hydrogeological parameters.

The entire problem stems from the fact that the notions of porosity and permeability, which concern points in an equation with partial derivatives, for instance, cannot be defined or measured at single points, since a porous medium is a conglomeration of solid grains and voids. Below a certain scale of volume, porosity and permeability have no physical significance.

The REV method consists of saying that we give to one mathematical point in space the porosity or permeability of a certain volume of material surrounding it. This volume is the REV. It will be used to define and possibly measure the 'mean' property of the volume in question. Consequently, this concept involves an integration in space. It is obviously the first method that comes to mind. Behind it lies the idea of a sample, which is collected and from which the relevant property is estimated by measurement. More exactly, the size of the REV is defined by saying that it is

(a) sufficiently large so as to contain a large number of pores, thus allowing us to define a global property, while ensuring that the effect of the fluctuations from one pore to another are negligible. One may take, for example, 1 cm^3, or 1 dm^3.

(b) sufficiently small so that the parameter variations from one domain to the next may be approximated by continuous functions, in order that we may use infinitesimal calculus, without introducing any error that may be picked up by the measuring instruments at the macroscopic scale, where meters and hectometers are the usual dimensions.

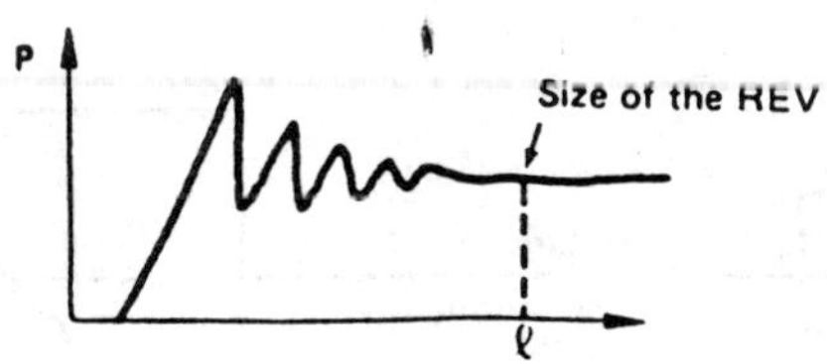

Figure 6.1.1: Definition of the REV.

This is, incidentally, someqhat like the problem in fluid mechanics of passing from the 'corpuscular' scale to that of the 'particle of matter'. It should be noted that in a fractured medium, the size of the REV may be quite astonishingly large, thus not satisfying the second hypothesis of *continuous functions* on the scale of the measuring instruments.

The size of the REV (measured, for example, by one of its characteristic dimensions, l, such as the radius of a sphere or the side of a cube) is, generally, linked to the existence of a flattening of the curve that connects the studied integral property, P, with the dimension l (Fig. 6.1.1). However, nothing allows us to assert that such a flattening always exists. The size of the REV may thus stay quite arbitrary.

Other important objections that can be made to this porous medium concept are of two kinds.

- First, it is very poorly suited for the treatment of discontinuities in the medium. When, in a thought experiment, the REV is moved across a discontinuity, the studied property is subjected to a continuous variation (Fig. 6.1.2). Sometimes, this poses problems of how to correctly represent boundaries, or limits between two media.

- The most important objection is that it gives no basis for studying the structure of the property in space. The most that can be said is that the spatial variations of the studied property must be smooth in accordance with the same thought process as above concerning the discontinuities.

Marle (1967) has suggested a more rigorous conceptualization of spatial integration. In order to achieve this, he proposes the use of an integrable nonnegative weighting function $m(x)$, such that its integral, when extended over the entire space, is equal to 1. This weighting function would not necessarily have a bounded support. The macroscopic magnitude $\langle a \rangle (x)$ is

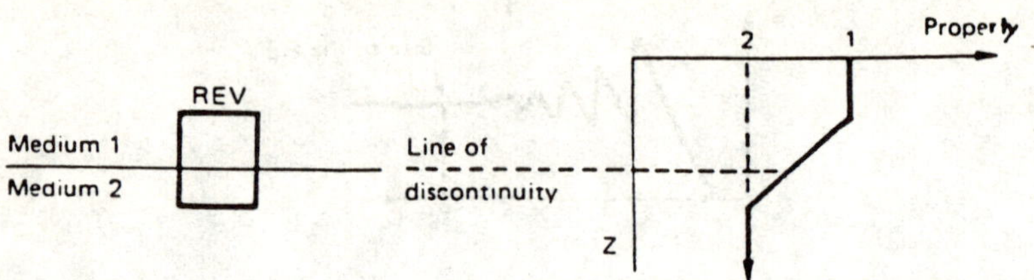

Figure 6.1.2: Definition of the properties of a discontinuous medium using the REV.

then defined from the local microscopic magnitude, $a(x)$, by a *convolution* extended over the whole space of a by m, in the form

$$\langle a \rangle (x) = \int a(x + x')m(x')dx',$$

where x stands for the coordinates in three-dimensional space (x_1, x_2, x_3). For the study of porosity, we choose an *indicator*, $a(x)$. If the point x is in a pore, $a(x) = 1$; if it is in a grain, $a(x) = 0$.

Furthermore, Marle suggests that this definition be generalized to the properties 'a' that are not continuous in the entire space* and that can be described by distributions. The convolution is then taken in the sense of distributions.

This method has the advantage of making the function $\langle a \rangle$ continuous and indefinitely differentiable, even if a is not, by a suitable choice of m. If the problem of the size of the REV is eliminated, that of the choice of the weighting function still remains arbitrary. However, with the help of this weighting function, it is possible to establish the connection between this method and a second one, examined below.

The *random functions* (RF) method is a more powerful is a more powerful concept. It consists of stating that the studied porous medium is a *realization* of a *random process*. Let us try to visualize this concept. Suppose we create in the laboratory several sand columns, each filled with the same type of sand. Each column represents the same porous medium, but is somewhat different from the others. Each column is a *realization* of the same porous medium, defined as the ensemble of all possible realizations (infinite in number) of the same process.

⁰* For example, a surface density of the adsorbed matter on the fluid-solid interface.

A property like porosity can then be defined, at a given geometrical point in space, as the average over all possible realizations of its point value (defined as 0 in a grain and 1 in a pore). One speaks of 'ensemble averages' instead of 'space averages'. For the sand columns just described, it is obvious that the ensemble average (or *expected value*) of these point porosities will be identical to the space average defined by taking the column itself as the REV. Furthermore, this ensemble average will be the same for any point of the column. We will define later the conditions necessary for this to be true.

In more general terms, a property Z will be called a random function (RF), $Z(x,\xi)$, if it varies both with the spatial coordinate system, x, and with the *state variable*, ξ, in the ensemble of realizations. Then, $Z(x,\xi_1)$ is a *realization* of Z, while $Z(x_o,\xi)$ is a *random variable*, i.e., the ensemble of the realizations of the RF Z at x_o, while $Z(x_o,\xi_1)$ is the single value of Z at x_o for realization ξ_1. To simplify the notations, the variable ξ is generally omitted.

If we want to find a less abstract example of a random porous medium, we may consider a series of sand dunes in a desert. Each crescent-shaped dune can be seen as a *realization* of a *stochastic process*; several thousands of them can be found in a given desert, formed from the same eolian sediments, thus having basically the same properties, although each is different.

The immense advantage of the stochastic approach is that one can study statistical properties of the porous medium in the ensemble of realizations other than just the expected value. One very often uses the *variance* (called dispersion variance) of the property, which characterizes the magnitude of the fluctuations with respect to the mean, and the *autocovariance* (or simply *covariance*), which characterizes the correlation between the values taken by the property at two neighboring points in space.

However, when studying a given porous medium, there will be only one realization of the conceptual random medium. Some assumptions are necessary to make this concept useful. The most common are stationarity and ergodicity.

Stationarity assumes that any statistical property of the medium (mean, variance, covariance, higher-order moments) is stationary in space, i.e., does not vary with a translation. It will be the same at any point of the medium. *Weak stationarity* refers to a medium where only the first two moments are stationary. If $Z(x)$ is the studied property, x being the coordinates in one, two, or three dimensions, then the random function $Z(x)$ satisfies:

(a) *Expected value*:

$$E[Z(x)] = m \ \text{ not a function of } x.$$

(b) *Covariance*:

$$E[(Z(x) - m)(Z(x + h) - m)]$$ not a function of X, but a function only of the lag h, a vector in two or three dimensions.

By developing, and labeling the covariance $C(h)$, we obtain

$$C(h) = E[Z(x) \cdot Z(x + h)] - m^2.$$

By definition,

$$C(0) = E[(Z(x) - m)^2] = \sigma_Z^2$$

is the variance of Z.

In more rigorous terms, strong stationarity means that all the probability distribution functions (pdf) of the random function $Z(x)$ are invariant under translation, whether we consider one point $p(Z(x))$, or n points $p(Z(x_1), ..., Z(x))$.

Ergodicity implies that the unique realization available behaves in space with the same pdf as the ensemble of possible realizations. In other words, by observing the variation in space of the property, it is possible to determine the pdf of the random function for all realizations. This is called the *statistical inference* of the pdf of the RF $Z(x)$.

In the vocabulary of stochastic processes, a phenomenon that is *stationary* and *ergodic* is called *homogeneous*. We would then use 'uniform' to describe a medium in which some property does does not vary in space. Geologists traditionally call it 'homogeneous'.

Other less stringent hypotheses can also be defined, e.g., stationarity of increments of Z. These will be defined later.

Marle (1967) compares this method to the one based on spatial integration. He shows that the stochastic definition may be regarded as the limiting case of an integral definition when the porous medium is assumed to be infinite, ergodic, and stationary, and the weighting function does not have a bounded support. As a matter of fact, spatial integration in an infinite volume reproduces the mathematical expectation over all possible realizations, if the medium is indeed stationary and ergodic. We shall use these two methods for defining the properties of porous media.

Other approaches can also be used to define the properties of porous media. One is that of *composite materials* (Beran, 1968) and, has been applied to porous media by Dagan (1979, 1981, 1982b).

6.2   Stochastic Approach to Permeability and Spatial Variability

We have seen in Sec. 6.1 that a probabilistic definition of a property like porosity can be given in a porous medium. However, the definition of the permeability as a random function requires a change of scale, which was proposed by Matheron in 1967, referring to the works of Schwydler (1962). As a matter of fact, point permeability cannot be used in the same way as point porosity, because, on the microscopic scale, Darcy's law implied by the notion of permeability, does not apply to the flow: it is the Navier-Stokes law that governs the relationship between the hydraulic head and the velocity.

Matheron (1967) has shown that Darcy's law is simply a consequence of the linearity of Navier's equation in the very complex geometry of a porous medium that leads to Darcy's law and the definition of permeability. We can thus, conceptually at least, link permeability to the geometric description of a porous medium. Such a geometric description of a medium (e.g., size and shape of pores) can be made stochastically, exactly as we have done in Sec. 1 for porosity. For instance, the permeability of simple geometrical media (fractures, tubes) depends on the aperture of the fractures, or the diameter of the tubes. These can be given a stochastic definition at a point in space (probability distribution function, expected value, spatial covariance, etc). In a more complex medium, the number of descriptors of the geometry increases, but conceptually, each of them can be given a stochastic definition on the microscopic scale.

As a consequence, permeability on the macroscopic scale, depending on stochastic microscopic quantities, can be regarded as a stochastic property and can be defined conceptually as a random function. This will have a probability distribution function, expected values, spatial covariance, etc.

Quite a number of authors have studied the pdf of permeability, hydraulic conductivity or transmissivity in a given aquifer. Their analysis is biased most of the time because they assume that the measurements taken at different locations are statistically independent, whereas, in reality, permeability usually displays a strong spatial correlation. Nevertheless, following Law (1944), Walton and Neill (1963), Krumbein (1936), Farengolts and Kolyada (1969), Ilyin *et al.* (1971), Jetel (1974), Freeze (1975) and Rousselot (1976), we can admit that permeability usually has a log-normal probability distribution function, whatever the nature of the rock. The variance of this spatial variability of permeability is quite high: if $Y = \ln k$, σ_Y^2 is generally in the range between 1 and 2 but can reach 10 in some cases.

The spatial correlation of transmissivity has also been studied, e.g., by Delhomme (1974, 1978a,b, 1979). He found that, in general, the stationarity hypothesis did not hold, and that stationarity on the first increments (called the intrinsic hypothesis) should be used. Instead of the covariance $C(h)$, one must then use the variogram $\gamma(h)$, which we will define in Sec. 6.6. The spatial correlation is important over distances that can be short (e.g. 10 m) or very long (up to 100 km), depending on the type of aquifer. There is, however, very often a strong erratic component (spatially uncorrelated) in the transmissivity, which may cause two wells not very far apart to have quite different transmissivities.

The spatial variability of permeability (or hydraulic conductivity, or transmissivity) leads us to the question of how to compose local permeability values in order to obtain an average permeability. In a *deterministic approach*, it is easy to show that the composition of uniform 'blocks', placed side by side in space, gives

- a rule of harmonic composition, if the blocks are in series

$$\sum l_i / K_{\text{mean}} = \sum (l_i / K_i) \ , \quad l_i \text{ length in flow direction} \ ,$$

- a rule of arithmetic composition, if the blocks are in parallel

$$K_{\text{mean}} \sum e_i = \sum (e_i K_i) \ , \quad e_i \text{ thickness.}$$

Here, we recognize the same rules as those of the composition of resistances derived from Ohm's law in electricity.

In a probabilistic approach, where the permeability may vary in all directions of space, Matheron (1967) has obtained the following results:

- If the flow is uniform (parallel flow lines), whatever the spatial correlation of the permeability and whatever the number of dimensions of the

space, the average permeability always ranges between the harmonic[*]
mean and the arithmetic[*] mean of the local permeabilities.

- If the probability distribution function of the permeability is log-normal,
  and if the flow is two-dimensional, the average permeability is exactly
  equal to the geometric[*] mean of the local permeability in uniform flow.

- If the flow is not uniform (converging radial, for example), there is no
  law of composition, constant in time, that makes it possible to define
  a mean Darcian permeability. This problem is quite worrying from
  the conceptual viewpoint in so far as it is precisely through pumping
  tests in wells that the permeability (or transmissivity) of an aquifer is
  measured *in situ*.

Gelhar (1976), Bakr *et al.* (1978) and Gutjahr *et al.* (1978) also give
linearized approximations of the average permeability in uniform flow, for a
normal probability distribution function of permeability:

$$1\text{-}D: \quad k_M = k_G(1 - \sigma_Y^2/2)$$
$$2\text{-}D: \quad k_M = k_G$$
$$3\text{-}D: \quad k_M = k_G(1 + \sigma_Y^2/6)$$

where k_M is the average permeability, k_G is the geometric mean permeability,
and σ_Y^2 is the variance of $Y = \ln k$.

An example of how these results can be obtained is given in the next
section.

6.3   Stochastic Partial Differential Equations

The flow equation in a porous medium is based on mass balance concepts. It
is written as a partial differential equation, analogous to the heat equation:

If the properties of the porous medium (e.g., the permeability, K, or the
storativity, S_s) are considered as random functions, then such an equation is
called a *stochastic partial differential equation*. This means that the solution
of the equation for the head, h, is also a RF. Solving this equation means

---

[0][*] Harmonic mean : $1/K_M = E(1/K)$
   Arithmetic mean : $K_M = E(K)$
   Geometric mean : $\ln K_M = E(\ln K)$.

determining the pdf of h (in particular its first moments) from the prescribed values of the pdf of K and S_s.

We shall give here a brief outline of several methods for solving such equations.

6.3.1   Properties of stochastic partial differential equations

For equations in which the parameters are random functions, one must first define what the derivative of a stochastic process is. If K is a random function, the quadratic mean derivative K' is defined by

$$\lim_{\Delta x \to 0} E\left[\frac{K(x + \Delta x) - K(x)}{\Delta x} - K'(x)\right] = 0.$$

Also, K' is a random function.

The complete solution of a stochastic partial differential equation consists of obtaining all the probability distribution functions, at every location and at all times, of the unknown random function, e.g., the head. This is almost always impossible to achieve. Therefore, one generally looks for (1) the probability distribution function of the unknown at several particular locations, or (2) the moments of the unknown function: expected value, variance, covariance. These moments can sometimes only be evaluated approximately.

Even to obtain such an approximate and limited solutions, one must often make some hypotheses on the stochastic processes in question.

- If these hypotheses concern the input parameters (e.g., boundary conditions, source terms, coefficients), the corresponding solution is said to be 'honest'.

- If these hypotheses concern the unknown solution, whose form is a-priori unknown, the solution is said to be 'dishonest'. This does not mean that a dishonest solution is necessarily incorrect if these hypotheses (e.g., stationarity...) are based on valid physical reasoning. Dishonest solutions can, on the contrary, sometimes, be more precise than honest ones. It is only when the assumptions are not physically based that they may be invalid (see Keller, 1964; Lumley and Panafosky, 1964; and Schweppe, 1973).

Let us briefly examine some methods of solution.

6.3.2 Spectral methods

This method is applicable to second-order stationary stochastic processes for both inputs and outputs. If $Y(x)$ is second-order stationary, the *spectrum* (or *spectral density*) of Y is the *Fourier transform* of its autocovariance function:

$$\varphi(k) = \frac{1}{2\pi} \int_{-\infty}^{+\infty} e^{-iks} \operatorname{cov}\left[Y(x+s), Y(x)\right] ds. \qquad (6.3.1)$$

Using the inverse Fourier transform, one can also write

$$\operatorname{cov}\left[Y(x+s), Y(x)\right] = C(s) = \int_{-\infty}^{+\infty} e^{iks} \varphi(k) dk. \qquad (6.3.2)$$

The following *representation theorem* will be used: if the second-order stationary stochastic process $Y(x)$ is of zero mean $E(Y) = 0$ and of covariance $C(s)$, then one can define a complex associated process (i.e., $Z \in C$ if $Y \in R$) that satisfies

$$Y(x) = \int_{-\infty}^{+\infty} e^{ikx} dZ(k), \qquad (6.3.3)$$

$$
\begin{aligned}
E\left[dZ(k_1)dZ^*(k_2)\right] &= 0 & &\text{if } k_1 \neq k_2 \\
E\left[dZ(k_1)dZ^*(k_1)\right] &= \varphi(k_1) & &\text{i.e. if } k_1 = k_2.
\end{aligned}
\qquad (6.3.4)$$

Equation (6.3.3) is a *Fourier-Stieltjes integral* and the asterisk in (6.3.4) denotes the *complex conjugate*.

We shall give a simple example of the use of the spectral method, from Gelhar (1976), Bakr *et al.* (1978), and Gutjahr *et al.* (1978). Let us consider a one-dimensional steady-state flow in an infinite medium. The flow equation is written as

$$\frac{d}{dx}\left[K(x)\frac{dH}{dx}\right] = 0, \qquad (6.3.5)$$

where K is the permeability and H is the head.

We assume that $K(x)$ is a second-order stationary stochastic process. If we integrate (6.3.5) once, it gives

$$K(x)\frac{dH}{dx} = -q, \qquad (6.3.6)$$

where q is the *constant* flow rate in the flow tube. Dividing by K and defining $W = 1/K$, we obtain

$$\frac{dH}{dx} = -qW. \tag{6.3.7}$$

Let us define the expected value of H and W and their fluctuation around the average by

$$\bar{H} = E(H) \quad h = H - \bar{H}, \text{ thus } E(h) = 0$$

$$\bar{W} = E(W) \quad w = W - \bar{W}, \text{ thus } E(w) = 0.$$

By substituting in (6.3.7) and taking its expected value, we get

$$\frac{d\bar{H}}{dx} + \frac{dh}{dx} = -q(\bar{W} + w),$$

$$E\left(\frac{d\bar{H}}{dx}\right) = -qE(W), \qquad \frac{d\bar{H}}{dx} = -q\bar{W}, \tag{6.3.8}$$

$$\frac{dh}{dx} = -qw. \tag{6.3.9}$$

Assuming h to be second-order stationary ('dishonest' hypothesis) and using the 'representation theorem', we can define two complex stochastic processes such as

$$h(x) = \int\limits_{-\infty}^{+\infty} e^{ikx}\, dZ_h(k) \,, \quad w(x) = \int\limits_{-\infty}^{+\infty} e^{ikx}\, dZ_w(k).$$

We then take the first derivative of h and introduce it into (6.3.9)

$$\frac{dh}{dx} = \int\limits_{-\infty}^{+\infty} e^{ikx}\, ik\, dZ_h(k),$$

$$ik\, dZ_h(k) = -q\, dZ_w(k) \quad \text{thus} \quad dZ_h(k) = i\frac{q}{k}\, dZ_w(k).$$

From (6.3.4) we can calculate the spectrum of h:

$$\varphi_h(k) = E[dZ_h(k)\, dZ_h^*(k)]$$

$$= E\left\{\left[i\frac{q}{k}\, dZ_w(k)\right]\left[-i\frac{q}{k}\, dZ_w^*(k)\right]\right\} = \frac{q^2}{k^2}\, E\left[dZ_w(k)\, dZ_w^*(k)\right]$$

$$\varphi_h k = \frac{q^2}{k^2}\, \varphi_w(k). \tag{6.3.10}$$

We have now solved our problem. Equation (3.8) gives us the first moment of H, $\bar{H} = -q\bar{W}x + \text{constant}$, and (6.3.10) gives us the spectrum of H given the spectrum of W. Using (6.3.2), one can also determine the covariance and variance of h from the spectrum. For instance, if the following covariance is used for W, as suggested by Gutjahr *et al.* (1978), we obtain

$$\text{cov}\,[w(x+s), w(x)] = \sigma_W^2(1 - |s|/l)e^{-|s|/l},$$

where σ_W^2 is the variance of w and the distance l is called the *correlation length*, one obtains

$$\varphi_w(k) = \frac{2k^2\sigma_W^2 l^3}{\pi(1 + k^2 l^2)^2},$$

$$\text{cov}\,[h(x+s), h(x)] = q^2\sigma_W^2 l^2(1 + |s|/l)e^{-|s|/l},$$

$$\sigma_h^2 = C(0) = q^2 l^2 \sigma_W^2.$$

In this example the covariance and variance of h are constant all over the medium.

Gelhar *et al.* (1974, 1977, 1979a,b), Gelhar and Axness (1983), and Gelhar (1986) have used this spectral method extensively, mainly for the transport equation in their later articles.

6.3.3　The method of perturbations

We shall use the same example as before, i.e., (6.3.5). Let K be second-order stationary with $E(K) = \bar{K}$ and the 'fluctuation' $k = K - \bar{K}$, $E(k) = 0$. We shall also assume that the 'fluctuation' h of the solution H is second-order stationary ('dishonest' hypothesis), with $E(H) = \bar{H}$ and $h = H - \bar{H}$, $E(h) = 0$. We develop K and H to the first order, i.e., add to $\bar{K}$ and $\bar{H}$ a 'small perturbation', i.e., a fraction of their fluctuation

$$K = \bar{K} + \beta k \qquad H = \bar{H} + \beta h. \qquad \bullet \ (6.3.11)$$

Given k, we now look for h. We can introduce (6.3.11) into (6.3.5) and develop in β, disregarding the terms in β^2 (assumed to be small), viz.

$$\bar{K}\frac{d^2\bar{H}}{dx^2} + \beta\left(\bar{K}\frac{d^2 h}{dx^2} + \frac{dk}{dx}\frac{d\bar{H}}{dx} + k\frac{d^2\bar{H}}{dx^2}\right) = 0.$$

If this is to hold for any small β, each of these two terms must be equal to zero. Thus

$$\bar{K}\frac{d^2\bar{H}}{dx^2} = 0 \ \text{ or } \ \frac{d\bar{H}}{dx} = -q/\bar{K} \ \text{ and } \ \bar{H} = -qx/\bar{K} + \text{const.}$$

Substituting this result in the second term, we obtain

$$\frac{d^2h}{dx^2} = \frac{q}{\bar{K}^2} \frac{dk}{dx},$$

or

$$\frac{dh}{dx} = \frac{q}{\bar{K}^2} k + a, \tag{6.3.12}$$

where a is a constant. We take the expected value to be

$$E\left(\frac{dh}{dx}\right) = \frac{d}{dx} E(h) = \frac{q}{\bar{K}^2} E(k) + E(a).$$

As $E(h) = E(k) = 0$, we can see that $E(a) = 0$. Then, (6.3.12) gives directly

$$\mathrm{cov}\left(\frac{dh}{dx}\right) = \frac{q^2}{\bar{K}^4} \mathrm{cov}(k).$$

However, for a stationary random function with a differentiable covariance one can write

$$\mathrm{cov}\left(\frac{dh}{ds}\right) = -\frac{d^2}{ds^2} \mathrm{cov}(h).$$

Thus, if we can assume that

$$\frac{d}{ds} \mathrm{cov}\left[h(x), h(x+s)\right]\Big|_{s\to-\infty} = 0 \quad \text{and} \quad \mathrm{cov}\left[k(x), k(x+s)\right]\big|_{s\to-\infty} = 0 \,,$$

with two integration we find

$$\mathrm{cov}[h(x+s), h(x)] = -\frac{q^2}{\bar{K}^4} \int\limits_{-\infty}^{s} \int\limits_{-\infty}^{y} \mathrm{cov}[k(x), k(x+u)]\,du\,dy.$$

Again, we have found the expected value and covariance of the head. However, this time we must assume that σ_K^2 is small, otherwise the first-order development in β is not valid. To overcome this difficulty in the case of a permeability, where σ_K^2 is generally rather large, Gelhar has suggested the use of the logarithm of K. Equation (3.5) is written as

$$K \frac{d^2H}{dx^2} + \frac{dK}{dx} \frac{dH}{dx} = 0 \quad \text{or} \quad \frac{d^2H}{dx^2} + \frac{d}{dx}(\ln K) \frac{dH}{dx} = 0.$$

If $F = \ln K$ is second-order stationary, one again writes $F = \bar{F} + \beta f$, and h is expressed as a function of the covariance of f. Gelhar has shown that in one dimension the error involved in the method of perturbations is less than 10% if $\sigma_F^2 \leq 1$ (by comparing it to the exact spectral method).

Tang and Pinder (1977) have used the method of perturbations for the transport equation. Sagar (1978) applied it to the flow equation. Gelhar and Axness (1983) have used it for the same equations together with the spectral method. Winter *et al.* (1984) applied it in the second order to the transport equation.

6.3.4　Simulation method (Monte-Carlo)

This is probably the most powerful method, where fewer assumptions are required. However, it is a numerical method, which may require much central processing unit (CPU) time and a careful examination of the results. The principle of the method is very simple. Let $Z(x,\xi)$ be a *stochastic process*, x being the coordinates in space and ξ the state variable. Remember that $Z(x,\xi_1)$ is called a realization of Z. One first generates 'simulations' of Z in the probabilistic sense, i.e., a large number of realizations of Z. To do so, we must know the probability distribution function of Z and its covariance (or variogram) if Z is spatially correlated.

Note that the knowledge of the probability distribution function of Z was not necessary in the two previous methods.

Then, for each of these realizations, the parameter represented by $Z(x,\xi_i)$ is completely determined and known (e.g., the permeability or the source term or the boundary conditions). Thus, the flow equation can be solved numerically for each realization, giving the value of the dependent variable, e.g., $h(x,\xi_i)$. It is then possible to statistically analyze the ensemble of calculated solutions $h(x,\xi_i)$ for $i = 1,...,N$: expected value, variance, histogram, and distribution function for each location x. It is no longer necessary to assume that h is stationary; these statistics can be calculated at each point. The covariance or variogram can also be determined if h is found to be stationary, or intrinsic.

There are some difficulties associated with the simulation method. First, a large number of realizations, N, is necessary in order to get meaningful statistics: from 50 to several hundreds or thousands. Secondly, as N is necessarily finite, one can always calculate an experimental variance or co-variance, even for a phenomenon where they do not exist. It is preferable to check that when N increases, these statistics indeed become constant. Finally, the solution can be a function of the mesh size. Because the numerical solution requires an estimate of the average of $Z(x,\xi_i)$ over a mesh, this estimate becomes less variable as the mesh becomes larger, simply because of the integration. Thus, the variability of the solution, $h(x,\xi)$, will also be affected. Furthermore, one must realize that if C (or γ) is the correlation structure of Z in space, then the correlation structure of the average of Z over a mesh will be the integrated covariance or variogram. This has not always been recognized in the past.

The main difficulty with the simulation method is how to generate the realizations, $Z(x,\xi_i)$. Freeze (1975) assumed that Z (in this case, the hydraulic conductivity in a one-dimensional flow problem) was not spatially correlated. When the probability distribution of Z was known, independent values were drawn randomly in each mesh. In two dimensions, Smith and Freeze (1979) and Smith and Schwartz (1980, 1981a,b) imposed a correlation structure on Z (the hydraulic conductivity) using the method of the 'nearest neighbour'. The correlation is imposed by a kind of 'moving average' of the value of Z, taken in adjacent meshes. Binsariti (1980) generated the complex covariance matrix of Z and took a vector of independent random numbers. He solved for the correlated $Z(x,\xi_i)$ by triangulation of the covariance matrix using Cholesky's method (see also Neuman, 1984). Meija and Rodriguez-Iturbe (1974) used spectral methods.

Delhomme (1979) used the method of the *turning bands*, developed by Matheron (1973), which is a very powerful tool in two dimensions (see also Chiles, 1977; Mantoglu and Wilson, 1982). Delhomme also used *conditional simulations* of Z instead of simple simulations. This is a great improvement on the Monte Carlo method for practical problems. Indeed, the stochastic process, Z, is then said to be *conditioned* by the measurements $Z(x_j)$ in space: all the realizations $Z(x,\xi_i)$ must have the measured values $Z(x_j)$ at each point x_j, where a measurement has been made. The method used to generate these conditional simulations is based on *kriging*.

Nonconditional simulations are suitable for studying the theoretical variability of a process: the statistics of Z are assumed to be known, but no measured values are available. On the contrary, conditional variability is only that which stems from the uncertainty in the estimation of Z between measurement points. *Conditional simulations* are thus a logical follow-up to *kriging*. Delhomme (1979) used them for transmissivities and mentioned that the transmissivity could be further conditioned by the inverse problem. Such conditioning is also discussed by Neuman and Yakowitz (1979), Neuman (1984), and Dagan (1982a,b).

6.4 Example of stochastic solution to the transport equation

Let us consider a perfectly stratified infinite medium, where the permeability $K(z)$ is a stationary RF. Let the flow velocity be parallel to the stratification, so that the velocity vector has only one component, $U(z)$ in the x direction, U being also a RF.

The classical dispersion equation is

$$D_L \frac{\partial^2 C}{\partial x^2} + D_T \frac{\partial^2 C}{\partial z^2} - U(z) \frac{\partial C}{\partial x} = \frac{\partial C}{\partial t},$$

where C is the concentration, and D_L, D_T are the longitudinal and transversal dispersion coefficients.

Another approach can be used to represent the same transport. Consider a slug injection of tracer at a point (x_o, z_o). Let us follow the movement of a *tracer particle* located at point (x_o, z_o) at time $t = 0$, and at point (X_t, Z_t) at some later time t. We can write:

$$X_t = x_o + \xi_t + \int_0^t U(z_\tau)d\tau$$

$$Z_t = z_o + \zeta_t,$$

where ξ_t and ζ_t represent the *dispersion process* defined by a *Brownian motion*, i.e., *Gaussian stochastic process* of zero mean having a variance proportional to time:

$$\sigma_\xi^2 = 2D_L t \qquad \sigma_\zeta^2 = 2D_T t.$$

This representation of transport is equivalent to the *convection-diffusion equation* if D_L and D_T are constant (Kolmogorov, 1931). The expectation of the concentration at location (x_1, z_1) and time t_1 is equal to the probability density of the particle at (x_1, z_1, t_1).

The symbols X_t and Z_t denote stochastic processes. Let us calculate only the first two moments of X_t. This will us (i) the average velocity of the tracer, and (ii) the variance of its position, hence the macroscopic longitudinal dispersion coefficient.

One finds (Matheron and Marsily, 1980)

$$E(X_t) = E(U)t,$$

$$\sigma^2_{X_t} = 2D_L t + 2 \int_0^t (t - \tau) \frac{1}{2(\pi D_T \tau)^{1/2}} \cdot$$

$$\cdot \int_{-\infty}^{+\infty} \exp(-s^2/4D_T \tau)\, \mathrm{cov}(s)\, ds\, d\tau,$$

where $\mathrm{cov}(s)$ is the covariance function of $U(z)$. If we choose a Gaussian covariance function

$$\mathrm{cov}(s) = \sigma^2_U \exp\left(- \frac{m^2}{2} \left(\frac{s}{l}\right)^2\right)$$

we find

$$\sigma^2_{X_t} = 2D_L t + 2\sigma^2_U \frac{l}{m} \left[\frac{1}{3D_T^2} \left(\frac{l^2}{m^2} + 2D_T t\right)^{3/2} \right.$$

$$\left. - \frac{1}{3D_T^2} \left(\frac{l}{m}\right)^3 - \frac{1}{mD_T} t\right].$$

If transport was diffusive, one should have $\sigma^2_{X_t} = 2D_A t$, where D_A would be a constant macroscopic longitudinal dispersion coefficient. This is clearly not the case for the Gaussian covariance function of the velocity, even for $t \to \infty$.

Only for a very special type of covariance function of U (with a 'hole effect', i.e., a zero integral and a linear behaviour at the origin of the Laplace transform of the covariance) can transport be shown to be diffusive for large times (see Matheron and Marsily, 1980).

6.5   The problem of estimation of a RF by kriging

Let us assume that the RF function, $Z(x)$, is stationary, with known expected value, m, and covariance, $C(h)$. The problem of *estimation* of the RF Z can be stated in the following way. Let $Z_1, ..., Z_n$ be n measurements of Z in a realization, at locations $x_1, ..., x_n$. We want to estimate Z_o at an unmeasured location x_o. To simplify, we define

$$Y(x) = Z(x) - m.$$

Thus, $E(Y) = 0$, and we will estimate Y_o.

The kriging estimation, Y_o^* of Y_o, will be taken as a linear combination of all the available measurements

$$Y_o^* = \sum_{i=1}^{n} \lambda_o^i Y_i. \tag{6.5.1}$$

The estimation will be said optimal by imposing

$$E\left[(Y_o^* - Y_o)^2\right] \text{ minimum.} \tag{6.5.2}$$

This is a minimum variance estimation.

If we substitute (6.5.1) in (6.5.2) we get

$$E\left[(Y_o^* - Y_o)^2\right] =$$

$$= E\left[\left(\sum_i \lambda_o^i Y_i - Y_o\right)^2\right] = E\left[\left(\sum_i \lambda_o^i Y_i\right)\left(\sum_j \lambda_o^j Y_j\right)\right]$$

$$- 2E\left[\sum_i \lambda_o^i Y_i Y_o\right] + E(Y_o^2)$$

$$= \sum_i \sum_j \lambda_o^i \lambda_o^j E(Y_i Y_j) - 2\sum_i \lambda_o^i E(Y_i Y_o) + E(Y_o^2).$$

However, by definition

$$E(Y_i Y_j) = C(x_i - x_j) + m^2 = C(x_i - x_j),$$

since

$$m = E(Y) = 0,$$

and

$$E(Y_o^2) = C(0).$$

Then

$$E\left[Y_o^* - Y_o)^2\right] = \sum_i \sum_j \lambda_o^i \lambda_o^j C(x_i - x_j)$$

$$- 2\sum_i \lambda_o^i C(x_i - x_o) + C(0). \tag{6.5.3}$$

The solution of (6.5.2) is obtained by equating to zero the partial derivatives of (6.5.3) with respect to the unknown λ_o^i's. This results in a linear system of n equations with n unknowns

$$\sum_j \lambda_o^j C(x_i - x_j) = C(x_i - x_o), \quad i = 1, \dots, n.$$

This system has only one solution if C is a positive definite function and the x_i's are distinct.

Once the λ_o^i's have been calculated, one can calculate this estimation variance

$$\text{var}(Y_o^* - Y_o) = E\left[(Y_o^* - Y_o)^2\right] - [E(Y_o^* - Y_o)]^2,$$

to obtain

$$E(Y_o^* - Y_o) = E(Y_o^*) - E(Y_o) = \sum_i \lambda_o^i E(Y_i) - E(Y_o) = 0.$$

Then

$$\text{var}(Y_o^* - Y_o) = E\left[(Y_o^* - Y_o)^2\right] = C(0) - \sum_i \lambda_o^i C(x_i - x_o).$$

We have thus, explicitly, given the *variance* of the *estimation error* of Y_o, or Z_o.

This is the simplest example of kriging, often called 'simple kriging'. More details can be found in Matheron (1971), Journel and Huijbregts (1978), Marsily (1986). Nonstationary cases are also considered.

6.6　The intrinsic hypothesis: definition of the variogram

In the mining industry (estimation of ore grades), it has been shown that, in certain cases, the hypothesis of second-order stationarity with a finite variance, $C(0)$, is not satisfied by the data. This is frequently the case also in hydrology. The experimental variance increases with the size of the area under consideration. A less stringent hypothesis, called the *intrinsic hypothesis*, has been developed to make the estimation possible.

6.6.1 The intrinsic hypothesis

The intrinsic hypothesis consists in assuming that even if the variance of Z is not finite, the variance of the first-order increments of Z is finite and these increments are themselves second-order stationary, i.e., that $Z(x+h) - Z(x)$ satisfies

$$\left. \begin{array}{rcl} E\left[Z(x+h) - Z(x)\right] & = & m(h) \\[2mm] \mathrm{var}\left[Z(x+h) - Z(x)\right] & = & 2\gamma(h) \end{array} \right\} \text{ function of } h, \text{ not of } x,$$

where h is a vector in the one-, two-, or three-dimensional space and $\gamma(h)$ is generally only a function of the distance h.

Although this is not absolutely necessary, it is usually assumed that $m = 0$. If this were not the case, but $m(x+h) - m(x) = m(h)$, the function $Z(x) - m(x)$ would satisfy this condition.

The variance of the increment then defines a new function called the *variogram*, $\gamma(h)$

$$E\left[Z(x+h) - Z(x)\right] = 0, \tag{6.6.1}$$

$$\gamma(h) = \frac{1}{2}\,\mathrm{var}\left[Z(x+h) - Z(x)\right]. \tag{6.6.2}$$

Equations (6.6.1) and (6.6.2) make it possible to write

$$\gamma(h) = \frac{1}{2}\,E\left\{[Z(x+h) - Z(x)]^2\right\}, \tag{6.6.3}$$

where $\gamma(h)$ is the mean quadratic increment of Z between two points separated by the distance h.

If we compare the intrinsic hypothesis with the hypothesis of second-order stationarity, we see that (6.6.1) is equivalent to $E\left[Z(x)\right] = m$ (constant mathematical expectation), but

$$C(h) = E\left[Z(x+h)Z(x)\right] - m^2.$$

Is there a relation between the covariance and the variogram? In the case where both exist, i.e., in the stationary hypothesis, we can write

$$\gamma(h) = \frac{1}{2}\,E\left[Z(x+h)^2\right] - E\left[Z(x+h)Z(x)\right] + \frac{1}{2}\,E\left[Z(x)^2\right],$$

where we can see that

$$\gamma(h) = C(0) - C(h).$$

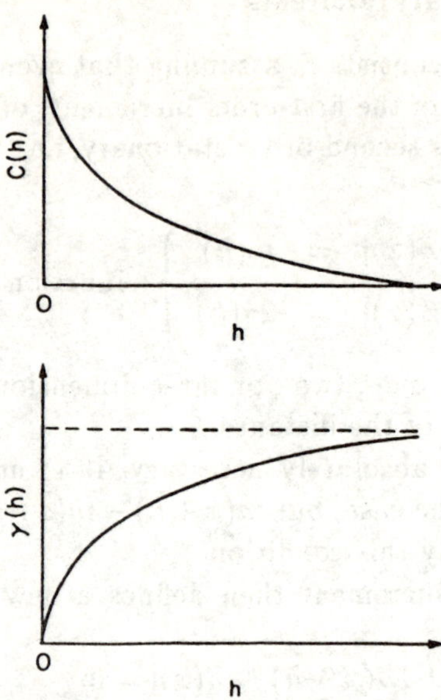

Figure 6.6.1: Covariance and variogram.

If we know the covariance, the variogram is simply its reflection with respect to the horizontal axis and with a vertical shift (Fig. 6.6.1).

When var(Z) is finite, the variogram tends towards an asymptotic value equal to this variance, which is also called the *sill* of the variogram (the distance at which the variogram reaches its asymptotic value is called the *range*). However, if the phenomenon under consideration does not have a finite variance, the variogram will never have a horizontal asymptotic value (Fig. 6.6.2). Not just any function $\gamma(h)$ can be a variogram, just as the covariance must be positive definite. It is indeed possible to show that:

- Minus γ must be conditionally positive definite, i.e., for all $x_1, ..., x_n \in R^m$ ($m = 1, 2$ or 3) and for all $\lambda_1, ..., \lambda_n \in R$, n coefficients satisfying $\sum_i \lambda_i = 0$, then

$$- \sum_i \sum_j \lambda_i \lambda_j \gamma(x_i - x_j) \geq 0.$$

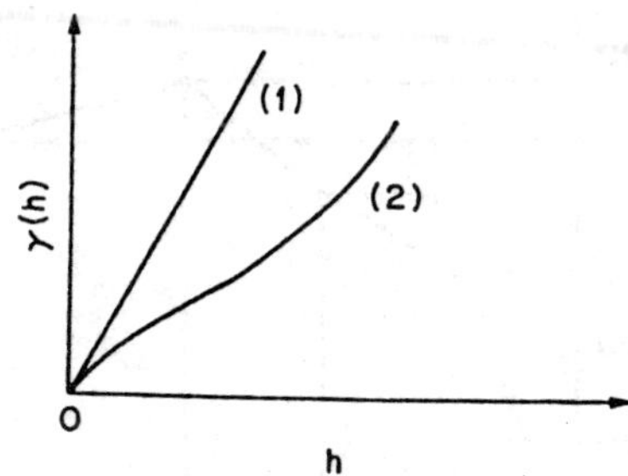

Figure 6.6.2: Variogram of a phenomenon with infinite variance.

- $\gamma(h)$ for $(h) \to \infty$ must necessarily increase less rapidly than $|h|^2$, i.e.,

$$\lim_{h \to \infty} \frac{\gamma(h)}{|h|^2} \to 0.$$

In practice, only a limited class of functions is used to describe variograms. We shall present a few of them in connection with *statistical inference* (determination of the variogram from the data).

Note that a slightly different set of kriging equations is used when the phenomenon is intrinsic and not stationary: the covariance is replaced by the variogram, and an additional condition of unbiasness is added (see Journel and Huijbregts, 1978), Marsily, 1986).

6.6.2   Determination of the variogram

We have defined the variogram in the case where the mean is constant by

$$\gamma(h) = \frac{1}{2} E\left\{ [Z(x+h) - Z(x)]^2 \right\}.$$

To estimate the variogram, we use the measurement points Z_i and assume *ergodicity* on the increments (i.e., that space averages can be used to estimate the averages in the whole set of realizations).

First we define a certain number of classes of distances between the measurement points, e.g.

$$0 < d_1 < 1 \text{ km} \qquad 1 < d_2 < 2 \text{ km} \qquad 2 < d_3 < 3 \text{ km}$$
$$3 < d_4 < 5 \text{ km} \qquad 5 < d_5.$$

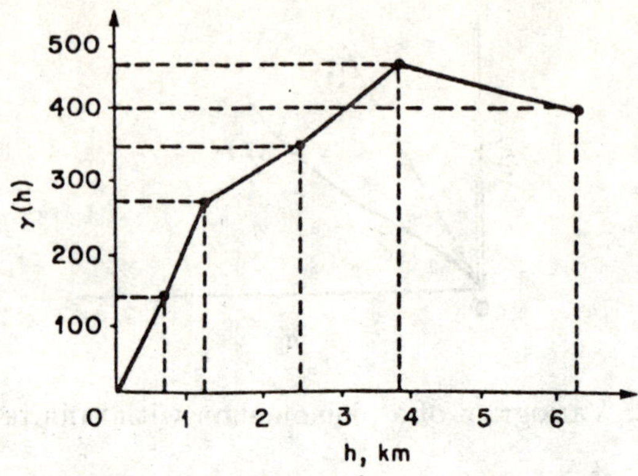

Figure 6.6.3: Experimental variogram.

Then, taking into account all possible pairs of points i and j for each class of distance, we calculate:

(1) the number of pairs present in the class;

(2) the average distance in the class;

(3) the average square increment $\frac{1}{2}(Z_i - Z_j)^2$.

With a set of 50 measurements we obtain, for example:

class	d_1	d_2	d_3	d_4	d_5
Number of elements	500	350	250	100	25
Average distance	0.7	1.3	2.4	3.8	6.2
$\frac{1}{2}(Z_i - Z_j)^2$ average	130	275	350	570	400

Note that the number of pairs that can be formed from a set of n points is $n(n-1)/2$; for 50 points this gives 1225 pairs. However, generally they are not evenly distributed. There are more pairs at short than at long distances. The variogram becomes more and more uncertain as $|h|$ increases. At large distances certain points may play a privileged role and introduce errors into the estimation. It may be necessary to eliminate a few measurements when calculating the variogram (Fig. 6.6.3).

However, we have seen that all functions cannot be variograms. In a class of acceptable analytical functions, we choose a given form and fit the parameters of this function to the observed points. The main types of variograms commonly used are: linear; in $|h|^\lambda$, $\lambda < 2$; spherical: exponential;

Gaussian; cubic. The forms and equations of these variograms are given in Fig. 6.6.4, adapted from Delhomme (1976). For example, the variogram in Fig. 6.6.3 would be interpreted as a spherical one and the two parameters, w and a, would be fitted by hand on the data (Fig. 6.6.5). Note that a piecewise linear variogram (i.e., made of segments of straight lines) is not acceptable; it is not in general a positive definite function (see Armstrong and Jabin, 1981).

6.6.3 Behaviour of the variogram for large h

Note that an unbounded variogram, e.g., a linear one, suggests that the field has infinite variance and that there is no covariance function; the intrinsic hypothesis is the only acceptable one here. But if the variogram reaches a sill, as for example in Fig. 6.6.5, then the covariance function exists for the phenomenon in question.

6.6.4 Behaviour close to the origin

Theoretically, for $h = 0$, $\gamma(h) = 0$ regardless of the variogram. However, very often variograms exhibit a jump at the origin, as in Fig. 6.6.6. This apparent jump at the origin is called the *nugget effect*, as it originated in the mining industry. Indeed, if a core contains a nugget, the concentration will be very high, whereas neighboring cores even with high mineral concentration will never be as rich: there is an 'erratic' component in the behaviour.

Such behaviour is very frequently found when data are analyzed (e.g., transmissivities). To take it into account, one just adds the quantity C to the variogram fitted on the data as if C were the origin

$$\gamma(h) = C\left[1 - \delta(h)\right] + \gamma'(h),$$

where $\delta(h)$ is the Kronecker δ ($\delta = 1$ if $h = 0$, $\delta = 0$ if $h \neq 0$) and $\gamma'(h)$ is the variogram fitted on the data with C as origin.

This *nugget effect* can also be attributed to measurement errors or to the fact that the data have not been collected with a sufficiently small interval to show the underlying continuous behaviour of the phenomenon (Fig. 6.6.7).

A horizontal variogram, i.e., $\gamma(h) = C$, $\forall h > 0$, is called a variogram with pure nugget effect. It expresses a purely random phenomenon without spatial structure.

Much work is at present being done to establish procedures that improve the quality and the robustness of the determination of the variogram (see, for instance, Armstrong, 1984, and Diamond and Armstrong, 1984).

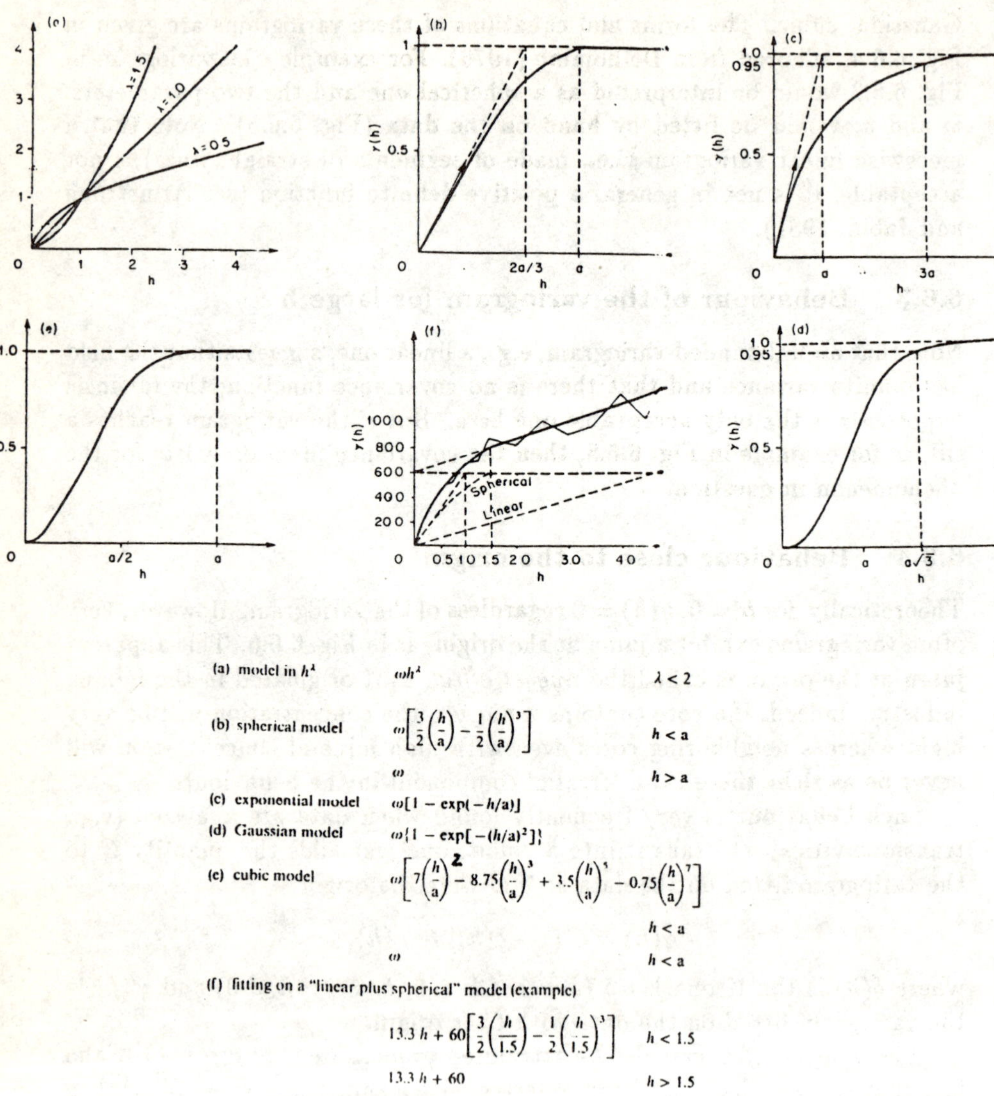

(a) model in h^λ ωh^λ $\lambda < 2$

(b) spherical model $\omega\left[\dfrac{3}{2}\left(\dfrac{h}{a}\right) - \dfrac{1}{2}\left(\dfrac{h}{a}\right)^3\right]$ $h < a$

    ω $h > a$

(c) exponential model $\omega[1 - \exp(-h/a)]$

(d) Gaussian model $\omega\{1 - \exp[-(h/a)^2]\}$

(e) cubic model $\omega\left[7\left(\dfrac{h}{a}\right)^2 - 8.75\left(\dfrac{h}{a}\right)^3 + 3.5\left(\dfrac{h}{a}\right)^5 - 0.75\left(\dfrac{h}{a}\right)^7\right]$ $h < a$

    ω $h < a$

(f) fitting on a "linear plus spherical" model (example)

    $13.3\,h + 60\left[\dfrac{3}{2}\left(\dfrac{h}{1.5}\right) - \dfrac{1}{2}\left(\dfrac{h}{1.5}\right)^3\right]$ $h < 1.5$

    $13.3\,h + 60$ $h > 1.5$

Figure 6.6.4: Common variogram models.

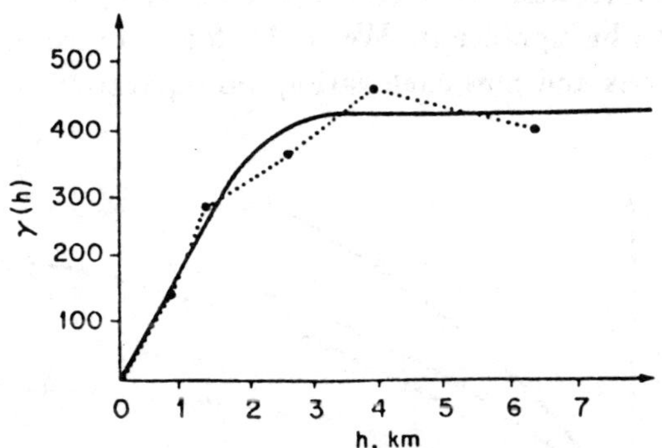

Figure 6.6.5: Spherical variogram fitted to the experimental one, $w = 430$, $a = 3$ km.

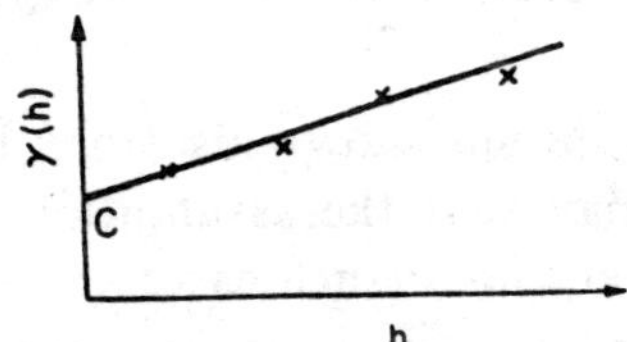

Figure 6.6.6: The nugget effect.

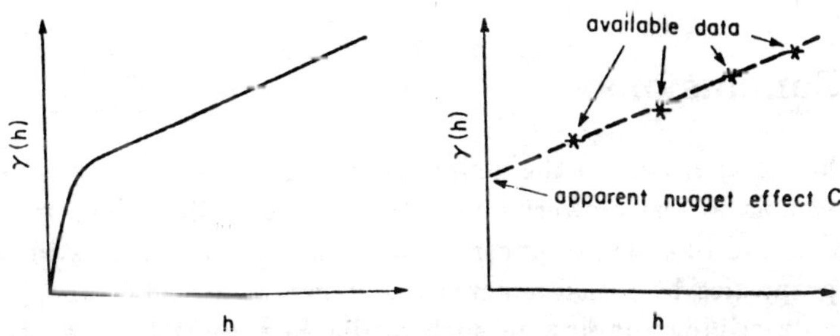

Figure 6.6.7: Underlying continuous behaviour of a variogram with a nugget effect.

6.6.5   Anisotropy in the variogram

It may be useful to compute the variogram while assuming that $\gamma(h)$ is also a function of the direction of the vector h. Of course, this requires more data points in order to be significant. We could, for instance, use four (or eight) classes of directions and plot each variogram separately, as in Fig. 6.6.8.

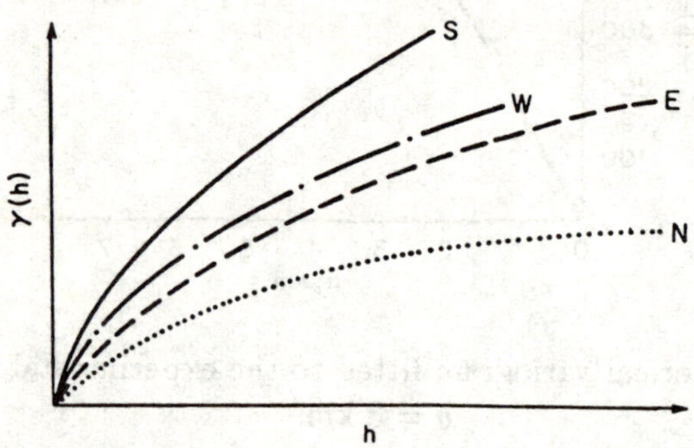

Figure 6.6.8: Directional variogram.

Generally, variograms do not show anisotropy like Fig. 10. If they do, then (1) this may be a sign that the assumption of stationarity (or even intrinsic behaviour) does not hold (such cases are dealt with among others, by Marsily, 1986), or (2) if the intrinsic or the stationary hypothesis is valid, this anisotropy can be eliminated by an appropriate linear transformation in the coordinate system. This will permit us to krige as usual in the new system.

6.7   Conclusions

The stochastic approach to the definition of the properties of porous media can be seen as a very powerful tool for understanding the variability of these media, and to assess to uncertainty associated with the estimation: (i) of these properties by measurements (as shown in Sec. 6.5, and (ii) of the quantities describing the flow in such media (e.g., head, concentration) as shown in Secs. 6.3 and 6.4.

The work along these lines is at present very active, both theoretically and on practical applications.

References

Armstrong, M. and Jabin, R. Variogram models must be positive definite, *J. Int. Assoc. Math. Geol.*,

Armstrong, M. Common problems seen on variograms, *J. Int. Assoc. Math. Geol.*, 16(3):305-313, 1984.

Baker, A. A., Gelhar, L. W., Gutjahr, A. L. and McMillan, J. R. Stochastic analysis of spatial variability in subsurface flow. Part I: comparison of one- and three-dimensional flows, *Water Resour. Res.*, 14(2):263-271, 1978.

Beran, M. J. *Statistical Continuum Theories*. Wiley (Interscience), New York, 1968.

Binsariti, A. A. *Statistical Analysis and Stochastic Modeling of the Cortaro Aquifer in Southern Arizona*. Ph. D. dissertation, Dept. of Hydrology and Water Resources, Univ. of Arizona, Tucson, Arizona, 1980.

Chiles, J. P. *Statistique des Phénomènes Non Stationnaires*. Thèse, Univ. of Nancy 1, France, 1977.

Dagan, G. Models of groundwater flow in statistically homogeneous porous formations, *Water Resour. Res.*, 15(1):47-63, 1979.

Dagan, G. Analysis of flow through heterogeneous random aquifers by the method of embedding matrix. 1. Steady flow. *Water Resour. Res.*, 17(1):107-121, 1981.

Dagan, G. Stochastic modelling of groundwater flow by unconditional and conditional probabilities. 2. The solute transport. *Water. Resour. Res.*, 18(4):835-848, 1982a.

Dagan, G. Analysis of flow through heterogeneous random aquifers. 2. Unsteady flow in confined formations. *Water Resour. Res.*, 18(5):1571-1585, 1982b.

Delhomme, J. P. *La Cartographie d'Une Grandeur Physique à Partir de Données de Différentes Qualités*. Meet. Int. Assoc. Hydrogeol., 10(1):185-194, 1974.

Delhomme, J. P. *Application de la Théorie des Variables Régionalisées Dans les Sciences de l'Eau*. Thèse, Univ. Paris VI, 1976.

Delhomme, J. P. Kriging in hydroscience. *Adv. Water Resour.*, 1(5):252-266, 1978a.

Delhomme, J. P. *Application de la Théorie des Variables Régionalisées Dans les Sciences de l'Eau*. Bull. Bur. Rech. Géol. Min., Ser. 2, Sect. III, No. 4-1978, 341-375, 1978b.

Delhomme, J. P. Spatial variability and uncertainty in groundwater flow parameters: a geostatistical approach. *Water Resour. Res.*, 15(2):269-280, 1979.

De Smedt, F., Belhadi, M. and Nurul, K. *Study of the Spatial Variability of the Hydraulic Conductivity with Different Measurements Techniques*. Proc. Int. Symp. Stochastic Approach Subsurface Flow. Montvillargenne, Int. Assoc. Hydraul. Res. (Marsily, G. de, ed.) Fontainebleau, France, 1985.

Diamond, P. and Armstrong M. Robustness of variograms and conditioning of kriging matrices. *J. Int. Assoc. Math. Geol.*, **16**(8):809-822, 1984.

Farengolts, Z. D. and Kolyada, M. N. Opyt primeneiyaé metodov matematicheskoy statistiki slya izutcheniyazakonov respredeleniya gidrogeologichestikh parametrov. *Trudy Vsegingeo*, **17**:76-112, 1969.

Freeze, R. A. A stochastic-conceptual analysis of one-dimensional groundwater flow in non-uniform, homogeneous media. *Water Resour. Res.*, **11**(5):725-741, 1975.

Gelhar, L. W. *Effects of Hydraulic Conductivity Variations on Groundwaters Flows*. Proc. Int. Symp. Stochastic Hydraulics (1976), 2nd Int. Assoc. Hydraulics Res., Lund, Sweden. (Hjort, P., Jönssen, L. and Larsen, P., eds.) Water Res. Pub., Fort Collins, Colorado, 409-428, 1977.

Gelhar, L. W. Stochastic subsurface hydrology: from theory to applications. *Water Resour. Res.*, **22**(9):135S-145S, 1986.

Gelhar, L. W. and Axness, C. L. Three-dimensional stochastic analysis of macrodispersion in aquifers. *Water Resour. Res.*,**19**(1):161-180, 1983.

Gelhar, L. W. Baker, A. A. Gutjahr, A. L. and McMillan, J. R. Comments on stochastic conceptual analysis of one dimensional groundwater flow in a non-uniform homogeneous medium, by Freeze, R.A., and reply by Freeze, R.A. *Water Resour. Res.*, **13**(2):477-480, 1977.

Gelhar, L. W. Gutjahr, A. L. and Naff, R. L. Stochastic analysis of macrodispersion in a stratified aquifer. *Water Resour. Res.*, **15**(6):1387-1397, 1979a.

Gelhar, L. W. KO, P. Y., KWAI, H. H. and WILSON, J. L. *Stochastic Modelling of Groundwater Systems*. R. M. Parsons Lab. for Water Resources and Hydrodynamics, Rep. 189, M.I.T., Cambridge, Mass., 1974.

Gelhar, L. W. Wilson, J. L. and Gutjahr, A. L. Comments on 'Simulation of groundwater flow and mass transport under uncertainty', by Tang, D.II., Pinder, G.F. – Reply. *Adv. Water Resour.*, 2(2):101-102, 1979b.

Gutjahr, A. L. Gelhar, L. W. BAKR, A.A. and McMILLAN, J.R. Stochastic analysis of spatial variability in subsurface flows. Part II: Evaluation and application. *Water Resour. Res.*, 14(5):953-960, 1978.

Ilyan, N. L., Chernychev, S. N., Dzekster, E. S. and Zilberg, V. S. *Ostenka tochnosti opredeleniya vodopronitsayemosti gronykh porod.* Nedra, Moscow, 1971.

Journel, A. G. and Huijbregts, C. *Mining Geostatics.* Academic Press, New York, 1978.

Jetel, J. *Complément régional de l'information sur les paramètres pétrophysiques en vue de l'élaboration des modèles de systèmes aquifères.* Proc. Meet. Int. Assoc. Hydrogeol., Mém., 10(1):199-203, 1974.

Keller, J. B. *Stochastic equations and wave propagation in random media.* Proc. Symp. Appl. Math., 16th, Am. Math. Soc., 145-170. Providence, R.I., 1964.

Kolmogorov, A. N. Uber die analytischen Methoden in der Wahrscheinlichkeitsrechnung. *Math. Ann.*, 104:415-458, 1931.

Krumbein, W. C. Application of the logarithmic moments to size frequency distribution of sediments. *J. Sediment. Petrol.*, 6(1):35-47, 1936.

Law, J. A statistical approach to the intersticial heterogeneity of sand reservoirs. *Trans. Am. Inst. Min. Metall. Pet. Eng.*, 155:202-222, 1944.

Lumley, J. L. and Panofsky, H. A. *The Structure of Atmospheric Turbulence.* Wiley, New York, 1964.

Mantoglou, A. and Wilson, J. L. The turning methods for simulation of random fields using line generation by a spectral method. *Water Resour. Res.*, 18(5):1379-1304, 1982.

Marle, C. Ecoulements monophasiques en milieu poreux. *Rev. Inst. Fr. Pet.*, 22(10):1471-1500, 1967.

Marsily, G. de. *Quantitative Hydrogeology. Groundwater Hydrology for Engineers.* Academic Press, New York, 440 p., 1986.

Matheron, G. *Les variables régionalisées et leur estimation.* Masson, Paris, 1965.

Matheron, G. The Theory of Regionalized Variables and Its Applications. Paris School of Mines. Cath. Cent. Morphologie Math., 5. Fontainebleau, 1975.

Matheron, G. The intrinsic random functions and their applications. *Adv. Appl. Prob.*, 5:438-468, 1973.

Matheron, G. and Marsily, G. de. Is transportation in porous media always diffusive ? A counter example. *Water Resour. Res.*, 16(5):901-917, 1980.

Meija, J. M. and Rodriguez-Iturbe, I. On the synthesis of random field sampling from the spectrum: an application to the generation of hydrologic spatial proceses. *Water Resour. Res.*, 10(4):705-712, 1974.

Neuman, S. P. Role of geostatistics in subsurface hydrology. In *Geostatistics for Nature Resources Characterization. Proc. NATO-ASI*, (Verly, G., David, M., Journel, A.G. and Maréchal, A., eds.), Part I, 787-816, Reidel, Dordrecht, The Netherlands, 1984.

Neumann, S. P. and Yakowitz, S. A statistical approach to the inverse problem of aquifer hydrology. 1. Theory. *Water Resour. Res.*, 15(4):845-860, 1979.

Rousselor, D. *Proposition Pour Une Loi de Distribution des Perméabilités Ou Transmissivités.* Rep. Bur. Rech. Géol. Min., Serv. Géol. Jura-Alpes-Lyon, 1976.

Sagar, B. Galerkin finite elements procedure for analyzing flow through random media. *Water Resour. Res.*, 14(6):1035-1044, 1978.

Schweppe, F. *Uncertain Dynamic Systems.* Prentice Hall, New York, 1973.

Schwydler, M. I. Les courants d'écoulement dans les milieux hétérogènes. *Izd. Akad. Nauk. SSSR*, 3:185-190, 1862.

Smith, L. and Freeze, R.A. Stochastic analysis of steady state groundwater flow in a bounded domain. 1. One-dimensional simulations. *Water Resour. Res.*, 15(3):521-528, 1979.

Smith, L. and Freeze, R.A. Stochastic analysis of steady state groundwater flow in a bounded domain. 2. Two-dimensional simulations. *Water Resour. Res.*, 15(6):1543-1559, 1979.

Smith, L. and Schwartz, F.W. Mass transport. 1. A stochastic analysis of macrodispersion. *Water Resour. Res.*, 16(2): 303-313, 1980.

Smith, L. and Schwartz, F.W. Mass transport. 2. Analysis of uncertainty in prediction. *Water Resour. Res.*, **17**(2):351-369, 1981a.

Smith, L. and Schwartz, F.W. Mass transport. 3. Role of hydraulic conductivity data in prediction. *Water Resour. Res.*, **17**(5):1463-1479, 1981b.

Tang, D. H. and Pinder, G. F. Simulation of groundwater flow and mass transport under uncertainty. *Adv. Water Resour.*, **1**(1):25-30, 1977.

Walton, W.C. and Neill, I. C. Statistical analysis of specific capacity data for a dolomite aquifer. *J. Geophys. Res.*, **68**(8):2251-2262, 1963.

Winter, C. L., Newman, C. M. and Neuman, S. P. A perturbation expansion for diffusion in a random field. *SIAM J. Appl. Math.*, **44**(2):411-424, 1984.

List of Main Symbols

C	Covariance.
D	Dispersion coefficient.
d_i	Class of distances, $i = 1, 2....$
E	Expected value.
h	Space interval.
K	Permeability.
l	Characteristic or correlation length.
m	Mean value.
N	Number of realizations of the stochastic process.
n	number of samples.
$p(Z)$	Probability distribution function of Z.
q	Flow rate.
S_s	Storativity.
t	Time.
U	Velocity.
x	Spatial coordinate.
z	Spatial coordinate.
Z	Random function or stochastic process.

Greek Letters

γ	Variogram.
δ	Kronecker delta function.
σ	Variance.
φ	Spectral density.

Subscripts

G Geometric mean value.
L Longitudinal.
M Average value.
T Transversal.

Special Symbols

$*$ As superscript, complex conjugate or kriging estimation.
$\overline{(..)}$ Average value of (..).

Index